D0351560

A-Level

Mathematics

for Edexcel Core 2

The Complete Course for Edexcel C2

Contents

Chapter 5

Sequences and Series

Chapter 6

Differentiation

Chapter 7

Integration

Reference

About this book

In this book you'll find...

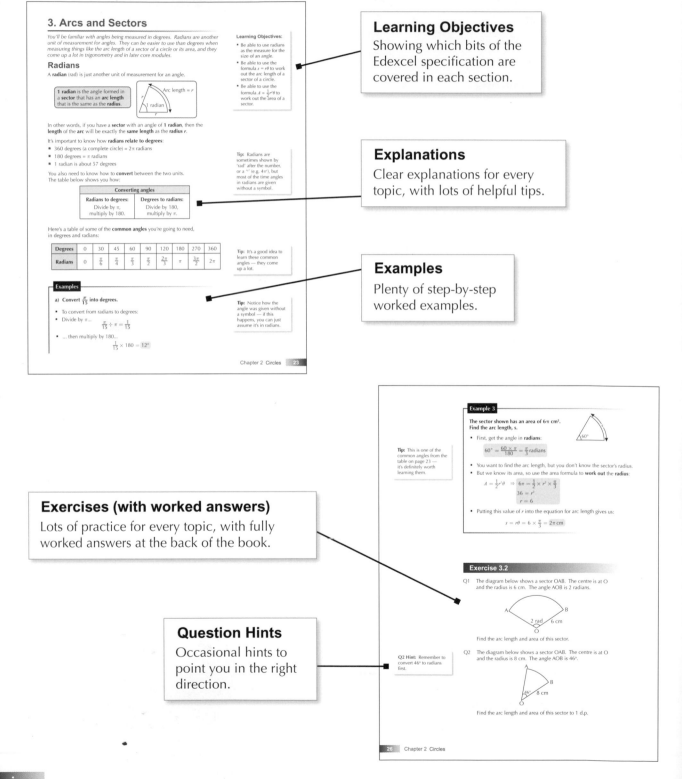

Learning Objectives
Showing which bits of the Edexcel specification are covered in each section.

Explanations
Clear explanations for every topic, with lots of helpful tips.

Examples
Plenty of step-by-step worked examples.

Exercises (with worked answers)
Lots of practice for every topic, with fully worked answers at the back of the book.

Question Hints
Occasional hints to point you in the right direction.

Review Exercise — Chapter 2

Q1 Give the radius and the coordinates of the centre of the circles
with the following equations:
a) $x^2 + y^2 = 9$
b) $(x - 2)^2 + (y + 4)^2 = 4$
c) $x(x + 6) = y(8 - y)$

Q2 Find the equations for the circles with the following properties:
a) centre (3, 2), radius 6
b) centre (−4, −8), radius 8
c) centre (0, −3), radius $\sqrt{14}$

Q3 A circle has the equation $x^2 + y^2 - 4x + 6y - 68 = 0$.
Find the coordinates of the centre of the circle and its radius.

Q4 The circle C has equation $(x - 2)^2 + (y - 1)^2 = 100$.
The point A (10, 7) lies on the circle.
Find an equation of the tangent at A.

Q5 The circle C has equation $x^2 + y^2 - 12x + 2y + 11 = 0$.
The point A (1, −2) lies on the circle.
Find an equation of the tangent at A.

Q6 The diagram below shows a sector ABC of a circle, with centre A
and a radius of 10 cm. The angle BAC is 0.7 radians.

Find the arc length BC and area of this sector.

Q7 The sector ABC below is part of a circle, where the angle BAC is 50°.

Given that the area of the sector is 20π cm², find the arc length BC.
Give your answer in terms of π.

28   Chapter 2 Circles

Review Exercises
Mixed questions covering the whole chapter, with fully worked answers.

Exam-Style Questions — Chapter 2

1   C is a circle with the equation: $x^2 + y^2 - 2x - 10y + 21 = 0$.

a) Find the centre and radius of C.   (5 marks)

The line joining P(3, 6) and Q(q, 4) is a diameter of C.

b) Show that $q = -1$.   (3 marks)

c) Find the equation of the tangent to C at Q, giving your answer in the form
$ax + by + c = 0$, where a, b and c are integers.   (5 marks)

2   A circle C is shown here.
M is the centre of C, and J and K both lie on C.

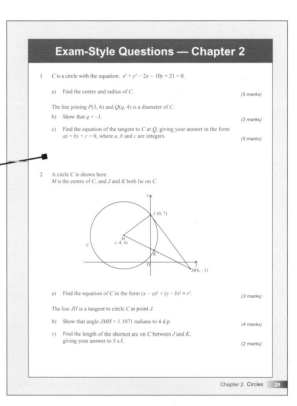

a) Find the equation of C in the form $(x - a)^2 + (y - b)^2 = r^2$.   (3 marks)

The line JH is a tangent to circle C at point J.

b) Show that angle JMH = 1.1071 radians to 4 d.p.   (4 marks)

c) Find the length of the shortest arc on C between J and K,
giving your answer to 3 s.f.   (2 marks)

Chapter 2 Circles   29

Exam-Style Questions
Questions in the same style as the ones you'll get in the exam, with worked solutions and mark schemes.

Formula Sheet
Contains all the formulas you'll be given in the C2 exam.

Glossary
All the definitions you need to know for the exam, plus other useful words.

Practice Exam Papers (on CD-ROM)
Two printable exam papers, with fully worked answers and mark schemes.

Published by CGP

Editors:
Helena Hayes, Paul Jordin, Simon Little, Kirstie McHale, Matteo Orsini Jones, Caley Simpson,
Charlotte Whiteley, Dawn Wright.

Contributors:
Katharine Brown, Michael Coe, Josephine Gibbons, Andy Pierson, Janet West.

ISBN: 978 1 84762 812 1

With thanks to Janet Dickinson and Jonathan Wray for the proofreading.
With thanks to Helen Greaves for the reviewing.

Printed by Elanders Ltd, Newcastle upon Tyne.
Clipart from Corel®

Text, design, layout and original illustrations © Coordination Group Publications Ltd. (CGP) 2012
All rights reserved.

Photocopying more than one chapter of this book is not permitted. Extra copies are available from CGP.
0870 750 1242 • www.cgpbooks.co.uk

1. Algebraic Division

Algebraic division means dividing one expression by another. It's just one way to simplify a fraction that has letters in it. Below we will look at one method you can use — long division.

Algebraic division

Important terms

There are a few words that come up a lot in algebraic division, so make sure you know what they all mean.

- **Divisor** — this is the thing you're dividing by. For example, if you divide $x^2 + 4x - 3$ by $x + 2$, the divisor is $x + 2$.
- **Quotient** — the bit that you get when you divide by the divisor (not including the **remainder**).
- **Remainder** — the bit that's **left over** after the division (it can't be divided by the divisor).

For algebraic division to work, the **degree** of the **divisor** has to be less than (or equal to) the **degree** of the **original polynomial** (for example, you couldn't divide $x^2 + 2x + 3$ by $x^3 + 4$ as $3 > 2$, but you could do it the other way around).

In C2, you will only be expected to divide by algebraic expressions of the form **(x + a)** or **(x − a)** — these are **linear divisors**.

Algebraic long division

You can use **long division** to divide two algebraic expressions, using the same method as you'd use for numbers.

Learning Objectives:

- Know what the terms quotient and remainder mean.
- Be able to carry out simple algebraic division of polynomials by $(x + a)$ or $(x - a)$.

Tip: The degree is the highest power of the variable. A polynomial is an algebraic expression made up of the sum of constant terms and variables raised to positive integer powers.
For example $x^3 - 2x + \frac{1}{2}$ is a polynomial, but $x^{-3} - 2x^{\frac{3}{2}}$ is not as it has a negative power and a fractional power of x.

Example 1

Divide $(x^3 + 3x^2 - 5x + 2)$ by $(x + 2)$.

- Start by dividing the first term in the polynomial by the first term of the divisor: $x^3 \div x = x^2$. Write this answer above the polynomial:

$$\begin{array}{r} x^2 \\ x + 2 \overline{)x^3 + 3x^2 - 5x + 2} \end{array}$$

- Multiply the divisor $(x + 2)$ by this answer (x^2) to get $x^3 + 2x^2$ and write this under the first two terms of the polynomial:

$$\begin{array}{r} x^2 \\ x + 2 \overline{)x^3 + 3x^2 - 5x + 2} \\ x^3 + 2x^2 \end{array}$$

Tip: If you do C3, you'll meet algebraic long division again there.

Tip: Note that we only divide each term by the 'x' term, not the '$x + 2$'. The +2 bit is dealt with in the steps in between.

Tip: The remaining polynomial at this stage is $x^2 - 5x + 2$.

- Subtract this from the main expression to get x^2. Bring down the $-5x$ term just to make things clearer for the next subtraction.

$$
\begin{array}{r}
x^2 \\
x + 2 \overline{) x^3 + 3x^2 - 5x + 2} \\
-\ \underline{x^3 + 2x^2} \\
x^2 - 5x
\end{array}
$$

- Now divide the first term of the remaining polynomial (x^2) by the first term of the divisor (x) to get x (the second term in the answer).

$$
\begin{array}{r}
x^2 + x \\
x + 2 \overline{) x^3 + 3x^2 - 5x + 2} \\
-\ \underline{x^3 + 2x^2} \\
x^2 - 5x
\end{array}
$$

- Multiply $(x + 2)$ by x to get $x^2 + 2x$, then subtract again and bring down the +2 term.

$$
\begin{array}{r}
x^2 + x \\
x + 2 \overline{) x^3 + 3x^2 - 5x + 2} \\
-\ \underline{x^3 + 2x^2} \\
x^2 - 5x \\
-\ \underline{x^2 + 2x} \\
-\ 7x + 2
\end{array}
$$

- Divide $-7x$ by x to get -7 (the third term in the answer). Then multiply $(x + 2)$ by -7 to get $-7x - 14$.

$$
\begin{array}{r}
x^2 + x - 7 \\
x + 2 \overline{) x^3 + 3x^2 - 5x + 2} \\
-\ \underline{x^3 + 2x^2} \\
x^2 - 5x \\
-\ \underline{x^2 + 2x} \\
-\ 7x + 2 \\
-\ \underline{-7x - 14} \\
16
\end{array}
$$

- After subtracting, this term (16) has a degree that's **less** than the degree of the divisor, $(x + 2)$, so it can't be divided. This is the **remainder**.

- So $(x^3 + 3x^2 - 5x + 2) \div (x + 2) = \boxed{x^2 + x - 7 \text{ remainder } 16.}$

Tip: So this means $x^3 + 3x^2 - 5x + 2 = (x^2 + x - 7)(x + 2) + 16$.

Example 2

Divide $(2x^3 - 2x + 3)$ by $(x - 3)$.

- This polynomial doesn't have an x^2 term, so add a $0x^2$ to where the x^2 term would be. This means you won't miss any terms out when doing the division.

$$x - 3\overline{)2x^3 + 0x^2 - 2x + 3}$$

- Then divide the first term in the polynomial by the first term of the divisor.

$$\begin{array}{r} 2x^2 \\ x - 3\overline{)2x^3 + 0x^2 - 2x + 3} \end{array}$$

- Multiply the divisor by this answer and subtract this from the original expression. Bring down the next term.

$$\begin{array}{r} 2x^2 \\ x - 3\overline{)2x^3 + 0x^2 - 2x + 3} \\ -\ 2x^3 - 6x^2 \\ \hline 6x^2 - 2x \end{array}$$

- Divide the first term of the remaining polynomial by the first term of the divisor.

$$\begin{array}{r} 2x^2 + 6x \\ x - 3\overline{)2x^3 + 0x^2 - 2x + 3} \\ -\ 2x^3 - 6x^2 \\ \hline 6x^2 - 2x \end{array}$$

- Multiply the divisor by this answer and subtract this from the remaining polynomial. Bring down the next term.

$$\begin{array}{r} 2x^2 + 6x \\ x - 3\overline{)2x^3 + 0x^2 - 2x + 3} \\ -\ 2x^3 - 6x^2 \\ \hline 6x^2 - 2x \\ -\ 6x^2 - 18x \\ \hline 16x + 3 \end{array}$$

- Divide the first term of the remaining polynomial by the first term of the divisor. Multiply the divisor by this answer and subtract this from the remaining polynomial.

$$
\begin{array}{r}
2x^2 + 6x + 16 \\
x - 3{\overline{\smash{\big)}\,2x^3 + 0x^2 - 2x + 3}} \\
-\ \underline{2x^3 - 6x^2} \\
6x^2 - 2x \\
-\ \underline{6x^2 - 18x} \\
16x + 3 \\
-\ \underline{16x - 48} \\
51
\end{array}
$$

Tip: So this means
$2x^3 - 2x + 3 =$
$(2x^2 + 6x + 16)(x - 3) + 51$.

- So we get a remainder of 51.
- Therefore $(2x^3 - 2x + 3) \div (x - 3) = 2x^2 + 6x + 16$ remainder 51.

Exercise 1.1

Q1 Hint: If the remainder is 0, it means the polynomial divides exactly by the divisor. So the divisor is a factor of the polynomial — there's more about this in the next section.

Q1 Use long division to solve the following.
In each case state the quotient and remainder.
a) $(x^3 - 2x^2 - 4x + 8) \div (x - 3)$
b) $(x^3 - x^2 - 11x - 10) \div (x + 2)$
c) $(x^3 - x^2 - 3x + 3) \div (x - 2)$
d) $(x^3 - 2x^2 - 5x + 6) \div (x + 3)$
e) $(x^3 - 9x^2 + 7x + 33) \div (x + 2)$

Q2, 3 Hint: Remember, you can add in any missing terms, just give them a coefficient of 0.

Q2 Divide $(x^3 - 5x + 4)$ by $(x - 1)$ using long division.

Q3 Divide $(x^3 - x^2 - 11)$ by $(x - 3)$ using long division.

Q4 Divide the cubic $(2x^3 + 3x^2 - 11x - 6)$ by the expressions below:
a) $x + 1$
b) $x + 2$
c) $x - 1$

Q5 $f(x) = x^3 + 2x^2 - 7x - 2$, express $f(x)$ in the form $(x - 2)g(x)$, where $g(x)$ is a quadratic.

2. The Remainder and Factor Theorems

The Remainder Theorem and Factor Theorem are useful tools. The Remainder Theorem gives an easy way to work out remainders and the Factor Theorem is an extension of the Remainder Theorem that helps you to factorise polynomials.

Learning Objectives:

- Be able to use the Remainder Theorem and Factor Theorem.
- Be able to determine the remainder when the polynomial f(x) is divided by (ax + b).
- Know that if f(x) = 0 when x = a, then (x − a) is a factor of f(x).
- Be able to factorise cubic expressions, e.g. $6x^3 + 11x^2 - x - 6$.

The Remainder Theorem

The Remainder Theorem gives you a quick way of **working out** the **remainder** from an algebraic division, but **without** actually having to do the division.

The Remainder Theorem says:

> When you divide **f(x)** by **(x − a)**, the remainder is **f(a)**.

So just stick **x = a** into the polynomial.

Dividing a polynomial by divisors that are multiples of each other, e.g. (x + 2) and (3x + 6), will produce the same remainder. This means you can simplify the divisor to get it in the form x − a and the Remainder Theorem still holds.

Example

Use the Remainder Theorem to work out the remainder when $(2x^3 - 3x^2 - 3x + 7)$ is divided by (x − 2).

- So: $f(x) = 2x^3 - 3x^2 - 3x + 7$

- You're dividing by (x − 2), so $a = 2$

- So the remainder must be:

$$f(a) = f(2) = (2 \times 8) - (3 \times 4) - (3 \times 2) + 7 = \boxed{5}$$

Tip: Be careful if you're dividing by something like (x + 7), as *a* will be negative. In this case, you'd get a = −7.

If you want the remainder after dividing by something like **(ax − b)**, there's an extension to the Remainder Theorem:

> When you divide **f(x)** by **(ax − b)**, the remainder is $f\left(\frac{b}{a}\right)$.

Tip: Note, $\frac{b}{a}$ is just the value of *x* that would make the bracket 0.

Example

Find the remainder when you divide $(2x^3 - 3x^2 - 3x + 7)$ by (2x − 1).

- So: $f(x) = 2x^3 - 3x^2 - 3x + 7$

- You're dividing by (2x − 1), so comparing it to (ax − b) we get:

$$a = 2 \quad \text{and} \quad b = 1$$

- So the remainder must be:

$$f\left(\frac{b}{a}\right) = f\left(\frac{1}{2}\right) = 2\left(\frac{1}{8}\right) - 3\left(\frac{1}{4}\right) - 3\left(\frac{1}{2}\right) + 7 = \boxed{5}$$

Fact: This page breaks the UK record for the most appearances of the word "remainder" on a single page. The title had previously been held by page 12 of "The Remainder Theorem Annual 1983".

If you're given the remainder when a polynomial is divided by something, you can use the Remainder Theorem to work **backwards** to find an **unknown coefficient** in the original polynomial.

Example

When $(x^3 + cx^2 - 7x + 2)$ is divided by $(x + 2)$, the remainder is -4. Use the Remainder Theorem to find the value of c.

- The polynomial was divided by $(x + 2)$, so: $\boxed{a = -2}$

- When you divide f(x) by $(x - a)$, the remainder is f(a), so:

$$f(-2) = -4 \quad \Rightarrow \quad (-2)^3 + c(-2)^2 - 7(-2) + 2 = -4$$
$$-8 + 4c + 14 + 2 = -4$$
$$4c = -12$$
$$\boxed{c = -3}$$

Exercise 2.1

Q1 Use the Remainder Theorem to work out the remainder in each of the following divisions:

a) $2x^3 - 3x^2 - 39x + 20$ divided by $(x - 1)$

b) $x^3 - 3x^2 + 2x$ divided by $(x + 1)$

c) $6x^3 + x^2 - 5x - 2$ divided by $(x + 1)$

d) $x^3 + 2x^2 - 7x - 2$ divided by $(x + 3)$

e) $4x^3 - 6x^2 - 12x - 6$ divided by $(2x + 1)$

f) $x^3 - 3x^2 - 6x + 8$ divided by $(2x - 1)$

Q2 The remainder when $x^3 + px^2 - 10x - 19$ is divided by $(x + 2)$ is 5. Use the Remainder Theorem to find the value of p.

Q3 When $x^3 - dx^2 + dx + 1$ is divided by $(x + 2)$ the remainder is -25. Use the Remainder Theorem to find the value of d.

Q4 When $x^3 - 2x^2 + 7x + k$ is divided by $(x + 1)$ the remainder is -8. Find the value of k.

Q5 Hint: Find the remainder from both divisions in terms of p.

Q5 $f(x) = x^4 + 5x^3 + px + 156$. The remainder when f$(x)$ is divided by $(x - 2)$ is the same as the remainder when f(x) is divided by $(x + 1)$. Use the Remainder Theorem to find the value of p.

The Factor Theorem

If you get a **remainder of zero** when you divide f(x) by ($x - a$), then ($x - a$) must be a **factor** of f(x). This is the **Factor Theorem**.

The **Factor Theorem** states:

> If **f(x)** is a polynomial, and **f(a) = 0**, then **($x - a$)** is a factor of **f(x)**.

This also works the other way round — if ($x - a$) is a factor of f(x), then you know without having to do calculations that f(a) = 0.

Tip: Remember, a root is just a value of x that makes f(x) = 0. So if you know the roots of f(x), you also know the factors of f(x) — and vice versa.

Example 1

Use the Factor Theorem to show that ($x - 2$) is a factor of ($x^3 - 5x^2 + x + 10$).

- So: $a = 2$

- Work out f(a): f(a) = f(2) = $8 - (5 \times 4) + 2 + 10 = 0$

- The remainder is 0, so that means ($x - 2$) divides into $x^3 - 5x^2 + x + 10$ **exactly**. So ($x - 2$) must be a **factor** of $x^3 - 5x^2 + x + 10$.

Example 2

Show that ($2x + 1$) is a factor of f(x) = $2x^3 - 3x^2 + 4x + 3$.

- The question's giving you a big hint here. Notice that $2x + 1 = 0$ when $x = -\frac{1}{2}$. So plug this value of x into f(x).

- If you show that f($-\frac{1}{2}$) = 0, then the Factor Theorem says that ($x + \frac{1}{2}$) is a factor — which means that $2 \times (x + \frac{1}{2}) = (2x + 1)$ is also a factor.

$$f(x) = 2x^3 - 3x^2 + 4x + 3$$

Tip: Remember from page 5, if ($x - a$) is a factor of f(x) then any multiple of ($x - a$) will also be a factor.

- And so: f$\left(-\frac{1}{2}\right) = 2\left(-\frac{1}{8}\right) - 3\left(\frac{1}{4}\right) + 4\left(-\frac{1}{2}\right) + 3 = 0$

- So, by the Factor Theorem, ($x + \frac{1}{2}$) is a factor of f(x), and so ($2x + 1$) is also a factor.

Just one more useful thing to mention about polynomials and factors:

> If the **coefficients** in a polynomial **add up to 0**, then **($x - 1$)** is a **factor**.

This works for all polynomials — there are no exceptions.

Tip: If you put $x = 1$ into a polynomial f(x), x^2, x^3 etc. are all just 1, so f(1) is the sum of the coefficients.

Example

Factorise the polynomial f(x) = $6x^2 - 7x + 1$.

- The coefficients (6, –7 and 1) add up to 0.

- That means f(1) = 0, and so ($x - 1$) is a factor.

- Then just factorise it like any quadratic to get this:

$$f(x) = 6x^2 - 7x + 1 = (6x - 1)(x - 1)$$

Q1 Use the Factor Theorem to show that:

a) $(x - 1)$ is a factor of $x^3 - x^2 - 3x + 3$

b) $(x + 1)$ is a factor of $x^3 + 2x^2 + 3x + 2$

c) $(x + 2)$ is a factor of $x^3 + 3x^2 - 10x - 24$

Q2 Use the Factor Theorem to show that:

a) $(2x - 1)$ is a factor of $2x^3 - x^2 - 8x + 4$

b) $(3x - 2)$ is a factor of $3x^3 - 5x^2 - 16x + 12$

Q2 Hint: Be careful with the fractions here.

Q3 a) Use the Factor Theorem to show that $(x - 3)$ is a factor of $x^3 - 2x^2 - 5x + 6$.

b) Show, by adding the coefficients, that $(x - 1)$ is also a factor of this cubic.

Q4 $f(x) = 3x^3 - 5x^2 - 58x + 40$. Use the Factor Theorem to show that the following are factors of $f(x)$:

a) $(x + 4)$ b) $(3x - 2)$

Q5 The cubic $qx^3 - 4x^2 - 7qx + 12$ has a factor of $(x - 3)$. Find the value of q.

Q6 The polynomial $f(x) = x^3 + cx^2 + dx - 2$ has factors $(x - 1)$ and $(x - 2)$. Using the Factor Theorem, find the values of c and d.

Factorising a cubic

You may be asked to **factorise** a cubic **without** being given any **factors**. If you are, the best way to find a factor of the cubic is to use trial and error:

> - First, **add up** the **coefficients** to check if $(x - 1)$ is a factor.
> - If that doesn't work, keep trying small numbers (find $f(-1)$, $f(2)$, $f(-2)$, $f(3)$, $f(-3)$ and so on) until you find a number that gives you **zero** when you put it in the **cubic**. Call that number k. $(x - k)$ is a **factor of the cubic**.

Once you have a factor, here's how to **factorise the cubic**:

> 1) Write down the **factor** you know **$(x - k)$**, and another set of brackets: $(x - k)($ $)$.
> 2) In the brackets, put the **x^2 term** needed to get the right x^3 term.
> 3) In the brackets, put in the **constant** (that when multiplied by k gives the constant in the cubic).
> 4) Put in the **x term** by comparing the number of x's on both sides.
> 5) Check there are the **same number** of x^2's on both sides.
> 6) **Factorise** the quadratic you've found — if that's possible.

Tip: If you can't factorise the quadratic, just leave it as it is.

Examples

a) Factorise $x^3 + 6x^2 + 5x - 12$.

- Check to see if the coefficients add up to 0: $1 + 6 + 5 - 12 = 0$.
- They do, so: $(x - 1)$ is a factor.
- Then factorise your cubic to get:

  Work out the term that gives x^3 when multiplied by x — this is x^2.

  The term that gives -12 when multiplied by -1 is 12.

  $(x - 1)(x^2 \qquad)$

  $(x - 1)(x^2 \qquad + 12)$

  $(x - 1)(x^2 + 7x + 12)$

  Then work out the x-term.

- And then... $(x - 1)(x + 3)(x + 4)$

Tip: Finding the middle term (nx) is the tricky bit: $(x - 1)(x^2 + nx + 12)$. These brackets give two x-terms: '$-nx$' and '$12x$'. You want these to add up to $5x$ as in the original expression, so $n = 7$. Check this works by making sure that it also gives the original $6x^2$ term.

b) $f(x) = 2x^3 - 3x^2 - 12x + 20$.
Factorise $f(x)$ and find all the solutions of $f(x) = 0$.

- Check to see if the coefficients add up to 0: $2 - 3 - 12 + 20 = 7$.
- They don't — so use trial and error for values of x, until you find a number that gives you a value of zero...

  $f(-1) = (2 \times -1) - (3 \times 1) - (12 \times -1) + 20 = 27$ ✗

  $f(2) = (2 \times 8) - (3 \times 4) - (12 \times 2) + 20 = 0$ ✓

- So: $(x - 2)$ is a factor.
- Then factorise your cubic to get: $(x - 2)(2x^2 \qquad)$

  $\Rightarrow (x - 2)(2x^2 \qquad - 10)$

  $\Rightarrow (x - 2)(2x^2 + x - 10)$

- And then... $(x - 2)(x - 2)(2x + 5) = (x - 2)^2(2x + 5)$

- So the solutions of $f(x) = 0$ are: $x = 2$ and $x = -\dfrac{5}{2}$

Tip: If you've added the coefficients and they don't add up to zero, then you don't need to do $f(1)$, as you already know that $(x - 1)$ isn't a factor.

Exercise 2.3

Q1 Factorise the following:
- a) $x^3 - 3x^2 + 2x$
- b) $2x^3 + 3x^2 - 11x - 6$
- c) $x^3 - 3x^2 + 3x - 1$
- d) $x^3 - 3x^2 + 4$
- e) $x^3 - 3x^2 - 33x + 35$
- f) $x^3 - 28x + 48$

Q2 $f(x) = x^3 - 2x^2 - 4x + 8$
- a) Factorise $f(x)$.
- b) Find the solutions of $f(x) = 0$.

Q3 Find the roots of the cubic equation $x^3 - x^2 - 3x + 3 = 0$.

Q3 Hint: "Find the roots" just means the same as "Find the solutions".

Q4 $f(x) = x^3 - px^2 + 17x - 10$, where $(x - 5)$ is a factor of $f(x)$.
- a) Find the value of p.
- b) Factorise $f(x)$.
- c) Find all the solutions to $f(x) = 0$.

Chapter 1 Algebra and Functions   9

Review Exercise — Chapter 1

Q1 Use long division to divide the cubics below.
In each case show your working, and state the quotient and remainder.
a) $x^3 - x^2 - 3x + 3$ by $(x + 3)$
b) $x^3 - 3x^2 - 5x + 6$ by $(x - 2)$
c) $x^3 + 2x^2 + 3x + 2$ by $(x + 2)$

Q2 Write the following functions f(x) in the form
$f(x) = (x + 2)g(x) + $ remainder (where $g(x)$ is a quadratic):
a) $f(x) = 3x^3 - 4x^2 - 5x - 6$
b) $f(x) = x^3 + 2x^2 - 3x + 4$
c) $f(x) = 2x^3 + 6x - 3$

Q3 Find the remainder when the following are divided by:
(i) $(x + 1)$, (ii) $(x - 1)$, (iii) $(x - 2)$
a) $f(x) = 6x^3 - x^2 - 3x - 12$
b) $f(x) = x^4 + 2x^3 - x^2 + 3x + 4$
c) $f(x) = x^5 + 2x^2 - 3$

Q4 Find the remainder when $f(x) = x^4 - 3x^3 + 7x^2 - 12x + 14$
is divided by:
a) $x + 2$ b) $2x + 4$ c) $x - 3$ d) $2x - 6$

Q5 The remainder when $x^3 + cx^2 + 17x - 10$ is divided by $(x + 3)$ is -16.
Use the Remainder Theorem to find the value of c.

Q6 Which of the following are factors of $f(x) = x^5 - 4x^4 + 3x^3 + 2x^2 - 2$?
a) $x - 1$ b) $x + 1$ c) $x - 2$ d) $2x - 2$

Q7 $f(x) = (x + 5)(x - 2)(x - 1) + k$.
If $(x + 2)$ is a factor of f(x), find the value of k.

Q8 Find the values of c and d so that $2x^4 + 3x^3 + 5x^2 + cx + d$
is exactly divisible by $(x - 2)(x + 3)$.

Q9 Given that $(x - 3)$ is a factor of the cubic $f(x) = x^3 - 9x^2 + 7x + 33$,
find the exact solutions of $f(x) = 0$.

> **Q9 Hint:** The quadratic formula (from C1) will come in handy here.

Q10 Find the roots of $f(x) = 0$ where $f(x) = x^3 + 6x^2 + 11x + 6$.

1 $f(x) = 2x^3 - 5x^2 - 4x + 3$

    a)   Find the remainder when $f(x)$ is divided by

        (i) $(x - 1)$

(2 marks)

        (ii) $(2x + 1)$

(2 marks)

    b)   Show using the Factor Theorem that $(x + 1)$ is a factor of $f(x)$.

(2 marks)

    c)   Factorise $f(x)$ completely.

(4 marks)

2 $f(x) = (4x^2 + 3x + 1)(x - p) + 5$, where p is a constant.

    a)   Find the value of $f(p)$.

(1 mark)

    b)   Find the value of p, given that when $f(x)$ is divided by $(x + 1)$, the remainder is -1.

(2 marks)

    c)   Find the remainder when $f(x)$ is divided by $(x - 1)$.

(1 mark)

3 $f(x) = 3x^3 + 8x^2 + 3x - 2$

    a)   Use the Factor Theorem to show that $(x + 2)$ is a factor of $f(x)$.

(2 marks)

    b)   Factorise $f(x)$ completely.

(4 marks)

    c)   Write down all the solutions to the equation:

$$3x^3 + 8x^2 + 3x - 2 = 0$$

(1 mark)

1. Equation of a Circle

Learning Objectives:

- Know that the equation of a circle with radius r and centre (a, b), is given by: $(x - a)^2 + (y - b)^2 = r^2$.

- Be able to find the equation of a circle, given its radius and the coordinates of its centre.

- Be able to find the radius and coordinates of the centre of a circle, given the equation of the circle.

If you've survived maths this long, you'll be very familiar with circles — you're probably a dab hand at working out their circumference and area. Now we're going to have a look at using equations to describe a circle. An equation of a circle can tell you its radius and where its centre is.

Equation of a circle with centre (0, 0)

The diagram to the right shows a circle centred on the origin $(0, 0)$ and with radius r.

You can describe a circle centred on the origin, with radius r, using the equation:

$$x^2 + y^2 = r^2$$

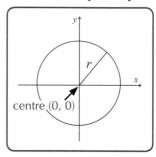

Even though you're dealing with circles, you get the equation above using **Pythagoras' theorem**. Here's how:

- The **centre** of the circle is at the origin, labelled **C**.

- **A** is any point on the circle, and has the coordinates **(x, y)**.

- **B** lies on the x-axis and has the same x-coordinate as A.

- So the length of line **CB** = x, and **AB** = y.

- Therefore, using Pythagoras' theorem to find the radius r, we get:

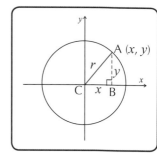

$$CB^2 + AB^2 = r^2$$

Which is... $\quad x^2 + y^2 = r^2$

If you're given the equation of a circle in the form $x^2 + y^2 = r^2$, you can work out the radius of the circle in the following way:

Example

A circle has the equation $x^2 + y^2 = 4$. Find the radius of the circle.

- **Compare** the equation $x^2 + y^2 = 4$ to $x^2 + y^2 = r^2$.
- Equating the two equations gives: $\quad r^2 = 4$

$$r = 2$$

Tip: Ignore the negative square root as the radius will be positive.

Equation of a circle with centre (a, b)

Unfortunately, circles aren't always centred at the origin. This means we need a general equation for circles that have a centre somewhere else — the point (a, b).

The general equation for circles with **radius r** and **centre (a, b)** is:

$$(x - a)^2 + (y - b)^2 = r^2$$

Notice that if the circle had a centre at $(0, 0)$, then you'd get $a = 0$ and $b = 0$, so you'd just get $x^2 + y^2 = r^2$ (the equation for a circle centred at the origin).

The example below shows how you get the equation of a circle when the coordinates of the centre and the value of the radius are given.

Example 1

Find the equation of the circle with centre (6, 4) and radius 3.

- If we draw a point P (x, y) on the **circumference** of the circle and join it to the centre $(6, 4)$, we can create a **right-angled triangle**.

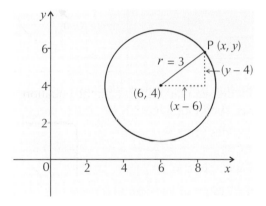

- The **sides** of this right-angled triangle are made up of the radius r (the hypotenuse), and sides of length $(x - 6)$ and $(y - 4)$
- Now let's see what happens if we use **Pythagoras' theorem**:

$$(x - 6)^2 + (y - 4)^2 = 3^2$$
$$\text{or:} \quad (x - 6)^2 + (y - 4)^2 = 9$$

This is the equation for the circle.

Tip: For the circle centred at the origin, the side lengths were r, x and y. Make sure you're happy with why the side lengths here are r, $(x - 6)$ and $(y - 4)$.

Example 2

What is the centre and radius of the circle with equation $(x - 2)^2 + (y + 3)^2 = 16$?

- **Compare** $(x - 2)^2 + (y + 3)^2 = 16$ with the general form:

$$(x - a)^2 + (y - b)^2 = r^2$$

- So, $a = 2$, $b = -3$ and $r = 4$.

  So the centre (a, b) is $(2, -3)$ and the radius (r) is 4.

- On a set of axes, the circle would look like this:

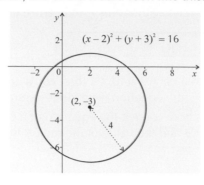

If you're given the centre and radius of a circle and you're asked to find the equation of the circle, just put the values of a, b and r into the equation $(x - a)^2 + (y - b)^2 = r^2$.

Example 3

Write down the equation of the circle with centre (–4, 2) and radius 6.

- The question says, 'Write down...', so you know you **don't** need to do any working.
- The **centre** of the circle is $(-4, 2)$, so $a = -4$ and $b = 2$.
- The **radius** is 6, so $r = 6$.
- Using the **general equation** for a circle $(x - a)^2 + (y - b)^2 = r^2$ you can write:

  $$(x + 4)^2 + (y - 2)^2 = 36$$

- On a set of axes, the circle would look like this:

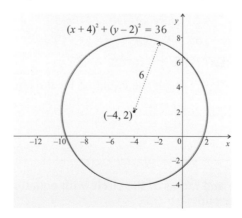

Q1 A circle has centre (0, 0) and radius 5.
Find an equation for the circle.

Q2 A circle C has radius 7 and centre (0, 0).
Find an equation for C.

Q3 Find the equation for each of the following circles:
   a) centre (2, 5), radius 3
   b) centre (–3, 2), radius 5
   c) centre (–2, –3), radius 7
   d) centre (3, 0), radius 4

> **Q3 Hint:** Be careful here with the +/ – signs.

Q4 Find the centre and radius of the circles
with the following equations:
   a) $(x - 1)^2 + (y - 5)^2 = 4$
   b) $(x - 3)^2 + (y - 5)^2 = 64$
   c) $(x - 3)^2 + (y + 2)^2 = 25$

Q5 A circle has centre (5, 3) and radius 8.
Find an equation for the circle.

Q6 A circle has centre (3, 1) and a radius of $\sqrt{31}$.
Find an equation for the circle.

Q7 The equation of the circle C is $(x - 6)^2 + (y - 4)^2 = 20$
   a) Find the coordinates of the centre of the circle.
   b) Find the radius of the circle and give
      your answer in the form $p\sqrt{5}$.

Q8 A circle has radius $\sqrt{5}$ and centre (–3, –2).
Find an equation for the circle.

Rearranging circle equations

Sometimes you'll be given an equation for a circle that doesn't look much like
$(x - a)^2 + (y - b)^2 = r^2$, for example, $x^2 + y^2 + ax + by + c = 0$.

- This means you can't immediately tell what
  the **radius** is or where the **centre** is.
- All you'll need to do is a bit of **rearranging** to get
  the equation into the familiar form.
- To do this, you'll normally have to **complete the square**.

> **Tip:** Completing the square was covered in C1. If it doesn't come flooding back to you when you look at these examples, then it might be useful to take a look back at your C1 stuff.

Example 1

The equation of a circle is $x^2 + y^2 - 6x + 4y + 4 = 0$.
Find the centre of the circle and the radius.

- We need to get the equation $x^2 + y^2 - 6x + 4y + 4 = 0$
  into the form: $(x - a)^2 + (y - b)^2 = r^2$.

- To do this, we need to **complete the square**.

- So, first **rearrange** the equation to group the x's and the y's together:

$$x^2 + y^2 - 6x + 4y + 4 = 0$$
$$x^2 - 6x + y^2 + 4y + 4 = 0$$

- Then **complete the square** for the x-terms and the y-terms.

$$x^2 - 6x + y^2 + 4y + 4 = 0$$
$$(x - 3)^2 - 9 + (y + 2)^2 - 4 + 4 = 0$$

Tip: To complete the square, write down the squared bracket that will produce the x^2 and x terms, then take off the number term that the bracket will also produce.

- Then **rearrange** to get it into the form $(x - a)^2 + (y - b)^2 = r^2$.

$$(x - 3)^2 + (y + 2)^2 = 9$$

Tip: Collect all the number terms to find r^2.

- This is the recognisable form, so we can use this equation to find:

the centre is $(3, -2)$ and the radius is $\sqrt{9} = 3$

Example 2

The equation of a circle is $x^2 + y^2 - 4x + 6y + 10 = 0$.
Find the centre of the circle and the radius.

- Get the equation $x^2 + y^2 - 4x + 6y + 10 = 0$
  into the form: $(x - a)^2 + (y - b)^2 = r^2$.

- **Rearrange** the equation to group the x's and the y's together:

$$x^2 + y^2 - 4x + 6y + 10 = 0$$
$$x^2 - 4x + y^2 + 6y + 10 = 0$$

- Then **complete the square** for the x-terms and the y-terms.

$$x^2 - 4x + y^2 + 6y + 10 = 0$$
$$(x - 2)^2 - 4 + (y + 3)^2 - 9 + 10 = 0$$

- Then **rearrange** to get it into the form $(x - a)^2 + (y - b)^2 = r^2$.

$$(x - 2)^2 + (y + 3)^2 = 3$$

- So: the centre is $(2, -3)$ and the radius is $\sqrt{3}$

Example 3

The equation of a circle is $x^2 + y^2 - 5x - 5y + 10 = 0$.
Find the centre of the circle and the radius.

- Get the equation $x^2 + y^2 - 5x - 5y + 10 = 0$
  into the form: $(x - a)^2 + (y - b)^2 = r^2$.

- **Rearrange** the equation to group the x's and the y's together:

$$x^2 + y^2 - 5x - 5y + 10 = 0$$
$$x^2 - 5x + y^2 - 5y + 10 = 0$$

- Then **complete the square** for the x-terms and the y-terms.

$$x^2 - 5x + y^2 - 5y + 10 = 0$$
$$(x - \tfrac{5}{2})^2 - \tfrac{25}{4} + (y - \tfrac{5}{2})^2 - \tfrac{25}{4} + 10 = 0$$

- Then **rearrange** to get it into the form $(x - a)^2 + (y - b)^2 = r^2$.

$$(x - \tfrac{5}{2})^2 + (y - \tfrac{5}{2})^2 = \tfrac{5}{2}$$

- So: the centre is $(\tfrac{5}{2}, \tfrac{5}{2})$ and the radius is $\sqrt{\tfrac{5}{2}}$.

Exercise 1.2

Q1 For each of the following circles find the radius
   and the coordinates of the centre.
   a) $x^2 + y^2 + 2x - 6y - 6 = 0$
   b) $x^2 + y^2 - 2y - 4 = 0$
   c) $x^2 + y^2 - 6x - 4y = 12$
   d) $x^2 + y^2 - 10x + 6y + 13 = 0$

Q2 A circle has the equation $x^2 + y^2 + 2x - 4y - 3 = 0$.
   a) Find the centre of the circle.
   b) Find the radius of the circle. Give your answer in the form $k\sqrt{2}$.

Q3 A circle has the equation $x^2 + y^2 - 3x + 1 = 0$.
   a) Find the coordinates of the centre of the circle.
   b) Find the radius of the circle.
      Simplify your answer as much as possible.

2. Circle Properties

Learning Objectives:

- Know that the angle in a triangle in a semicircle is a right angle.
- Know that the perpendicular from the centre of a circle to a chord bisects the chord.
- Know that when a radius of a circle meets a tangent to the circle at a point, the two will be perpendicular.

You'll have seen the circle rules at GCSE. But, it's important to keep them fresh in your mind as some are useful at A-Level too. You need to be able to look at a circle question and work out which of the rules apply to it.

Circle properties

Here is a reminder of some of the most useful properties of circles. Although it might not be obvious when you first look at a question, these rules could help you answer some tricky-sounding circle questions.

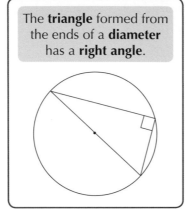

The **triangle** formed from the ends of a **diameter** has a **right angle**.

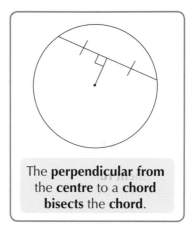

The **perpendicular from** the **centre** to a **chord** **bisects** the **chord**.

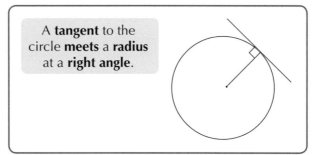

A **tangent** to the circle **meets** a **radius** at a **right angle**.

Tip: A chord is a line joining two points which lie on the circumference of a circle.

Tip: Bisecting means dividing into two equal parts.

Using circle properties

When using the rules above about perpendicularity, you've also got to remember the **gradient rule** for perpendicular lines from C1, which is:

> The gradients of perpendicular lines **multiply to give –1**.

Which means:

> Gradient of the perpendicular line =
> **–1 ÷ the gradient of the other one**.

On the next page are some examples which rely on knowing the circle rules above and the gradient rule.

Tip: Perpendicularity is a real word.

Example 1

The circle shown is centred at C. Points A and B lie on the circle.
Point B has coordinates (6, 3). The midpoint, M, of the line AB has
coordinates (4, 4). Line l passes through both C and M.

Find an equation for the line l.

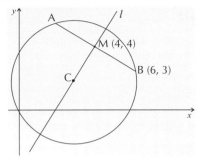

- AB is a **chord**. l goes through the circle's centre of the circle and **bisects** the chord. So we can say that the line l is **perpendicular** to the chord.

- We know two points on AB, so start by finding its **gradient**: $\dfrac{3-4}{6-4} = -\dfrac{1}{2}$

Tip: In the exam, after a question like this, they might then give you a bit more info and ask you to work out the equation for the circle.

- Use the **gradient rule** to work out the gradient of l:

$$\text{Gradient of } l = \dfrac{-1}{-\frac{1}{2}} = 2$$

- Then sub the gradient, 2, and the point on l that you know, (4, 4), into one of the equations for a **straight line** to work out the equation:

$$y - y_1 = m(x - x_1)$$
$$y - 4 = 2(x - 4)$$
$$y - 4 = 2x - 8$$
$$y = 2x - 4$$

Tip: $y - y_1 = m(x - x_1)$ is one of the ways of finding the equation of a straight line. It was covered in C1.

Example 2

Point A (6, 4) lies on a circle with the equation $x^2 + y^2 - 4x - 2y - 20 = 0$.

a) **Find the centre and radius of the circle.**

- Get the equation into the form: $(x - a)^2 + (y - b)^2 = r^2$.
  First, **rearrange** the equation to group the x's and the y's together:

$$x^2 + y^2 - 4x - 2y - 20 = 0$$
$$x^2 - 4x + y^2 - 2y - 20 = 0$$

- Then **complete the square** for the x-terms and the y-terms, and **rearrange** to get it into the form $(x - a)^2 + (y - b)^2 = r^2$.

$$(x - 2)^2 - 4 + (y - 1)^2 - 1 - 20 = 0$$
$$(x - 2)^2 + (y - 1)^2 = 25$$

- This shows: the centre is (2, 1) and the radius is 5.

b) Find the equation of the tangent to the circle at A.

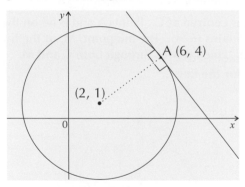

- The **tangent** is at **right angles** to the radius at (6, 4).
- Gradient of radius at (6, 4) = $\dfrac{4-1}{6-2} = \dfrac{3}{4}$

- Gradient of tangent = $\dfrac{-1}{\frac{3}{4}} = -\dfrac{4}{3}$

- Using $\quad y - y_1 = m(x - x_1)$

$$y - 4 = -\frac{4}{3}(x - 6)$$
$$3y - 12 = -4x + 24$$
$$\boxed{4x + 3y - 36 = 0}$$

Tip: The question doesn't ask for the equation in a particular form, so pick whatever's easiest.

Example 3

The points A (–2, 4), B (n, –2) and C (5, 5) all lie on the circle shown below. AB is a diameter of the circle.

Show that n = 6.

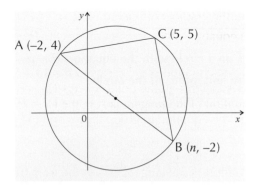

- The line AB is a **diameter** of the circle. So the angle ACB is an angle in a semicircle and must be a **right angle**.
- This means the lines AC and BC are **perpendicular** to each other.
- The **gradient rule** states that the gradients of two perpendicular lines multiply to give **–1**, so use this to work out n.

- First, find the gradient of AC: $m_1 = \dfrac{5 - 4}{5 - (-2)} = \dfrac{1}{7}$

- Then find the gradient of BC: $m_2 = \dfrac{-2 - 5}{n - 5} = \dfrac{-7}{n - 5}$

- Use the gradient rule: $m_1 \times m_2 = -1$

$$\dfrac{1}{7}\left(\dfrac{-7}{n - 5}\right) = -1 \quad \longleftarrow \boxed{\text{Cancel the 7s.}}$$

$$\dfrac{-1}{n - 5} = -1$$

$$1 = n - 5$$

$$\boxed{n = 6}$$

Exercise 2.1

Q1 The circle shown below has the equation $(x - 3)^2 + (y - 1)^2 = 10$.
The line shown is a tangent to the circle
and touches it at point A (4, 4).

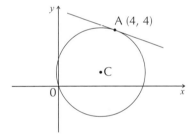

a) Find the centre of the circle, C.

b) Work out the gradient of the radius at (4, 4).

c) Find the equation of the tangent at A in the form $ax + by = c$.

Q2 A circle has the equation $(x + 1)^2 + (y - 2)^2 = 13$.
The circle passes through the point A (-3, -1).
Find the equation of the tangent at A in the form $ax + by + c = 0$.

> **Q2 Hint:** Do this in the same way as Q1. First find the centre of the circle, then the gradients of the radius and tangent.

Q3 A circle has the equation $(x - 3)^2 + (y - 4)^2 = 25$.
The circle passes through the point A (7, 1).
Find the equation of the tangent at A in the form $ax + by = c$.

Q4 The circle C has the equation $x^2 + y^2 + 2x - 7 = 0$.
Find an equation of the tangent to the circle at the point (-3, 2).

Q5 A circle has the equation $x^2 + y^2 + 2x + 4y = 5$.
The point A (0, -5) lies on the circle.
Find the tangent to the circle at A in the form $ax + by = c$.

Q6 The circle shown is centred at C. Points A and B lie on the circle. Point A has coordinates (–3, 7). The midpoint of the line AB, M, has coordinates (–1, 1). Line *l* passes through both C and M.

Q6 Hint: AB is a chord.

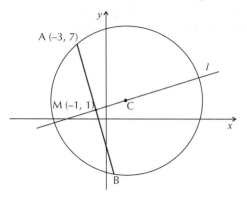

a) Use the information above to find an equation for the line *l*.

b) The coordinates of C are (2, 2).
   Find an equation for the circle.

Q7 The points A (–2, 12), B (4, 14) and C (8, 2) all lie on the circle shown below. Prove that the line AC is a **diameter** of the circle.

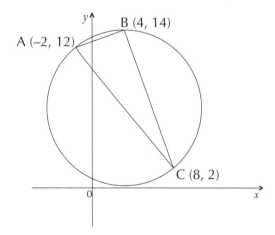

3. Arcs and Sectors

You'll be familiar with angles being measured in degrees. Radians are another unit of measurement for angles. They can be easier to use than degrees when measuring things like the arc length of a sector of a circle or its area, and they come up a lot in trigonometry and in later core modules.

Radians

A **radian** (rad) is just another unit of measurement for an angle.

> **1 radian** is the angle formed in a **sector** that has an **arc length** that is the same as the **radius**.

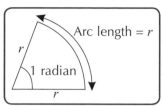

In other words, if you have a **sector** with an angle of **1 radian**, then the **length** of the **arc** will be exactly the **same length** as the **radius** r.

It's important to know how **radians relate to degrees**:
- 360 degrees (a complete circle) = 2π radians
- 180 degrees = π radians
- 1 radian is about 57 degrees

You also need to know how to **convert** between the two units. The table below shows you how:

Converting angles	
Radians to degrees:	**Degrees to radians:**
Divide by π, multiply by 180.	Divide by 180, multiply by π.

Here's a table of some of the **common angles** you're going to need, in degrees and radians:

Degrees	0	30	45	60	90	120	180	270	360
Radians	0	$\frac{\pi}{6}$	$\frac{\pi}{4}$	$\frac{\pi}{3}$	$\frac{\pi}{2}$	$\frac{2\pi}{3}$	π	$\frac{3\pi}{2}$	2π

Examples

a) Convert $\frac{\pi}{15}$ into degrees.

- To convert from radians to degrees:
- Divide by π...

$$\frac{\pi}{15} \div \pi = \frac{1}{15}$$

- ... then multiply by 180...

$$\frac{1}{15} \times 180 = \boxed{12°}$$

Learning Objectives:
- Be able to use radians as the measure for the size of an angle.
- Be able to use the formula $s = r\theta$ to work out the arc length of a sector of a circle.
- Be able to use the formula $A = \frac{1}{2}r^2\theta$ to work out the area of a sector.

Tip: Radians are sometimes shown by 'rad' after the number, or a '^{c}' (e.g. $4\pi^c$), but most of the time angles in radians are given without a symbol.

Tip: It's a good idea to learn these common angles — they come up a lot.

Tip: Notice how the angle was given without a symbol — if this happens, you can just assume it's in radians.

b) Convert 120 degrees into radians.

- You could smugly use the table on the last page to find the answer, but to show that the rule for converting from degrees to radians works...
- ... divide by 180 and then multiply by π.

$$\frac{120}{180} \times \pi = \boxed{\frac{2\pi}{3}}$$

c) Convert 297 degrees into radians.

- To convert from degrees to radians, divide by 180 and then multiply by π.

$$\frac{297}{180} \times \pi = \boxed{1.65\pi \ \text{ or } \ 5.18 \text{ rad (3 s.f.)}}$$

Tip: In the exam, you can give your answer to part c) in terms of π or as a rounded number, unless the question states otherwise. Generally though, it's better to just keep it in terms of π, and if the question asks for an exact answer you have to do this.

Exercise 3.1

Q1 Convert the angles below into radians. Give your answers in terms of π.

a) 180°   b) 135°   c) 270°

d) 70°   e) 150°   f) 75°

Q2 Convert the angles below into degrees.

a) $\frac{\pi}{4}$ b) $\frac{\pi}{2}$ c) $\frac{\pi}{3}$

d) $\frac{5\pi}{2}$ e) $\frac{3\pi}{4}$ f) $\frac{7\pi}{3}$

Arc length and sector area

A **sector** is part of a circle formed by **two radii** and part of the **circumference**. The **arc** of a sector is the **curved** edge of the sector. You can work out the **length** of the arc, or the **area** of the sector — as long as you know the **angle** at the **centre** (θ) and the **length** of the **radius** (r). When working out arc length and sector area you **always** work in radians.

Arc length

For a circle with **radius r**, a sector with **angle θ** (measured in **radians**) has **arc length s,** given by:

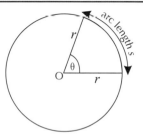

- If you put $\theta = 2\pi$ in this formula (and so make the sector equal to the whole circle), you find that the distance all the way round the outside of the circle is $s = 2\pi r$.
- This is just the normal **circumference** formula.

Sector area

For a circle with **radius r**, a sector with **angle θ** (measured in **radians**) has **area A**, given by:

$$A = \tfrac{1}{2}r^2\theta$$

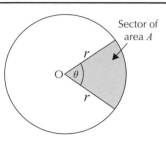

Sector of area A

- Again, if you put $\theta = 2\pi$ in the formula, you find that the area of the whole circle is $A = \tfrac{1}{2}r^2 \times 2\pi = \pi r^2$.
- This is just the normal '**area of a circle**' formula.

Example 1

Find the exact length L and area A in the diagram to the right.

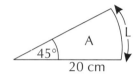

45°

A

L

20 cm

- It's asking for an arc length and sector area, so you need the angle in **radians**.

$$45° = \frac{45 \times \pi}{180} = \frac{\pi}{4} \text{ radians}$$

- Now put everything in your formulas:

$$L = r\theta = 20 \times \frac{\pi}{4} = 5\pi \text{ cm}$$

$$A = \frac{1}{2}r^2\theta = \frac{1}{2} \times 20^2 \times \frac{\pi}{4} = 50\pi \text{ cm}^2$$

Tip: Or you could just quote this if you've learnt the stuff on page 23 off by heart.

Example 2

Find the area of the shaded part of the symbol.

- Each 'leaf' has area:

$$\frac{1}{2} \times 20^2 \times \frac{\pi}{4} = 50\pi \text{ cm}^2$$

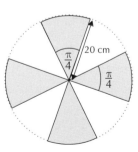

20 cm

$\frac{\pi}{4}$

$\frac{\pi}{4}$

- So the area of the whole symbol =

$$4 \times 50\pi = 200\pi \text{ cm}^2$$

Tip: Instead, you could use the total angle of all the shaded sectors (π).

Example 3

The sector shown has an area of 6π cm².
Find the arc length, s.

- First, get the angle in **radians**:

$$60° = \frac{60 \times \pi}{180} = \frac{\pi}{3} \text{ radians}$$

- You want to find the arc length, but you don't know the sector's radius.
- But we know its area, so use the area formula to **work out** the **radius**:

$$A = \frac{1}{2}r^2\theta \;\; \Rightarrow \;\; 6\pi = \frac{1}{2} \times r^2 \times \frac{\pi}{3}$$
$$36 = r^2$$
$$r = 6$$

- Putting this value of r into the equation for arc length gives us:

$$s = r\theta = 6 \times \frac{\pi}{3} = \boxed{2\pi \text{ cm}}$$

Tip: This is one of the common angles from the table on page 23 — it's definitely worth learning them.

Exercise 3.2

Q1 The diagram below shows a sector OAB. The centre is at O and the radius is 6 cm. The angle AOB is 2 radians.

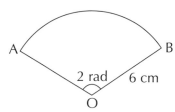

Find the arc length and area of this sector.

Q2 The diagram below shows a sector OAB. The centre is at O and the radius is 8 cm. The angle AOB is 46°.

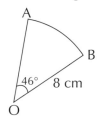

Find the arc length and area of this sector to 1 d.p.

Q2 Hint: Remember to convert 46° to radians first.

Q3   The diagram below shows a sector of a circle with a centre O
      and radius r cm. The angle AOB shown is θ.

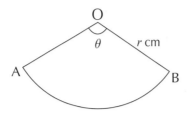

      For each of the following values of θ and r, give the arc length and
      the area of the sector. Where appropriate give your answers to 3 s.f.
      a) $\theta = 1.2$ radians, $r = 5$ cm
      b) $\theta = 0.6$ radians, $r = 4$ cm
      c) $\theta = 80°$, $r = 9$ cm
      d) $\theta = \frac{5\pi}{12}$, $r = 4$ cm

Q4   The diagram below shows a sector ABC of a circle, where the angle
      BAC is 0.9 radians.

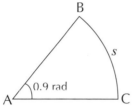

      Given that the area of the sector is 16.2 cm², find the arc length s.

Q5   A circle C has a radius of length 3 cm with centre O.
      A sector of this circle is given by angle AOB which is 20°.

      Find the length of the arc AB and the area of the sector.
      Give your answer in terms of π.

Q6   The sector shown has an arc length of 7 cm. The angle BAC
      is 1.4 rad. Find the area of the sector.

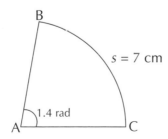

Review Exercise — Chapter 2

Q1 Give the radius and the coordinates of the centre of the circles with the following equations:

a) $x^2 + y^2 = 9$

b) $(x - 2)^2 + (y + 4)^2 = 4$

c) $x(x + 6) = y(8 - y)$

Q2 Find the equations for the circles with the following properties:

a) centre (3, 2), radius 6

b) centre (−4, −8), radius 8

c) centre (0, −3), radius $\sqrt{14}$

Q3 A circle has the equation $x^2 + y^2 - 4x + 6y - 68 = 0$.
Find the coordinates of the centre of the circle and its radius.

Q4 The circle C has equation $(x - 2)^2 + (y - 1)^2 = 100$.
The point A (10, 7) lies on the circle.
Find an equation of the tangent at A.

Q5 The circle C has equation $x^2 + y^2 - 12x + 2y + 11 = 0$.
The point A (1, −2) lies on the circle.
Find an equation of the tangent at A.

Q6 The diagram below shows a sector ABC of a circle, with centre A and a radius of 10 cm. The angle BAC is 0.7 radians.

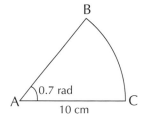

Find the arc length BC and area of this sector.

Q7 The sector ABC below is part of a circle, where the angle BAC is 50°.

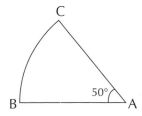

Given that the area of the sector is 20π cm², find the arc length BC.
Give your answer in terms of π.

Exam-Style Questions — Chapter 2

1   C is a circle with the equation: $x^2 + y^2 - 2x - 10y + 21 = 0$.

     a)   Find the centre and radius of C.

                (5 marks)

     The line joining $P(3, 6)$ and $Q(q, 4)$ is a diameter of C.

     b)   Show that $q = -1$.

                (3 marks)

     c)   Find the equation of the tangent to C at Q, giving your answer in the form
         $ax + by + c = 0$, where a, b and c are integers.

                (5 marks)

2   A circle C is shown here.
     M is the centre of C, and J and K both lie on C.

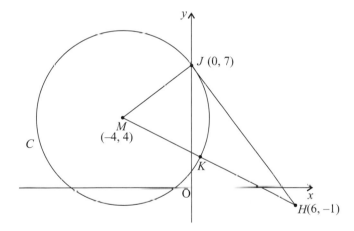

     a)   Find the equation of C in the form $(x - a)^2 + (y - b)^2 = r^2$.

                (3 marks)

     The line JH is a tangent to circle C at point J.

     b)   Show that angle $JMH = 1.1071$ radians to 4 d.p.

                (4 marks)

     c)   Find the length of the shortest arc on C between J and K,
         giving your answer to 3 s.f.

                (2 marks)

3   The diagram below shows the dimensions of a child's wooden toy.
     The toy is a prism with height 10 cm.

     Its cross-section is a sector of a circle with radius 20 cm and angle $\frac{\pi}{4}$ radians.

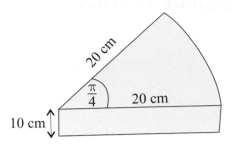

a)   Show that the volume of the toy, $V = 500\pi$ cm³.

(3 marks)

b)   Show that the surface area of the toy, $S = (150\pi + 400)$ cm².

(5 marks)

4   The diagram shows a circle C, with centre P. M is the midpoint of AB, a chord.

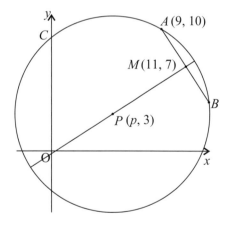

a)   Show that $p = 5$.

(5 marks)

b)   Find the equation of circle C.

(4 marks)

1. The Sine and Cosine Rules

In this section you'll see how SOHCAHTOA and the sine and cosine rules can be used to find the length of each side and the size of each angle in a triangle, as well as its area. You might have seen some of this at GCSE, so parts of this chapter will just be a recap.

Trig values from triangles

You need to know the values of **sin**, **cos** and **tan** at 30°, 60° and 45°, and there are two **triangles** that can help you remember them. It may seem like a long-winded way of doing it, but once you know how to do it, you'll always be able to work them out — even without a calculator.

The idea is you draw the triangles below, putting in their angles and side lengths. Then you can use them to work out special trig values like **sin 45°** or **cos 60°** with exact values instead of the decimals given by calculators.

First, make sure you can remember SOHCAHTOA:

$$\sin = \frac{\text{opp}}{\text{hyp}} \qquad \cos = \frac{\text{adj}}{\text{hyp}} \qquad \tan = \frac{\text{opp}}{\text{adj}}$$

Learning Objectives:

- Be able to find the values of sin, cos and tan of 30°, 60° and 45° (and the equivalent angles in radians) without a calculator.

- Know the sine and cosine rules and be able to use them on any triangle.

- Be able to work out the area of a triangle using the formula $\frac{1}{2}ab\sin C$.

These are the two triangles that you'll use:

Half an equilateral triangle with sides of length 2:

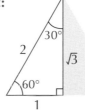

- You can work out the height using Pythagoras' theorem: height $= \sqrt{2^2 - 1^2} = \sqrt{3}$.

- Then you can use the triangle to work out sin, cos and tan of 30° and 60°.

$\sin 30° = \frac{1}{2}$	$\cos 30° = \frac{\sqrt{3}}{2}$	$\tan 30° = \frac{1}{\sqrt{3}}$
$\sin 60° = \frac{\sqrt{3}}{2}$	$\cos 60° = \frac{1}{2}$	$\tan 60° = \sqrt{3}$

Tip: If you're working in radians (see page 23), you just need to replace the angle in degrees with the equivalent angle in radians.
So $\sin \frac{\pi}{6} = \sin 30° = \frac{1}{2}$.

Tip: Make sure you're confident with these values as they'll come up a lot in C2-C4.

Right-angled triangle with two sides of length 1:

- You can work out the hypotenuse using Pythagoras' theorem: hypotenuse $= \sqrt{1^2 + 1^2} = \sqrt{2}$.

- Then you can use the triangle to work out sin, cos and tan of 45°.

$\sin 45° = \frac{1}{\sqrt{2}}$	$\cos 45° = \frac{1}{\sqrt{2}}$	$\tan 45° = 1$

The sine and cosine rules

There are three useful formulas you need to know for working out information about a triangle. There are **two** for finding the **angles** and **sides** (called the **sine rule** and **cosine rule**) and one for finding the **area**.

Tip: These rules work on any triangle, not just right-angled triangles.

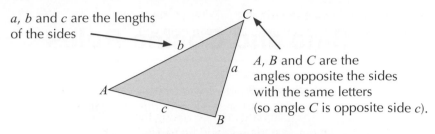

a, *b* and *c* are the lengths of the sides

A, *B* and *C* are the angles opposite the sides with the same letters (so angle *C* is opposite side *c*).

Tip: You can use any two bits of the sine rule to make a normal equation with just one = sign. The sine rule also works if you flip all the fractions upside down: $\frac{\sin A}{a} = \frac{\sin B}{b} = \frac{\sin C}{c}$.

The Sine Rule

$$\frac{a}{\sin A} = \frac{b}{\sin B} = \frac{c}{\sin C}$$

The Cosine Rule

$$a^2 = b^2 + c^2 - 2bc\cos A$$

Area of any triangle

$$\text{Area} = \frac{1}{2}ab\sin C$$

Tip: You can also rearrange the cosine rule into the form
$$\cos A = \frac{b^2 + c^2 - a^2}{2bc}$$
to find an angle.

To decide which rule to use, look at what you know about the triangle:

You can use the **sine rule** if:

- You know **any two angles** and a **side**.

Tip: Remember, if you know two angles you can work out the third by subtracting them from 180°.

You can **sometimes** use the **sine rule** if:

- You know **two sides** and an **angle that isn't between them**.

This doesn't always work though — sometimes there are **2 possible** triangles:

You can use the **cosine rule** if:

- You know **all three** sides... · ...or you know **two sides** and the **angle** that's **between** them.

Example 1

Find the missing sides and angles for $\triangle ABC$, in which $A = 40°$, $a = 27$ m and $B = 73°$. Then find the area of the triangle.

- Start by **sketching** the triangle. It doesn't have to be particularly accurate, but it can help you decide which rule(s) you need.

- Here you have 2 angles and a side, so you can use the **sine rule**.

- First, though, start by finding angle C (using the fact that the angles in a triangle add up to 180°).

$$\angle C = 180° - 73° - 40° = 67°$$

- Now use the sine rule to find the other sides one at a time:

$$\frac{a}{\sin A} = \frac{b}{\sin B} \Rightarrow \frac{27}{\sin 40°} = \frac{b}{\sin 73°} \qquad \frac{c}{\sin C} = \frac{a}{\sin A} \Rightarrow \frac{c}{\sin 67°} = \frac{27}{\sin 40°}$$

$$\Rightarrow b = \frac{27 \times \sin 73°}{\sin 40°} \qquad\qquad \Rightarrow c = \frac{27 \times \sin 67°}{\sin 40°}$$

$$= 40.2 \text{ m } (1 \text{ d.p.}) \qquad\qquad = 38.7 \text{ m } (1 \text{ d.p.})$$

- Now you've found the missing values, you can find the area using the formula:

$$\text{Area of } \triangle ABC = \frac{1}{2} ab \sin C$$

$$= \frac{1}{2} \times 27 \times 40.169... \times \sin 67°$$

$$= 499.2 \text{ m}^2 \ (1 \text{ d.p.})$$

Tip: Use the unrounded value for b here (rather than just 40.2).

Example 2

Find the values of X, Y and Z.

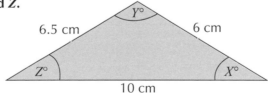

Tip: One of the hardest things about using the sine and cosine rules is matching the sides and angles given in the question to the ones in the formula. If it helps, label your triangle with A, B, C, a, b and c each time.

- You've been given all three sides but none of the angles, so start by using the **cosine rule** to find angle Z (you'll have to rearrange the formula a bit first).

$$a^2 = b^2 + c^2 - 2bc\cos A \Rightarrow \cos A = \frac{b^2 + c^2 - a^2}{2bc}$$

$$\Rightarrow \cos Z = \frac{10^2 + 6.5^2 - 6^2}{(2 \times 10 \times 6.5)}$$

$$\Rightarrow \cos Z = 0.817...$$

$$\Rightarrow Z = 35.2° \ (1 \text{ d.p.})$$

Tip: Just use the $\cos^{-1}$ button on your calculator to work out the value of Z from $\cos Z$.

- Use the cosine rule **again** to find the value of another angle. It doesn't matter which one you go for (using Y here).

$$a^2 = b^2 + c^2 - 2bc\cos A$$

$$\Rightarrow \cos Y = \frac{6^2 + 6.5^2 - 10^2}{2 \times 6 \times 6.5}$$

$$\Rightarrow \cos Y = -0.278...$$

$$\Rightarrow Y = 106.2° \text{ (1 d.p.)}$$

- Now that you have two of the angles, you can find the other by subtracting them from 180°:

$$X = 180° - 35.2° - 106.2°$$

$$= 38.6° \text{ (1 d.p.)}$$

Tip: You could try using the sine rule here, but it would give you a value of 73.8° for Y — and you can see from the diagram that angle Y is obtuse. So here you'd have to subtract 73.8° from 180° to get the actual value of Y — there's more on this later in the chapter.

- You can find the areas of more **complicated** shapes by turning them into **multiple triangles** stuck together, then using the **sine** and **cosine rules** on each individual triangle.

- This method can be used for working out angles and sides in real-life problems, such as calculating distances travelled or areas covered. Sometimes you'll see a problem that uses **bearings**.

Example 3

Rasmus the trawlerman has cast his nets between buoys in the North Sea (shown on the diagram below).

a) Find the area of sea his nets cover to 2 s.f.

Tip: Rasmus's boat is called the Sea Beast.

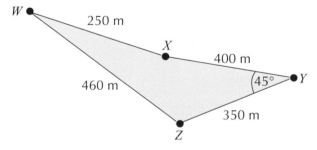

- You can start by finding the distance between X and Z (let's call it y) — this will split the area into 2 triangles.
  Do this by treating XYZ as a triangle and using the **cosine rule**:

$$a^2 = b^2 + c^2 - 2bc\cos A$$

$$\Rightarrow y^2 = 400^2 + 350^2 - 2(400)(350)\cos 45°$$

$$\Rightarrow y^2 = 84510.1...$$

$$\Rightarrow y = 290.7 \text{ m} \text{ (1 d.p.)}$$

- Now you have all three sides for the left-hand triangle, so you can find an angle (let's say W) with the **cosine rule**.

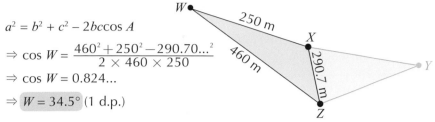

$a^2 = b^2 + c^2 - 2bc\cos A$

$\Rightarrow \cos W = \dfrac{460^2 + 250^2 - 290.70...^2}{2 \times 460 \times 250}$

$\Rightarrow \cos W = 0.824...$

$\Rightarrow \boxed{W = 34.5°}$ (1 d.p.)

Tip: You could have found the area of the right-hand triangle at the start as you had all the info you needed.

- Now you have enough information to find the **area** of each triangle with the formula on page 32.

  Left-hand triangle:

  $\text{Area} = \frac{1}{2}ab\sin C$

  $= \frac{1}{2} \times 250 \times 460 \times \sin 34.48...°$

  $= \boxed{32\,600\,\text{m}^2}$ (3 s.f.)

  Right-hand triangle:

  $\text{Area} = \frac{1}{2}ab\sin C$

  $= \frac{1}{2} \times 400 \times 350 \times \sin 45°$

  $= \boxed{49\,500\,\text{m}^2}$ (3 s.f.)

- So the total area of sea covered is:

  $32\,600 + 49\,500 = \boxed{82\,000\,\text{m}^2}$ (2 s.f.)

b) If X is on a bearing of 100° from W, on what bearing does Rasmus have to sail to get from X to Y (to 3 s.f.)?

- To find the bearing, find all the other angles round X and then subtract them from 360°. The unknown angle marked below is $180° - 100° = 80°$.

Tip: This comes from the rules of parallel lines — you did this at GCSE.

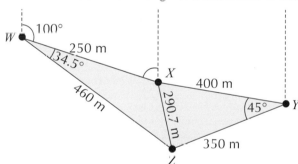

- To find the angle $\angle WXZ$, use the cosine rule on the left-hand triangle:

  $a^2 = b^2 + c^2 - 2bc\cos A \quad \Rightarrow \quad \cos \angle WXZ = \dfrac{250^2 + 290.70...^2 - 460^2}{2 \times 250 \times 290.70...}$

  $\Rightarrow \quad \angle WXZ = \cos^{-1} -0.444...$

  $\Rightarrow \quad \angle WXZ = 116.38° \,(2\,\text{d.p.})$

- Then do the same for angle $\angle YXZ$, using the right-hand triangle:

  $a^2 = b^2 + c^2 - 2bc\cos A \quad \Rightarrow \quad \cos \angle YXZ = \dfrac{400^2 + 290.70...^2 - 350^2}{2 \times 400 \times 290.70...}$

  $\Rightarrow \quad \cos \angle YXZ = 0.524...$

  $\Rightarrow \quad \angle YXZ = 58.36° \,(2\,\text{d.p.})$

- Now just subtract all the angles from 360° to find the bearing Rasmus should sail on to get from X to Y:

  $360° - 80° - 116.38° - 58.36° = 105.26° = \boxed{105°}$ (3 s.f.)

Give all answers to 3 significant figures unless otherwise stated.

Q1   Use the cosine rule to find
the length *QR*.

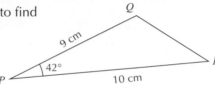

Q2   Use the sine rule to find
the length *TW*.

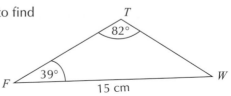

Q3   Find the length *AC*.

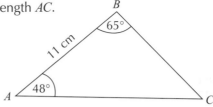

Q4   Find the size of angle *D*.

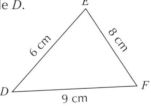

Q5   In triangle *JKL*: *JL* = 24 cm, *KL* = 29 cm and angle *L* = 62°.
Find the length *JK*.

Q6 Hint: Make sure
you set your calculator
to radians.

Q6   In triangle *GHI*: *HI* = 8.3 cm, *GH* = 6.4 cm and angle *H* = 2.3 rad.
Find the length *GI*.

Q7   In triangle *BCD*: *BC* = 14 cm, *CD* = 11 cm and *BD* = 23 cm.
Find the angle *C* (in degrees) to 1 d.p.

Q8   In triangle *PQR*: *PR* = 48 m, angle *P* = 0.66 rad and
angle *R* = 0.75 rad. Find the length *PQ*.

Q9 Hint: Sketching the
triangle first will make
it easier to see which
angle you need to find.
The smallest angle is
opposite the shortest
side, so it's easy to
identify.

Q9   In triangle *DEF*: *DE* = 8 cm, *EF* = 11 cm and *DF* = 16 cm.
Find the smallest angle (in degrees).

Q10   Find the area of this triangle.

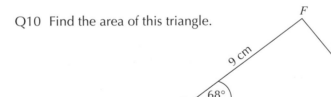

Q11 Find the area of this triangle.

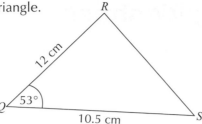

Q12 Find the area of this triangle to 2 d.p.

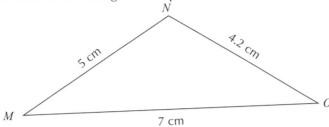

Q13 Two points, A and B, are both at sea level and on opposite sides of a mountain. The distance between them is 5 km. From A, the angle of elevation of the top of the mountain (M) is 21°, and from B, the angle of elevation is 17°.

a) Find the distance BM.

b) Hence find the height of the mountain to the nearest metre.

Q14 A ship sails 8 km on a bearing of 070° and then changes direction to sail 10 km on a bearing of 030°.

a) Draw a diagram to represent the situation.

b) What's the ship's distance from its starting position?

c) On what bearing must it now sail to get back to its starting position?

> **Q14 Hint:** Bearings are measured clockwise from the vertical (North).

Q15 a) Find the length AC.

b) Hence find the area of the quadrilateral $ABCD$.

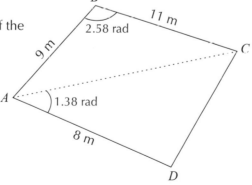

Q16 A children's pirate hat is to be made from 4 identical triangular cardboard faces with sides of 12 cm, 15 cm and 18 cm. What's the total area of cardboard needed?

2. Trig Identities

Learning Objectives:

- Know the trig identities $\tan x \equiv \dfrac{\sin x}{\cos x}$ and $\sin^2 x + \cos^2 x \equiv 1$.

- Be able to use trig identities to prove other relations.

- Be able to use trig identities to find the values of an expression.

Trig identities are really useful for simplifying expressions — they can make equations easier to solve by replacing one term with another.

Trig identities

There are **two trig identities** you need to know in C2. They're really useful in trigonometry and come up time and time again in C3 and C4, so make sure you learn them.

$$\tan x \equiv \frac{\sin x}{\cos x} \qquad \sin^2 x + \cos^2 x \equiv 1$$

Tip: The '$\equiv$' means that the relation is true for all values of x. An identity is just a relation which contains a '$\equiv$' sign.

- The second identity can also be rearranged into $\sin^2 x \equiv 1 - \cos^2 x$ or $\cos^2 x \equiv 1 - \sin^2 x$.

- You can use them to **prove** that two expressions are equivalent.

Example 1

Show that $\dfrac{\cos^2 \theta}{1 + \sin \theta} \equiv 1 - \sin \theta$.

- A good way of doing this kind of question is to play around with **one side** of the equation until it's the same as the other side.

- Start with the left-hand side: $\quad \dfrac{\cos^2 \theta}{1 + \sin \theta}$

- There are usually "clues" that you can pick up to help you decide what to do next. You know there's a trig identity for $\cos^2 \theta$, so start by **replacing** it in the fraction:

$$\frac{\cos^2 \theta}{1 + \sin \theta} \equiv \frac{1 - \sin^2 \theta}{1 + \sin \theta}$$

- The next step isn't quite as obvious, but if you look at the top of the fraction it might remind you of something — it's a **difference of two squares**.

$$\frac{1 - \sin^2 \theta}{1 + \sin \theta} \equiv \frac{(1 + \sin \theta)(1 - \sin \theta)}{1 + \sin \theta}$$

Tip: The replacements aren't always easy to spot — look out for things like differences of two squares, or 1's that can be replaced by $\sin^2 x + \cos^2 x$.

- Now you can just cancel the $1 + \sin \theta$ from the top and bottom of the fraction, and you get the answer you were looking for:

$$\frac{(1 + \sin \theta)(1 - \sin \theta)}{1 + \sin \theta} \equiv \boxed{1 - \sin \theta} \quad \longleftarrow \text{This is the right-hand side.}$$

Example 2

Show that $\cos \theta \tan \theta + 4 \sin \theta \equiv 5 \sin \theta$.

- Again, start with the left-hand side and see if there are any **replacements** you can easily make. Try putting in the trig identity for $\tan \theta$.

$$\cos \theta \tan \theta + 4 \sin \theta \equiv \cos \theta \frac{\sin \theta}{\cos \theta} + 4 \sin \theta$$

Tip: Always keep in mind what you're aiming for — here you want your answer in terms of just $\sin \theta$.

- Now you have a term with $\cos\theta$ on the top and bottom. **Cancel** these out and see what you're left with.

$$\cos\theta\,\frac{\sin\theta}{\cos\theta} + 4\sin\theta \equiv \sin\theta + 4\sin\theta$$

- Now it's just a case of adding the terms together, and you've arrived at the right-hand side.

$$\sin\theta + 4\sin\theta \equiv \boxed{5\sin\theta}$$

Example 3

Find $\sin\theta$ if $\cos\theta = \frac{2}{3}$, given that θ is an acute angle.

- The identity to use here is $\quad\sin^2\theta + \cos^2\theta \equiv 1$

- Rearranging this gives $\quad\sin^2\theta \equiv 1 - \cos^2\theta$

- Then put in the value of $\cos\theta$ in the question and square root each side:

$$\sin\theta = \sqrt{1 - \left(\frac{2}{3}\right)^2} = \sqrt{\frac{5}{9}} = \frac{\sqrt{5}}{3}$$

Tip: You're told that θ is acute here, which means that $\sin\theta$ is positive (this is covered on the next two pages), so you can ignore the negative square root.

Exercise 2.1

Q1 Use the identity $\tan\theta \equiv \dfrac{\sin\theta}{\cos\theta}$ to show that $\dfrac{\sin\theta}{\tan\theta} - \cos\theta \equiv 0$.

Q2 Use the identity $\sin^2\theta + \cos^2\theta \equiv 1$ to show that $\cos^2\theta \equiv (1 - \sin\theta)(1 + \sin\theta)$.

Q3 Given that x is acute, find $\cos x$ if $\sin x = \frac{1}{2}$.

Q3, 5 Hint: You're told that x is acute, so just ignore the negative roots.

Q4 Show that $4\sin^2 x - 3\cos x + 1 \equiv 5 - 3\cos x - 4\cos^2 x$.

Q5 Given that x is acute, find $\tan x$ if $\sin^2 x = \frac{3}{4}$.

Q4 Hint: If you're not told which identity to use, just play around with the ones you know until something works.

Q6 Show that $(\tan x + 1)(\tan x - 1) \equiv \dfrac{1}{\cos^2 x} - 2$.

Q7 Show that $(\sin\theta + \cos\theta)^2 + (\sin\theta - \cos\theta)^2 \equiv 2$.

Q8 Show that $\tan x + \dfrac{1}{\tan x} \equiv \dfrac{1}{\sin x \cos x}$.

Q8 Hint: $\dfrac{1}{\tan x} \equiv \dfrac{\cos x}{\sin x}$

Q9 Show that $4 + \sin x - 6\cos^2 x \equiv (2\sin x - 1)(3\sin x + 2)$.

Q10 Show that $\sin^2 x\cos^2 y - \cos^2 x\sin^2 y \equiv \cos^2 y - \cos^2 x$.

3. Trig Functions

Learning Objectives:

- Be able to sketch the graphs of sin x, cos x and tan x.

- Be able to sketch the common transformations of the graphs of sin x, cos x and tan x.

Being able to sketch the graphs of trig functions and their transformations is really useful — it'll come in handy later in the chapter when you have to solve equations within a given interval.

Graphs of trig functions

You should be able to draw the graphs of **sin x**, **cos x** and **tan x** without looking them up — including all the important points, like where the graphs cross the **axes** and their **maximum** and **minimum** points.

---

sin x

- The graph of $y = \sin x$ is **periodic** — it repeats itself every 360° (or 2π radians). So $\sin x = \sin (x + 360°) = \sin (x + 720°)$ $= \sin (x + 360°n)$, where n is an integer.

- It bounces between $y = -1$ and 1, and it can **never** have a value outside this range.

- It goes through the **origin** (as $\sin 0° = 0$) and then crosses the x-axis every **180°**.

- $\sin (-x) = -\sin x$. The graph has **rotational symmetry** around the origin, so you could rotate it 180° about (0, 0) and it would look the same.

- The graph of sin x looks like this:

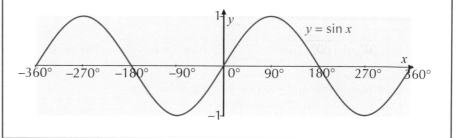

---

cos x

- The graph of $y = \cos x$ is also **periodic** with period 360° (or 2π radians). $\cos x = \cos (x + 360°) = \cos (x + 720°) = \cos (x + 360°n)$, where n is an integer.

- It also bounces between $y = -1$ and 1, and it can **never** have a value outside this range.

- It crosses the y-axis at $\boldsymbol{y = 1}$ (as $\cos 0° = 1$) and the x-axis at ±90°, ±270° etc.

- $\cos (-x) = \cos x$. The graph is **symmetrical** about the **y-axis**, so you could reflect it in the y-axis and it would look the same.

Tip: You say that sin x has a period of 360°.

Tip: Make sure you know the coordinates of the key points in radians as well.

Tip: The graphs of sin x and cos x are the same shape but shifted 90° along the x-axis. This makes them easier to remember, but make sure you don't get them mixed up.

- The graph of cos x looks like this:

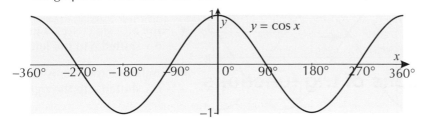

tan x

- The graph of $y = \tan x$ is also **periodic**, but this time it repeats itself every 180° (or π radians). So $\tan x = \tan(x + 180°) = \tan(x + 180°n)$, where n is an integer.

- It takes values between $-\infty$ and ∞ in each **180° interval**.

- It goes through the **origin** (as $\tan 0° = 0$).

- It's **undefined** at $\pm90°$, $\pm270°$, $\pm450°$...
  — at these points it **jumps** from ∞ to $-\infty$ or vice versa.

- The graph of $\tan x$ looks like this:

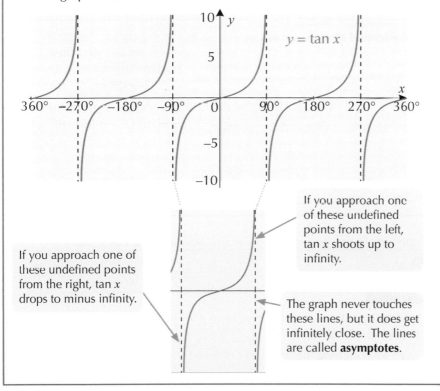

If you approach one of these undefined points from the right, tan x drops to minus infinity.

If you approach one of these undefined points from the left, tan x shoots up to infinity.

The graph never touches these lines, but it does get infinitely close. The lines are called **asymptotes**.

Tip: The best way to learn these functions is to practise sketching them and marking on the key points.

Tip: $y = \tan x$ is undefined at these points because you're dividing by zero. Remember from page 38 that $\tan x = \frac{\sin x}{\cos x}$, so when $\cos x = 0$, $\tan x$ is undefined, and $\cos x = 0$ when $x = 90°$, $270°$ etc. (see previous page).

Transformations

You came across different types of **transformations** in C1. A **translation** is a horizontal or vertical **shift** that doesn't change the shape of the graph. A **stretch** is exactly what it says — a horizontal or vertical **stretch** (or **squash**) of the graph. At C2, you'll need to be able to apply these types of transformation to **trig functions**.

Tip: For $y = \sin(x + c)$, the graph has a maximum where $x + c = 90°$, $450°$, etc. — i.e. when $x = 90° - c$, $450° - c$, ... It has a minimum where $x + c = 270°$, $630°$ ... and crosses the x-axis at $x + c = 0°$, $180°$, $360°$ ...

1. A translation along the x-axis: $y = \sin(x + c)$

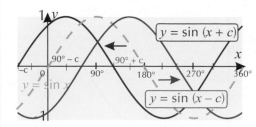

- For $c > 0$, $\sin(x + c)$ is just $\sin x$ **shifted c to the left**.

- Similarly, $\sin(x - c)$ is just $\sin x$ **shifted c to the right**.

Tip: In the diagram, n is about 2 for the blue curve and about 0.5 for the orange curve.

Tip: When $0 < n < 1$ it looks like a squash, but it's still called a stretch.

2. A vertical stretch: $y = n \sin x$

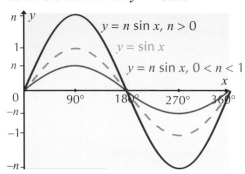

- For $y = n \sin x$, the graph of $y = \sin x$ is **stretched vertically** by a factor of n.

- If $n > 1$, the graph gets taller, and if $0 < n < 1$, the graph gets flatter.

- And if $n < 0$, the graph is also **reflected** in the x-axis.

Tip: Make sure you know which way the stretch goes. The larger n is, the more squashed the graph becomes. In the diagram, n is 3 for the blue curve and 0.5 for the orange curve.

3. A horizontal stretch: $y = \sin nx$

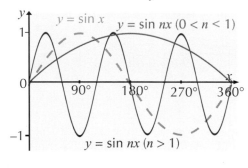

- For $y = \sin nx$, the graph of $y = \sin x$ is **stretched horizontally** by a factor of $\frac{1}{n}$.

- If $0 < n < 1$, the graph of $y = \sin x$ is **stretched horizontally outwards**, and if $n > 1$ the graph of $y = \sin x$ is **squashed inwards**.

- If $n < 0$, the graph is also **reflected** in the y-axis.

The same transformations will apply to the graphs of $y = \cos x$ and $y = \tan x$ as well.

Example 1

On the same axes, sketch the graphs of $y = \cos x$ and $y = -2 \cos x$ in the range $-2\pi \leq x \leq 2\pi$.

- Start by **sketching** the graph of $\cos x$ — have a look back at pages 40-41 if you need a reminder of how to do this.

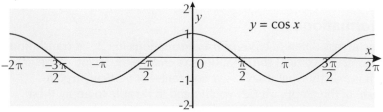

Tip: Don't worry that the angles are in radians here — the transformations work in the same way.

- Next, think about what the **transformed** graph will look like. It's in the form $y = n \cos x$, so it will be **stretched vertically**.

- $n = -2$, so it will be stretched by a factor of **2**. As n is negative, it will also be **reflected** in the x-axis. This is all the information you need to be able to sketch the graph.

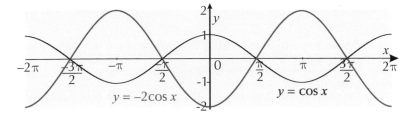

Example 2

On the same axes, sketch the graphs of $y = \tan x$ and $y = \tan 2x$ in the interval $-180° \le x \le 180°$.

- Again, start by sketching the graph of $y = \tan x$ (see p.41).

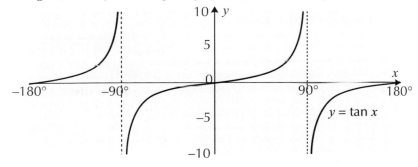

- This time it's in the form $y = \tan nx$, so it will be stretched horizontally. $n > 1$, so the graph will be stretched by a factor of $\frac{1}{2}$, which is the same as a squash by a factor of 2.

- To make it easier, draw dotted lines for the new asymptotes (divide the x-values of the old ones by 2) then draw the tan shape between them.

Tip: You'll have double the number of repetitions of the tan shape in the same interval because you've halved the period.

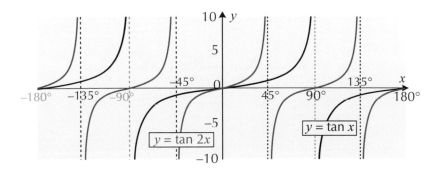

Q1 On the same set of axes, sketch the graphs of $y = \cos x$ and $y = \cos (x + 90°)$ in the interval $-180° \le x \le 180°$.

Q2 On the same set of axes, sketch the graphs of $y = \sin x$ and $y = \frac{1}{3}\sin x$ in the interval $-\pi \le x \le \pi$.

Q3 On the same set of axes, sketch the graphs of $y = \sin x$ and $y = \sin 3x$ in the interval $0° \le x \le 360°$.

Q4 On the same set of axes, sketch the graphs of $y = \cos x$ and $y = -\cos x$ in the interval $0 \le x \le 2\pi$.

Q5 c) Hint: Just look at what's happened to the graph, then think about which type of transformation is needed to achieve it.

Q5 a) Sketch the graph of $f(x) = \tan x$ in the interval $-90° \le x \le 270°$.
  b) Translate this graph 90° to the left and sketch it on the same set of axes as part a).
  c) Write down the equation of the transformed graph.

Q6 a) Sketch the graph of $y = \sin x$ in the interval $-2\pi \le x \le 2\pi$.
  b) Stretch the graph horizontally by a factor of 2 and sketch it on the same set of axes as part a).
  c) Write down the equation of the transformed graph.

Q7 The diagram shows the graph of $y = \sin x$ and a transformed graph.

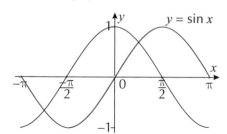

  a) Describe the transformation.
  b) Write down the equation of the transformed graph.

Q8 The diagram shows the graph of $y = \cos x$ and a transformed graph.

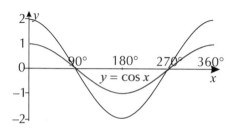

  a) Describe the transformation.
  b) Write down the equation of the transformed graph.

4. Solving Trig Equations

Once you know how to sketch the graphs of trig functions, you can use them to solve trig equations. Solving a trig equation just means finding the value (or values) of x that satisfies the given equation.

Learning Objectives:

- Be able to solve trig equations by sketching a graph.
- Be able to solve trig equations using a CAST diagram.
- Be able to solve trig equations of the form $\sin kx = n$ and $\sin (x + c) = n$.
- Be able to solve trig equations using trig identities.

Sketching a graph

To solve trig equations in a **given interval** you can use one of two methods. The first is drawing a **graph** of the function and reading solutions off the graph. You'll often find that there's **more than one** solution to the equation — in every **360° interval**, there are usually **two** solutions to an equation, and if the interval is bigger (see example 2 below), there'll be even more solutions.

Example 1

Solve $\cos x = \frac{1}{2}$ for $0° \leq x \leq 360°$.

- Start by using your **calculator** to work out the first value.
  For $\cos x = \frac{1}{2}$, $x = 60°$.

Tip: This is actually one of the common trig angles from page 31.

- Then **sketch a graph** of $\cos x$ in the interval you're interested in, and draw a horizontal line across for $y = \frac{1}{2}$. The points where the line and curve **meet** are the solutions of the equation.

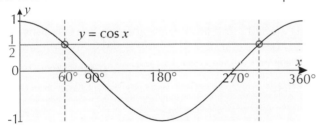

- Each 360° interval of the graph is symmetrical, so the second solution will be the same distance from 360° as the first is from 0°. You know one solution is 60°, so the other solution is $360° - 60° = 300°$.

Tip: If you are dealing with sin or tan instead, use the graphs to work out the solutions in a similar way.

- You now know all of the solutions in this interval: $x = 60°, 300°$

If you had an interval that was **larger** than **one repetition** of the graph (i.e. 360° or 2π for sin and cos, and 180° or π for tan), you'd just add or subtract **multiples** of 360° (for sin and cos) or 180° (for tan) onto the solutions you've found until you have **all** the solutions in the **given interval**.

Example 2

Solve $\sin x = -0.3$ for $0 \leq x \leq 4\pi$. Give your answers to 3 s.f.

- Again, use your **calculator** to work out the first value. For $\sin x = -0.3$, $x = -0.3046...$ However, this is **outside** the given interval for x, so **add on** 2π to find the first solution in the interval: $-0.3046... + 2\pi = 5.978...$

- Now **sketch a graph** of $\sin x$ in the interval you're interested in, and draw a horizontal line across at $y = -0.3$. This time, you'll need to draw the graph between $x = 0$ and $x = 4\pi$, so there'll be **2 repetitions** of the sin wave.

Tip: You add on 2π because the curve repeats every 2π radians. If you were working in degrees, you'd add on 360° instead (for tan, it would be π or 180°).

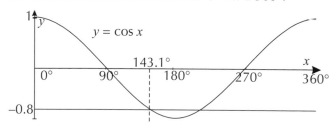

- You can see from the graph that there are **4 solutions** in the given interval (because the horizontal line crosses the curve 4 times). You've already found the one at $x = 5.978...$ — to find the next one, you use the **symmetry** of the graph. This first solution is $2\pi - 5.978... = 0.3046...$ away from 2π. Now, looking at the graph, the other solution between 0 and 2π will be $0.3046...$ away from π — i.e. $\pi + 0.3046... = 3.446...$

- For the other solutions (the ones between 2π and 4π), just **add 2π** onto the values you've already found: $3.4462... + 2\pi = 9.729...$ and $5.978... + 2\pi = 12.261...$

- So the solutions to $\sin x = -0.3$ for $0 \leq x \leq 4\pi$, to 3 s.f., are:

$$x = 3.45, 5.98, 9.73 \text{ and } 12.3.$$

Tip: If the interval was bigger, just keep on adding (or subtracting) lots of 2π until you have all the answers within the required interval.

Exercise 4.1

Q1 By sketching a graph, find all the solutions to the equations below in the interval $0° \leq x \leq 360°$. Give your answers to 1 decimal place.

a) $\sin x = 0.75$ b) $\cos x = 0.31$ c) $\tan x = -1.5$

d) $\sin x = -0.42$ e) $\cos x = -0.56$ f) $\tan x = -0.67$

Q2 By sketching a graph, find all the solutions to the equations below in the interval $0 \leq x \leq 2\pi$. Give your answers as exact values.

a) $\cos x = \dfrac{1}{\sqrt{2}}$ b) $\tan x = \sqrt{3}$ c) $\sin x = \dfrac{1}{2}$

d) $\tan x = \dfrac{1}{\sqrt{3}}$ e) $\tan x = 1$ f) $\cos x = \dfrac{\sqrt{3}}{2}$

Q2 Hint: All of these values relate to the common angles on p.31.

Q3 One solution of $\cos x = -0.8$ is $143.1°$ (1 d.p.). Use the graph below to find all the solutions in the interval $0° \leq x \leq 360°$.

Q4 Find all the solutions of the equation $\tan x = 2.5$ in the interval $0° \leq x \leq 1080°$. Give your answers to 1 decimal place.

Q4-5 Hint: Be careful with the intervals here — and remember that tan repeats every 180°.

Q5 Find all the solutions of the equation $\sin x = 0.81$ in the interval $-2\pi \leq x \leq 2\pi$. Give your answers to 3 significant figures.

Chapter 3 Trigonometry

Using a CAST diagram

The second way of finding the solutions to a trig equation is by using a **CAST diagram**. CAST stands for Cos, All, Sin, Tan, and it shows you where each of these functions is **positive** by splitting a 360° period into **quadrants**.

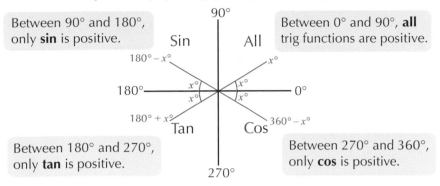

Between 90° and 180°, only **sin** is positive.

Between 0° and 90°, **all** trig functions are positive.

Between 180° and 270°, only **tan** is positive.

Between 270° and 360°, only **cos** is positive.

Tip: You can use a CAST diagram if you're working in radians too — the axes are just labelled with 0, $\frac{\pi}{2}$, π and $\frac{3\pi}{2}$ radians instead.

- To use a CAST diagram, you need to use your **calculator** to find the first solution of the trig function (or, if it's a common angle, you might just be able to recognise it).

- You then make the **same angle** from the **horizontal** in each of the four quadrants (shown in the diagram above), then measure each angle **anticlockwise** from 0°. So if the first solution was 45°, the solution in the sin quadrant would be 135° (180° − 45°) measured anticlockwise from 0°, and so on.

- **Ignore** the ones that give a **negative** result (unless the given value is negative — in which case you want the two quadrants in which the trig function is **negative**).

Tip: The angle you put into the CAST diagram should be acute (i.e. between 0° and 90°) — if you get a negative value, just put the positive value into the diagram. Or you can measure **clockwise** from 0° for negative angles.

Example 1

Find all the solutions of $\sin x = \frac{1}{2}$ for $0° \leq x \leq 360°$.

- Use a **calculator** to find the **first solution** (30°), and put this in your CAST diagram.

- Then add the **same angle** to each **quadrant**, measuring from the **horizontal** in each case.

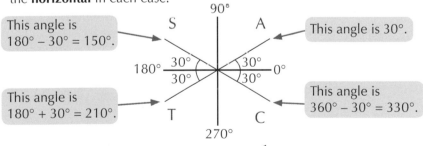

This angle is 180° − 30° = 150°.

This angle is 30°.

This angle is 180° + 30° = 210°.

This angle is 360° − 30° = 330°.

Tip: This value is also a common angle — you don't need to use a calculator if you can remember it.

Tip: Remember, each line you mark on the diagram represents the angle being measured anticlockwise from 0°.

- You need a **positive** value of sin x (because $\frac{1}{2}$ is positive), so you're only interested in the quadrants where sin x is positive — i.e. the first and second quadrants. There are two solutions: 30° and 150°.

- For values **outside** the interval $0° \leq x \leq 360°$, just find solutions between 0° and 360° and then **add** or **subtract multiples of 360°** to find solutions in the correct interval (there's an example of this on the next page). If the interval is in **radians**, add or subtract multiples of **2π** instead.

Example 2

Find all the solutions of tan x = –4 for 0 ≤ x ≤ 4π.
Give your answers to 3 s.f.

Tip: Don't be put off by the fact that this example uses radians — the method is just the same.

- Using a **calculator**, you'll find that the first solution is $x = -1.33$ (3 s.f.). **Ignore** the negative and just put the value 1.33 into the CAST diagram.

- Add the **same angle** to each **quadrant**, measuring from the **horizontal** in each case. Remember that you're working in **radians** here.

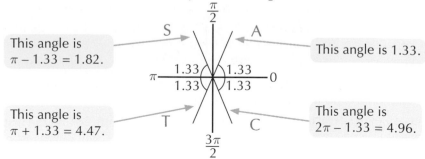

This angle is $\pi - 1.33 = 1.82$.

This angle is 1.33.

This angle is $\pi + 1.33 = 4.47$.

This angle is $2\pi - 1.33 = 4.96$.

Tip: Adding or subtracting 2π (or 360°) is the same as going round a full circle on the CAST diagram (so you end up back at the same position).

- You need a **negative** value of tan x (as –4 is negative), so just look at the quadrants where tan x is negative (i.e. the 'S' and 'C' quadrants). There are two solutions: 1.82 and 4.96 (as the first value was **negative**, you could have measure **clockwise** from 0 to find the solution 4.96).

- To find the solutions between 2π and 4π, just **add** 2π to the solutions you've already found.

- The solutions of tan x = –4 for $0 \le x \le 4\pi$ are
  $x = 1.82, 4.96, 8.10$ and 11.2 (3 s.f.).

Exercise 4.2

Q1 One solution of sin x = 0.45 is x = 26.7° (1 d.p.). Use a CAST diagram to find all the solutions in the interval $0° \le x \le 360°$.

Q2 Use a CAST diagram to find the solutions of the following equations in the interval $0° \le x \le 360°$. Give your answers to 1 d.p.

Q2 Hint: You might have to rearrange some of the equations first.

a) cos x = 0.8   b) tan x = 2.7   c) sin x = –0.15

d) tan x = 0.3   e) tan x = –0.6   f) sin x = –0.29

g) 4sin x – 1 = 0   h) 4cos x – 3 = 0   i) 5tan x + 7 = 0

Q3 Use a CAST diagram to find all the solutions to tan x = –8.4 in the interval $0 \le x \le 2\pi$. Give your answers to 3 s.f.

Q4 Use a CAST diagram to find all the solutions to sin x = 0.75 in the interval $0° \le x \le 720°$. Give your answers to 1 d.p.

Q5 Use a CAST diagram to find all the solutions to cos x = 0.31 in the interval $-180° \le x \le 180°$. Give your answers to 1 d.p.

Q6 Use a CAST diagram to find all the solutions to sin x = 0.82 in the interval $0 \le x \le 4\pi$. Give your answers to 3 s.f.

Changing the interval

Sometimes you'll have to solve equations of the form **sin kx = n** or
sin (x + c) = n (where n, k and c are numbers). In these situations it's usually
easiest to **change the interval** you're solving for, then **solve as normal** for kx
or $x + c$. You'll then need to remember to get the final solutions either by
dividing by k or **subtracting c** at the end.

Solving equations of the form sin kx = n

Let's start by looking at how to solve equations of the form **sin kx = n**.

- First, **multiply** the **interval** you're looking for solutions in by k. E.g. for the
  equation sin 2x = n in the interval $0 \leq x \leq 2\pi$, you'd look for solutions in the
  interval $0 \leq 2x \leq 4\pi$. Then **solve** the equation over the new interval.

- However, this gives you solutions for kx, so you then need to **divide** each
  solution by k to find the values of x.

You can either **sketch the graph** over the new interval (this will show you **how
many** solutions there are) or you can use the **CAST method** to find solutions
between 0° and 360° then add on multiples of 360° until you have all the
solutions in the new interval — use whichever method you prefer.

Example 1

Solve cos 4x = 0.6 for 0° ≤ x ≤ 360°. Give your answers to 1 d.p.

- First, **change** the **interval**. The interval is $0° \leq x \leq 360°$, and the value
  of k is 4, so **multiply** the whole interval by **4**: $0° \leq 4x \leq 1440°$.

- Then **solve** the equation to find the solutions for **4x**. I'm going to use a
  CAST diagram, but you could sketch a graph if you prefer.

- Find the **first solution** using a calculator: cos 4x = 0.6
  $$\Rightarrow 4x = 53.13° \text{ (2 d.p.)}$$

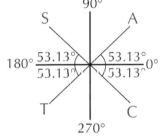

  You want the quadrants where cos is
  positive, so the other solution between
  0° and 360° is:
  $$4x = 360° - 53.13° = 306.87° \text{ (2 d.p.)}$$

- Now **add on** multiples of 360° to find **all** the solutions in the interval
  $0° \leq 4x \leq 1440°$ (to 2 d.p.):

  53.13°, 306.87°, 413.13°, 666.87°, 773.13°, 1026.87°, 1133.13°, 1386.87°.

- Remember, these are solutions for **4x**. To find the solutions for x,
  divide through by **4**. So the solutions to cos 4x = 0.6 in the interval
  $0° \leq x \leq 360°$ (to 1 d.p.) are:

  13.3°, 76.7°, 103.3°, 166.7°, 193.3°, 256.7°, 283.3°, 346.7°.

- It's a good idea to **check** your answers — just put your values of x into
  cos 4x and check that they give you 0.6. You can make sure you've got
  the **right number** of solutions too — there are 2 solutions to cos x = 0.6
  in the interval $0° \leq x \leq 360°$, so there'll be 8 solutions to cos 4x = 0.6
  in the same interval.

Tip: If you'd sketched
a graph, you could just
see how many times the
graph and line crossed
— there's an example of
this on the next page.

Example 2

Solve $\sin 3x = -\dfrac{1}{\sqrt{2}}$ for $0° \leq x \leq 360°$.

- This time you've got **$3x$** instead of x, which means the **interval** you need to find solutions in is $0° \leq 3x \leq 1080°$. So sketch the graph of $y = \sin x$ between 0° and 1080°.

Tip: This is actually one of the common angles from p.31 — so you could have found it without using a calculator.

- Use your **calculator** to find the **first solution**. You'll get $3x = -45°$, but this is outside the interval for $3x$, so use the pattern of the graph to find a solution in the interval. As the sin curve **repeats every 360°**, there'll be a solution at $-45° + 360° = 315°$.

- Now you can use the **symmetry** of the graph to find the other solution between 0° and 360° — the graph is symmetrical in each interval of 180°, so the other solution is at $180° + 45° = 225°$.

- You know the graph repeats every 360°, so **add on** lots of 360° to the answers you've just found to find the other solutions between 0° and 1080°: $3x = 585°, 675°, 945°, 1035°$.

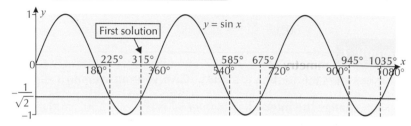

Tip: Again, you can check your answers using a calculator — and make sure they're in the right interval.

- Now you have 6 solutions for $3x$, so **divide them all by 3** to get the solutions for x: $x = 75°, 105°, 195°, 225°, 315°, 345°$.

Example 3

Find all the solutions of $\tan 2x = 1.4$ for $0 \leq x \leq 2\pi$. Give your answers to 3 s.f.

- Here, the interval needs to be multiplied by 2: $0 \leq 2x \leq 4\pi$.

- Use a calculator to work out the first solution: $2x = 0.9505$ (4 s.f.).

- Then use a CAST diagram to find the other solution between 0 and 2π:

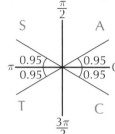

  You want the quadrants where tan is **positive**, so the other solution between 0 and 2π is: $2x = \pi + 0.9505 = 4.092$ (4 s.f.).

- To find the solutions between 2π and 4π, just **add 2π** to the solutions you've already found: $2x = 7.234, 10.38$ (4 s.f.).

- Finally, to find the solutions for x between 0 and 2π, just divide by 2: $x = 0.475, 2.05, 3.62, 5.19$ (3 s.f.)

Solving equations of the form sin (x + c) = n

The method for solving equations of the form **sin (x + c) = n** is similar — but instead of multiplying the interval, you have to add or subtract the value of c.

- **Add** (or **subtract**) the value of c to the **whole interval** — so the interval $0° \leq x \leq 360°$ becomes $c \leq x + c \leq 360° + c$ (you add c onto each bit of the interval).

- Now **solve** the equation over the **new interval** — again, you can either sketch a graph or use a CAST diagram.

- Finally, **subtract** (or **add**) c from your solutions to give the values for x.

Example 1

Solve cos (x + 60°) = $\frac{3}{4}$ for −360° ≤ x ≤ 360°, giving your answers to 1 d.p.

- You've got cos (x + 60°) instead of cos x — so **add 60°** to each bit of the interval. The new interval is: $-300° \leq x + 60° \leq 420°$.

- Use your **calculator** to get the **first solution**:
$$\cos (x + 60°) = \frac{3}{4} \quad \Rightarrow \quad x + 60° = 41.4° \text{ (1 d.p.)}$$

- Use the **symmetry** of the graph to find the other solution between 0° and 360°. The first solution is 41.4° away from 0°, so the next solution will be 41.4° away from 360° — i.e. at $360° - 41.4° = 318.6°$ (1 d.p.).

- The cos graph **repeats** every 360°, so to find the other solutions, **add and subtract 360°** from the answers you've just found, making sure they're still within the interval you want:
$x + 60° = 401.4°, 678.6°$ (not in interval), $-41.4°, -318.6°$ (not in interval).

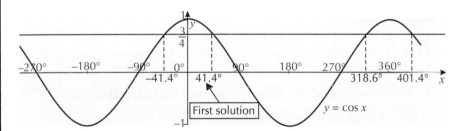

- These solutions are for cos (x + 60°) so you need to **subtract 60° from each value** to find the solutions for x (to 1 d.p.):
$x = -101.4°, -18.6°, 258.6°$ and $341.4°$

- So there are **4 solutions**, and they're all in the required interval (−360° ≤ x ≤ 360°). Again, you can **check** your answers by putting them back into cos (x + 60°) and making sure you get $\frac{3}{4}$.

Tip: Be careful with the interval here — the original interval isn't $0° \leq x \leq 360°$.

Tip: You can see from the graph that there should be 4 solutions in this interval, but be careful — if your graph wasn't accurate, it would be easy to miss a solution.

Example 2

Solve tan $(x - 75°) = 2$ for $0° \leq x \leq 360°$. Give your answers to 1 d.p.

- First, find the new interval. This time you want to **subtract 75°** from each bit of the interval, so the new interval is $-75° \leq x - 75° \leq 285°$. You'll need to sketch the graph of tan x over this interval.

- Use your **calculator** to find the **first solution**:
  tan $(x - 75°) = 2 \Rightarrow$ $x - 75° = 63.4°$ (1 d.p.)

Tip: To see if there are any other solutions within the interval, add and subtract 180° to the values you've just found: $x - 75° = -116.6°$, 423.4°. Both of these values are outside the required interval, so there are only 2 solutions.

- Now you can use the **pattern** of the graph to find the other solution in the interval — the tan graph **repeats** every **180°**, so add on 180° to the solution you've already found: $63.4° + 180° = 243.4°$ (1 d.p.)

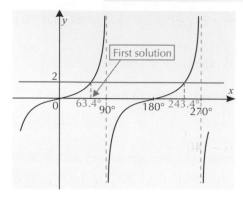

- Finally, **add on 75°** to find the solutions in the interval $0° \leq x \leq 360°$ (to 1 d.p.): $x = 138.4°, 318.4°$.

Example 3

Solve sin $(x + \frac{\pi}{3}) = -\frac{\sqrt{3}}{2}$ for $0 \leq x \leq 4\pi$.

- This time, **add $\frac{\pi}{3}$** to each bit of the interval: $\frac{\pi}{3} \leq x + \frac{\pi}{3} \leq \frac{13\pi}{3}$

- $\frac{\sqrt{3}}{2}$ is a **common trig value** — from p.31 you know that: sin $\frac{\pi}{3} = \frac{\sqrt{3}}{2}$, so put $\frac{\pi}{3}$ into the **CAST diagram**:

Tip: Have a look back at the example on p.48 to see how to deal with negative values.

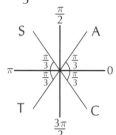

As you're finding solutions for $-\frac{\sqrt{3}}{2}$, you want the quadrants where sin is **negative** (the 3rd and 4th quadrants), so the solutions between 0 and 2π are:

$x + \frac{\pi}{3} = \pi + \frac{\pi}{3} = \frac{4\pi}{3}$ and

$x + \frac{\pi}{3} = 2\pi - \frac{\pi}{3} = \frac{5\pi}{3}$

- To find the other solutions (between 2π and 4π), just **add 2π** to the solutions you've already found: $x + \frac{\pi}{3} = \frac{10\pi}{3}, \frac{11\pi}{3}$

- Finally, **subtract $\frac{\pi}{3}$** from each solution to find the values of x:

$x = \pi, \frac{4\pi}{3}, 3\pi, \frac{10\pi}{3}$

In this exercise, give all answers in degrees to 1 d.p.
and all answers in radians to 3 s.f.

Q1 Solve $\sin 2x = 0.6$ in the interval $0° \leq x \leq 360°$.

Q2 Solve $\tan 4x = 4.6$ in the interval $0° \leq x \leq 360°$.

Q3 Solve $\cos 3x = -0.24$ in the interval $0° \leq x \leq 360°$.

Q4 Find all the solutions to $\cos 2x = 0.72$ in the interval $0 \leq x \leq 2\pi$.

Q5 Find all the solutions to $\sin 3x = -0.91$ in the interval $0 \leq x \leq 2\pi$.

Q6 Solve $\tan \frac{x}{2} = 2.1$ in the interval $0° \leq x \leq 360°$.

Q6 Hint: Don't let the ½ throw you — it's exactly the same method as before, except you divide the interval by 2 instead of multiplying.

Q7 Find all the solutions to $\cos (x - 27°) = 0.64$ in the interval $0° \leq x \leq 360°$.

Q8 Solve $\tan (x - 140°) = -0.76$ in the interval $0° \leq x \leq 360°$.

Q9 Find all the solutions to $\sin (x + 36°) = 0.45$ in the interval $0° \leq x \leq 360°$

Q10 Find all the solutions to $\tan (x + 73°) = 1.84$ in the interval $0° \leq x \leq 360°$.

Q11 Find all the solutions to $\sin (x - \frac{\pi}{4}) = -0.25$ in the interval $0 \leq x \leq 2\pi$.

Q12 Solve $\cos (x + \frac{\pi}{8}) = 0.13$ in the interval $0 \leq x \leq 2\pi$.

Using trig identities to solve equations

Sometimes you'll be asked to find solutions to an equation that has
a **tan** term as well as a sin or cos term in it.
In these situations you might need to use the **trig identity** for tan x (p.38):

$$\tan x \equiv \frac{\sin x}{\cos x}$$

This will **eliminate** the tan term, and you'll be left with an equation
just in terms of sin or cos.

Example

Solve $3 \sin x - \tan x = 0$ for $0 \leq x \leq 2\pi$.

- The equation has both $\sin x$ and $\tan x$ in it, so writing $\tan x$ as $\frac{\sin x}{\cos x}$ would be a good place to start.

$$3 \sin x - \tan x = 0 \quad \Rightarrow \quad 3 \sin x - \frac{\sin x}{\cos x} = 0$$

Tip: Don't make the mistake of cancelling $\sin x$ from the equation instead of factorising — you'll lose the solutions to $\sin x = 0$ if you do.

- Next get rid of the $\cos x$ on the bottom by multiplying the whole equation by $\cos x$.

$$3 \sin x - \frac{\sin x}{\cos x} = 0 \quad \Rightarrow \quad 3 \sin x \cos x - \sin x = 0$$

- You now have a common factor of $\sin x$ so factorise:

$$3 \sin x \cos x - \sin x = 0 \quad \Rightarrow \quad \sin x (3 \cos x - 1) = 0$$

- Now you've got two things multiplying together to make zero. That means **one** of them must be equal to zero.

$$\sin x (3 \cos x - 1) = 0 \quad \Rightarrow \quad \boxed{\sin x = 0 \text{ or } 3 \cos x - 1 = 0}$$

Tip: Remember $\sin x = 0$ every π radians. Have a look back at the graph on p.40 to see why.

- If $\sin x = 0$, then the solutions in the required interval are $x = 0$, π and 2π.

- If $3 \cos x - 1 = 0$, start by rearranging into $\cos x = \frac{1}{3}$. Use your calculator to find the first solution ($x = 1.2309...$).

- You can then use a CAST diagram to find the other solutions:

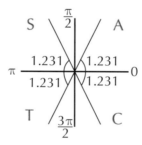

- You need a positive solution for $\cos x$, so look at the 1st and 4th quadrants: 1.2309... and $2\pi - 1.2309... = 5.0522...$ radians.

- You now have all the solutions:
  $x = 0$, 1.231 (4 s.f.), π, 5.052 (4 s.f.) and 2π radians

Similarly, if you have a **$\sin^2 x$** or **$\cos^2 x$**, you can use the other identity from p.38 to rewrite one trig function in terms of the other.

$$\boxed{\sin^2 x + \cos^2 x \equiv 1}$$

Example

Solve $2\sin^2 x + 5\cos x = 4$ for $0° \le x \le 360°$.

- You can't do much while the equation's got both $\sin x$ and $\cos x$ in it. So **replace** the $\sin^2 x$ with $1 - \cos^2 x$:

  $$2\sin^2 x + 5\cos x = 4 \quad \Rightarrow \quad 2(1 - \cos^2 x) + 5\cos x = 4$$

  Tip: Now $\cos x$ is the only trig function you need to deal with.

- **Multiply out** the bracket and **rearrange** it so that you've got zero on one side — you get a **quadratic** in $\cos x$:

  $$2(1 - \cos^2 x) + 5\cos x = 4 \quad \Rightarrow \quad 2\cos^2 x - 5\cos x + 2 = 0$$

- It's easier to **factorise** the quadratic if you make the **substitution** $y = \cos x$:

  $$2y^2 - 5y + 2 = 0 \Rightarrow (2y - 1)(y - 2) = 0 \Rightarrow (2\cos x - 1)(\cos x - 2) = 0$$

- One of the brackets must be 0. So you get 2 equations:

  $$2\cos x - 1 = 0 \Rightarrow \cos x = \frac{1}{2} \text{ or } \cos x - 2 = 0 \Rightarrow \cos x = 2$$

- $\cos x$ is always between -1 and 1, so $\cos x = 2$ has **no solutions**. This means the only solutions are those for $\cos x = \frac{1}{2}$: $x = 60°$ and $300°$.

  Tip: The solutions to $\cos x = \frac{1}{2}$ were found on page 45.

Exercise 4.4

In this exercise, give all non-exact answers in degrees to 1 d.p. and all non-exact answers in radians to 3 s.f.

Q1 Solve each of the following equations for values of x in the interval $0° \le x \le 360°$:

a) $(\tan x - 5)(3\sin x - 1) = 0$
b) $5\sin x \tan x - 4\tan x = 0$
c) $\tan^2 x - 9$
d) $4\cos^3 x = 3\cos x$
e) $3\sin x = 5\cos x$
f) $5\tan^2 x - 2\tan x = 0$
g) $6\cos^2 x - \cos x - 2 = 0$
h) $7\sin x + 3\cos x = 0$

Q1 Hint: You might not need to use any trig identities to answer some of these. Think about whether or not each equation can be factorised.

Q2 Find the solutions to each of the following equations in the given interval:

a) $\tan x = \sin x \cos x$ $0 \le x \le 2\pi$
b) $5\cos^2 x - 9\sin x = 3$ $-2\pi \le x \le 4\pi$
c) $2\sin^2 x + \sin x - 1 = 0$ $-2\pi \le x \le 2\pi$

Q3 a) Show that the equation $4\sin^2 x = 3 - 3\cos x$ can be written as $4\cos^2 x - 3\cos x - 1 = 0$.

b) Hence solve the equation $4\sin^2 x = 3 - 3\cos x$ in the interval $0 \le x \le 2\pi$.

Q4 Find all the solutions of the equation $9\sin^2 2x + 3\cos 2x = 7$ in the interval $0° \le x \le 360°$.

Q4 Hint: Watch out — there are 8 solutions for this one.

Review Exercise — Chapter 3

Q1   Write down the exact values of cos 30°, sin 30°, tan 30°, cos 45°, sin 45°, tan 45°, cos 60°, sin 60° and tan 60°.

Q2   For triangle $\triangle ABC$, in which $A = 30°$, $C = 25°$ and $b = 6$ m:

    a) Find all the sides and angles of the triangle.

    b) Find the area of the triangle.

Q3   For triangle $\triangle PQR$, in which $p = 13$ km, $q = 23$ km and $R = 20°$:

    a) Find all the sides and angles of the triangle.

    b) Find the area of the triangle.

Q4   Find all the angles in the triangle below, in degrees to 1 d.p.

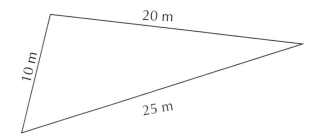

Q5   Find the missing sides and angles for the 2 possible triangles $\triangle ABC$ which satisfy $b = 5$, $a = 3$, $A = 35°$.

> **Q5 Hint:** This is tricky — sketch it first and try to think how you could make 2 triangles from the numbers given.

Q6   Show that $\tan x - \sin x \cos x \equiv \sin^2 x \tan x$.

Q7   Show that $\tan^2 x - \cos^2 x + 1 \equiv \tan^2 x(1 + \cos^2 x)$.

Q8   Simplify: $(\sin y + \cos y)^2 + (\cos y - \sin y)^2$.

Q9   Show that $\dfrac{\sin^4 x + \sin^2 x \cos^2 x}{\cos^2 x - 1} \equiv -1$.

Q10 Sketch the following graphs in the interval $-360° \le x \le 360°$, making sure you label all of the key points.

    a) $y = \cos x$ b) $y = \sin x$ c) $y = \tan x$

Q11 Below is the graph of $y = \cos x$ and a transformation of the graph.
What is the equation of the transformed graph?

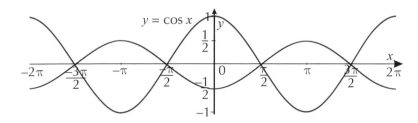

Q12 Below is a graph of $y = \sin x$ and a transformation of the graph.
What is the equation of the transformed graph?

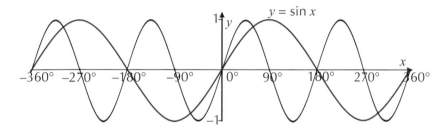

Q13 Sketch the following pairs of graphs on the same axes:

a) $y = \cos x$ and $y = \frac{1}{2}\cos x$ (for $0° \leq x \leq 360°$)

b) $y = \sin x$ and $y = \sin(x + 30°)$ (for $0° \leq x \leq 360°$)

c) $y = \tan x$ and $y = \tan 3x$ (for $0° \leq x \leq 180°$)

Q14 a) Solve each of these equations for $0° \leq \theta \leq 360°$:

(i) $\sin \theta = \frac{\sqrt{3}}{2}$ (ii) $\tan \theta = -1$ (iii) $\cos \theta = -\frac{1}{\sqrt{2}}$

b) Solve each of these equations for $-180° \leq \theta \leq 180°$ (giving your answers to 1 d.p.):

(i) $\cos 4\theta = -\frac{2}{3}$ (ii) $\sin(\theta + 35°) = 0.3$ (iii) $\tan \frac{\theta}{2} = 500$

Q15 Find all the solutions to $6\sin^2 x = \cos x + 5$ in the interval $0 \leq x \leq 2\pi$ (exact answers or to 3 s.f.).

Q16 Solve $3\tan x + 2\cos x = 0$ for $-90° \leq x \leq 90°$.

Q17 Find all the solutions of the equation $6\sin^2 x + \sin x - 1 = 0$ in the interval $0 \leq x \leq 2\pi$, giving your answers to 3 s.f. where appropriate.

Q18 Find all the solutions of the equation $\tan x - 3\sin x = 0$ in the interval $0° \leq x \leq 720°$, giving your answers to 1 d.p.

Exam-Style Questions — Chapter 3

1 The diagram shows the locations of two walkers, X and Y, after walking in different directions from the same start position.

   X walked due south for 150 m.
   Y walked 250 m on a bearing of 100°.

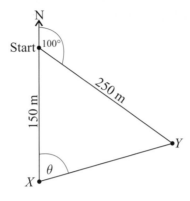

   a) Calculate the distance between the two walkers, in m to the nearest m.

   (2 marks)

   b) Show that $\dfrac{\sin\theta}{\sin 80°} = 0.93$ to 2 decimal places.

   (2 marks)

2 Find all the values of x, in the interval $0 \le x \le 2\pi$, for which:

   $$2 - \sin x = 2\cos^2 x$$

   Give your answers in terms of π.

   (6 marks)

3 Solve the following equations, for $-\pi \le x \le \pi$:

   a) $(1 + 2\cos x)(3\tan^2 x - 1) = 0$

   (6 marks)

   b) $3\cos^2 x = \sin^2 x$

   (4 marks)

4   For an angle x, $3 \cos x = 2 \sin x$.

a)   Find $\tan x$.

(2 marks)

b)   Hence, or otherwise, find all the values of x in the interval $0 \leq x \leq 360°$
for which $3 \cos x = 2 \sin x$, giving your answers to 1 d.p.

(2 marks)

5   a)   (i)   Sketch, for $0 \leq x \leq 360°$, the graph of $y = \cos (x + 60°)$.

(2 marks)

(ii)   Write down all the values of x, for $0 \leq x \leq 360°$, where $\cos (x + 60°) = 0$.

(2 marks)

b)   Sketch, for $0 \leq x \leq 180°$, the graph of $y = \sin 4x$.

(2 marks)

c)   Solve, for $0 \leq x \leq 180°$, the equation: $\sin 4x = 0.5$,
giving your answers in degrees.

(4 marks)

6   Solve, for $0 \leq x \leq 2\pi$:

a) $\tan \left(x + \dfrac{\pi}{6} \right) = \sqrt{3}$

(4 marks)

b) $2 \cos \left(x - \dfrac{\pi}{4} \right) = \sqrt{3}$

(4 marks)

c) $\sin 2x = -\dfrac{1}{2}$

(6 marks)

7   a)   Show that the equation:
$$2(1 - \cos x) = 3 \sin^2 x$$
can be written as
$$3 \cos^2 x - 2 \cos x - 1 = 0$$

(2 marks)

b)   Use this to solve the equation
$$2(1 - \cos x) = 3 \sin^2 x$$
for $0 \leq x \leq 360°$, giving your answers to 1 d.p.

(6 marks)

1. Logs

Learning Objectives:

- Be able to convert between index and log notation.
- Know and be able to use the laws of logs.
- Know and be able to use the change of base formula for logs.

Logarithm might sound scary, but it's just a fancy way of describing the power that something has been raised to. Once you know how to use logs, you can solve all sorts of equations that involve powers.

Logs

A **logarithm** is just the power that a number needs to be **raised to** to produce a given value.

Before now, you've used **index notation** to represent powers, but sometimes it's easier to work with **log notation**.

In **index notation**, 3^5 means that **5** lots of **3** are multiplied together. **3** is known as the **base**. You now need to be able to **switch** from index notation into **log notation**, and vice versa.

Tip: If you're struggling to get your head round this, try putting in some numbers — e.g. $\log_3 9 = 2$ is the same as $3^2 = 9$.

Log notation looks like this:

$$\log_a b = c$$ ... which means the **same** as the **index notation**... $$a^c = b$$

- The little number '*a*' after 'log' is the **base**.
- '*c*' is the **power** the base is being **raised** to.
- '*b*' is the answer you get when *a* is raised to the power *c*.
- Log means '**power**', so the log above really just means: "what is the power you need to raise *a* to if you want to end up with *b*?"

Tip: In index notation, this is saying that $a^1 = a$, and $a^0 = 1$ (you know this from C1).

As $\log_a b = c$ is the same as $a^c = b$, it means that:

$$\log_a a = 1$$ and $$\log_a 1 = 0$$

- The **base** of a log must always be a **positive integer** (otherwise the log isn't defined for some values).
- So you **can't** take a log of a **negative number** (there's no power you can raise a positive number to to make it negative — this is more obvious if you look at the graphs on p.65).

You can get logs to any base, but **base 10** is the **most common**. The button marked 'log' on your calculator uses base 10.

Tip: Some calculators allow you to find logs to any base.

Here's an example of a log with a base 10:

Index notation: $10^2 = 100$

or

log notation: $\log_{10} 100 = 2$

- So the **logarithm** of 100 to the **base 10** is 2, because 10 raised to the **power** of 2 is 100.
- The base goes here but it's usually left out if it's 10.

Example 1

Write down the values of the following:

a) $\log_2 8$

- **Compare** to $\log_a b = c$. Here the **base** (a) is 2. And the answer (b) is 8.
- So think about the **power** (c) that you'll need to raise 2 to to get 8.
- 8 is 2 raised to the power of 3, so $2^3 = 8$ and $\boxed{\log_2 8 = 3}$

b) $\log_9 3$

- Work out the **power** that 9 needs to be raised to to get 3.
- 3 is the square root of 9, or $9^{\frac{1}{2}} = 3$, so $\boxed{\log_9 3 = \frac{1}{2}}$

> **Tip:** Don't get caught out here — the power is actually a fraction.

c) $\log_5 5$

- Remember that anything to the power of 1 is itself ($a^1 = a$),

$$\text{so } \boxed{\log_5 5 = 1}$$

> **Tip:** So $\log_a a = 1$ for any value of a.

Example 2

Write the following using log notation:

a) $5^3 = 125$

- You just need to make sure you get things in the **right place**.
- 3 is the **power** (c) or logarithm that 5 (a, the **base**) is raised to to get 125 (b).
- So sub into $\log_a b = c$ to get: $\boxed{\log_5 125 = 3}$

b) $3^0 = 1$

- You'll need to remember this one: $\boxed{\log_3 1 = 0}$

> **Tip:** This is true for any base a. Any logarithm of 1 is always 0 because a^0 is always 1 (p.60):
> $$\log_a 1 = 0$$

Example 3

Find the value of p such that $\log_5 p = -2$.

- $a = 5$, $b = p$ and $c = -2$.
- So sub into $a^c = b$ to get: $5^{-2} = p \Rightarrow \frac{1}{5^2} = p \Rightarrow \boxed{p = \frac{1}{25} \text{ or } 0.04}$

Exercise 1.1

In this exercise, log means $\log_{10}$.

Q1 Write the following using log notation:

    a) $2^3 = 8$ b) $5^4 = 625$ c) $49^{\frac{1}{2}} = 7$

    d) $8^{\frac{2}{3}} = 4$ e) $10^{-2} = \frac{1}{100}$ f) $2^{-3} = 0.125$

> **Q1 Hint:** If you're finding these tricky to get your head around, work out what a, b and c are first, and then substitute them into $\log_a b = c$.

Q2 Write the following using log notation:

    a) $4^x = 9$ b) $x^3 = 40$ c) $8^{11} = x$

Q3 Write the following using index notation (you don't need to work out any unknowns):

a) $\log_5 125 = 3$ b) $\log 10\,000 = 4$ c) $\log_{\frac{1}{2}} 4 = -2$

d) $\log_7 a = 6$ e) $\log_5 t = 0.2$ f) $\log_4 m = 1$

g) $\log_{\frac{1}{4}} p = \frac{1}{2}$ h) $\log k = 5$ i) $\log_x a = m$

Q4 Find the value of the following.
Give your answer to 3 d.p. where appropriate.

a) $\log 1000$ b) $\log 0.01$ c) $\log 1$

d) $\log 2$ e) $\log 3$ f) $\log 6$

Q5 Find the value of:

a) $\log_2 4$ b) $\log_3 27$ c) $\log_5 0.2$

Q6 Find the value of x, where $x \geq 0$, by writing the following in index notation:

a) $\log_x 49 = 2$ b) $\log_x 8 = 3$ c) $\log_x 100\,000 = 5$

d) $\log_x 3 = \frac{1}{2}$ e) $\log_x 7 = \frac{1}{3}$ f) $\log_x 2 = \frac{1}{5}$

g) $\log_3 x = 4$ h) $\log_2 x = 6$ i) $\log_7 x = 1$

j) $\log_9 x = \frac{1}{2}$ k) $\log_{64} x = \frac{1}{3}$ l) $\log_{27} x = \frac{2}{3}$

Q7 In each part, use index notation to write y in terms of x, given that:

a) $\log_a x = 2$ and $\log_a y = 4$.   b) $\log_a x = 3$ and $\log_{2a} y = 3$.

c) $\log_a x = 5$ and $\log_a y = 20$.

Laws of logs

You'll need to be able to **simplify** expressions containing logs in order to answer trickier questions — e.g. to **add** or **subtract** two logs you can combine them into one log. To answer these questions, you'll need to use the **laws of logarithms**. These **only work** if the **base** of each log is the same:

Tip: These laws are really useful, so make sure you know them off by heart.

Laws of Logarithms

$$\log_a x + \log_a y = \log_a (xy) \qquad \log_a x - \log_a y = \log_a \left(\frac{x}{y}\right) \qquad \log_a x^k = k \log_a x$$

So $\log_a \frac{1}{x} = \log_a (x^{-1}) = -\log_a x$

Example 1

Simplify the following:

a) $\log_3 4 + \log_3 5$

- First check that the logs you're **adding** have the same base.
  They do (it's 3) so it's OK to combine them.

- You need to use the law $\log_a x + \log_a y = \log_a (xy)$.
- So here, $a = 3$ (the base), $x = 4$ and $y = 5$.

$$\log_3 4 + \log_3 5 = \log_3 (4 \times 5) \boxed{= \log_3 20}$$

b) $\log_3 4 - \log_3 5$

- You're **taking** one log from another here, and the bases are the same (both 3).
- You need to use the law $\log_a x - \log_a y = \log_a \left(\frac{x}{y}\right)$.
- So here, $a = 3$ (the base), $x = 4$ and $y = 5$.

$$\log_3 4 - \log_3 5 \boxed{= \log_3 \left(\frac{4}{5}\right)}$$

c) $3 \log_4 2 + 2 \log_4 5$

- The logs are being **added**, but first you must use the law $\log_a x^k = k \log_a x$ to get rid of the 3 and 2 in front of the logs.
- Using the law, the logs become:

$$3 \log_4 2 = \log_4 (2^3) \boxed{= \log_4 8}$$

$$2 \log_4 5 = \log_4 (5^2) \boxed{= \log_4 25}$$

- The logs both have the same base (4) so the expression becomes:

$$\log_4 8 + \log_4 25 = \log_4 (8 \times 25) \boxed{= \log_4 200}$$

Tip: It's a good idea to split this into separate steps so you don't make a mistake.

You might be asked to simplify expressions that contain unknown variables too.

Example 2

a) Write the expression $2 \log_a 6 - \log_a 9$ in the form $\log_a n$, where n is a number.

- Use $\log_a x^k = k \log_a x$ to **simplify** $2 \log_a 6$:

$$2 \log_a 6 = \log_a 6^2 = \log_a 36$$

- Then use $\log_a x - \log_a y = \log_a \left(\frac{x}{y}\right)$:

$$\log_a 36 - \log_a 9 = \log_a (36 \div 9) \boxed{= \log_a 4}$$

b) Write the expression $\log_{10}\left(\dfrac{100x^2}{y^3}\right)$ in terms of $\log_{10} x$ and $\log_{10} y$.

- Use the laws of logs to **break up** the expression:

$$\log_{10}\left(\frac{100x^2}{y^3}\right) = \log_{10} 100x^2 - \log_{10} y^3$$
$$= \log_{10} 100 + \log_{10} x^2 - \log_{10} y^3$$
$$= \log_{10} 10^2 + \log_{10} x^2 - \log_{10} y^3$$
$$= 2 \log_{10} 10 + 2 \log_{10} x - 3 \log_{10} y$$
$$\boxed{= 2 + 2 \log_{10} x - 3 \log_{10} y}$$

Tip: All three of the laws are used here. And if you're not sure where the 2 in the final answer has come from, remember that $\log_a a = 1$.

Changing the base

Your calculator can work out $\log_{10}$ for you — but it can't do any old log, so if you needed to calculate logs with a **different base**, you'd be stuck. Luckily you can **change the base** of any log to base 10 using this formula and then just pop it into your calculator:

$$\text{Change of Base: } \log_a x = \frac{\log_b x}{\log_b a}$$

Tip: You're trying to work out what power you'd need to raise 7 to to get 4.

Examples

a) **Find the value of $\log_7 4$ to 4 d.p.**

- This is too tricky to work out without a calculator. So **change** the **base** of the log to **10**.
- Here, $a = 7$ and $x = 4$. And we want $b = 10$. So:

$$\log_7 4 = \frac{\log_{10} 4}{\log_{10} 7} \quad = 0.7124 \quad (4 \text{ d.p.})$$

- You can check this on your calculator by doing:

$$7^{0.7124\ldots} = 4$$

b) **Find the value of $\log_3 2$ to 4 d.p.**

- **Change** the **base** of the log to **10**.
- Here, $a = 3$ and $x = 2$. And we want $b = 10$. So:

$$\log_3 2 = \frac{\log_{10} 2}{\log_{10} 3} \quad = 0.6309 \quad (4 \text{ d.p.})$$

Exercise 1.2

Q1 Simplify the following, where possible:

a) $\log_a 2 + \log_a 5$ b) $\log_m 8 + \log_m 7$ c) $\log_b 8 - \log_b 4$
d) $\log_m 15 - \log_m 5$ e) $3 \log_n 4$ f) $2 \log_a 7$
g) $\frac{1}{2} \log_b 16$ h) $\frac{2}{3} \log_a 125$ i) $\log_3 a - \log_2 a$

Q2 Write each of the following expressions as a single log:

a) $2 \log_a 5 + \log_a 4$ b) $3 \log_m 2 - \log_m 4$ c) $3 \log_n 4 - 2 \log_n 8$
d) $\frac{2}{3} \log_b 216 - 2 \log_b 3$ e) $1 + \log_a 6$ f) $2 - \log_b 5$

Q3 Hint: Rewrite the numbers as products of their prime factors.

Q3 If $\log_a 2 = x$, $\log_a 3 = y$ and $\log_a 5 = z$, write in terms of x, y and z:

a) $\log_a 6$ b) $\log_a 16$ c) $\log_a 60$

Q4 Simplify each of the following as much as possible:

a) $\log_b b^3$ b) $\log_a \sqrt{a}$
c) $\log_m 4m - 2 \log_m 2$ d) $\log_{2b} 4 + \log_{2b} b - \log_{2b} 2$

Q5 Prove that:

a) $\log_2 4^x = 2x$ b) $\dfrac{\log_a 54 - \log_a 6}{\log_a 3} = 2$

Q6 Find the value of the following logs to 3 s.f.:

a) $\log_6 3$ b) $\log_9 2$ c) $\log_3 13$ d) $\log_5 4$

2. Exponentials

Over the last few pages, you met logarithms — now it's time to deal with exponentials. An exponential has the effect of reversing a log (it's actually called an inverse — you'll do more on inverses in C3). You can use exponentials to solve equations involving logs and vice versa.

Learning Objectives:

- Know that an exponential function takes the form $y = a^x$.
- Know the shape of the graph $y = a^x$.
- Be able to solve equations of the form $a^x = b$.

Exponentials

Exponentials are functions of the form $y = a^x$ (or $f(x) = a^x$), where $a > 0$. **All** graphs of exponential functions have the **same basic shape**.

$y = a^x$ for $a > 1$

- All the graphs go through **1** at $x = 0$ since $a^0 = 1$ for any a.

- $a > 1$ — so y **increases as x increases**.

- The **bigger** a is, the **quicker** the graph increases (so the curve is **steeper**).

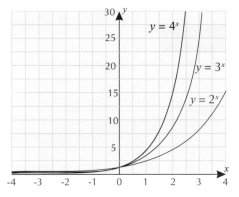

- As x **decreases**, y **decreases** at a **smaller and smaller rate** — y will approach zero, but never actually get there.

- This means as $x \to \infty$, $y \to \infty$ and as $x \to -\infty$, $y \to 0$.

Tip: Notice that the graph of $y = a^x$ is always positive. This is why you can't take the log of a negative number (remember, $y = a^x$ means $\log_a y = x$).

Tip: The notation $x \to \infty$ just means 'x tends to ∞' (i.e. x gets bigger and bigger).

Similarly $y \to 0$ means 'y tends to 0' (gets smaller and smaller).

$y = a^x$ for $0 < a < 1$

- All the graphs go through **1** at $x = 0$ since $a^0 = 1$ for any a.

- $0 < a < 1$ — so y **decreases as x increases**.

- The **smaller** a is, the **faster** the graphs decrease (so the curve is **steeper**).

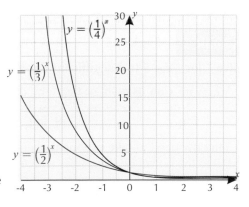

- As x **increases**, y **decreases** at a **smaller and smaller rate** — y will approach zero, but never actually get there.

- This means as $x \to \infty$, $y \to 0$ and as $x \to -\infty$, $y \to \infty$.

Tip: These graphs have the same shapes as the ones above but reflected in the y-axis. This makes sense since $\left(\frac{1}{a}\right)^x$ is the same as a^{-x} — remember your C1 graph transformations.

Exponentials and logs

Solving equations

Exponential functions are always in the form $a^x = b$. If you remember, in the first part of this chapter we said that:

$$a^x = b \quad \text{... is the same as...} \quad \log_a b = x$$

This means that exponentials and logs are the **inverses** of each other — so you can use logs to **get rid** of exponentials and vice versa. This is useful for solving equations. It also means that:

$$a^{\log_a x} = x = \log_a a^x$$

You can use this to get rid of logs easily when solving equations.

Tip: Laws of logs help prove this — i.e. $\log_a a^x = x \log_a a = x$.

Example 1

Solve $2^{4x} = 3$ to 3 significant figures.

- To solve the equation, you want **x on its own**.
- To do this you can **take logs** of both sides:

$$\log 2^{4x} = \log 3$$

- Now use one of the **laws of logs**: $\log x^k = k \log x$:

$$4x \log 2 = \log 3$$

- You can now **divide** both sides by $4 \log 2$ to get x on its own:

$$x = \frac{\log 3}{4 \log 2}$$

- But $\dfrac{\log 3}{4 \log 2}$ is just a number you can find using a **calculator**:

$$x = 0.396 \text{ (to 3 s.f.)}$$

Tip: Here, a log of base 10 has been used. There's more about this in the example on the next page.

Tip: Don't make the mistake of writing:
$$\frac{\log 3}{\log 2} = \log\left(\frac{3}{2}\right)$$
as you can't cancel terms like this.

Example 2

Solve $7 \log_{10} x = 5$ to 3 significant figures.

- You want **x on its own**, so begin by **dividing** both sides by 7:

$$\log_{10} x = \frac{5}{7}$$

- You now need to **take exponentials** of both sides by doing '10 to the power of' both sides (since the log is to base 10):

$$10^{\log_{10} x} = 10^{\frac{5}{7}}$$

- Logs and exponentials are **inverse** functions, so they **cancel out**:

$$x = 10^{\frac{5}{7}}$$

- Again, $10^{\frac{5}{7}}$ is just a number you can find using a **calculator**:

$$x = 5.18 \text{ (to 3 s.f.)}$$

Tip: If you're taking exponentials, make sure you use the same base as the log (here it's 10).

Tip: You could get from $\log_{10} x = \frac{5}{7}$ to $x = 10^{\frac{5}{7}}$ just by changing from log notation to index notation — see page 60.

Example 3

Use logarithms to solve the following for x, giving the answers to 3 s.f.

a) $10^{3x} = 4000$

- There's an 'unknown' in the power, so **take logs of both sides**.

- In theory, it doesn't matter what **base** you use, but your calculator has a '$\log_{10}$' button, so base 10 is usually a good idea. But whatever base you use, use the **same** one for **both sides**.

- So taking logs to base 10 of both sides of the above equation gives:

$$\log 10^{3x} = \log 4000$$

- On the previous page you saw that $\log_a a^x = x$, so $\log_{10} 10^{3x} = 3x$, so you can simplify the left-hand side:

$$3x = \log 4000, \text{ so } \boxed{x = 1.20 \text{ (to 3 s.f.)}}$$

Tip: If you'd used a different base here you would have had to do $\dfrac{\log_a 4000}{3 \log_a 10}$ to find x.

b) $7^x = 55$

- Again, **take logs** of both sides, and use the log rules:

$$\log_{10} 7^x = \log_{10} 55$$
$$x \log_{10} 7 = \log_{10} 55$$

- So:

$$x = \frac{\log_{10} 55}{\log_{10} 7} = \boxed{2.06 \text{ (to 3 s.f.)}}$$

c) $\log_2 x = 5$

- To get rid of a log, you **take exponentials** — so you do 2 (the base) to the power of each side.

$$2^{\log_2 x} = 2^5$$

- Think of 'taking logs' and 'taking exponentials' as opposite processes — one cancels the other out. So you get:

$$x = 2^5 = \boxed{32}$$

Tip: Remember, when taking exponentials you must always use the same base as the logarithm — here the base is 2.

In the exam, you might be asked to solve an equation where you have to use a **combination** of the methods covered in this chapter. It can be tricky to work out what's needed, but just remember that you're trying to get x **on its own** — and think about which laws will help you do that.

The next page has some examples.

Example 4

Solve $6^{x-2} = 3^x$, giving your answer to 3 s.f.

- Start by taking **logs** of both sides (you can use any base, so use 10).

$$\log 6^{x-2} = \log 3^x$$

- Now use $\log x^k = k \log x$ on both sides:

$$(x - 2) \log 6 = x \log 3$$

- **Multiply out** the brackets and **collect** all the x terms on one side:

$$x \log 6 - 2 \log 6 = x \log 3 \Rightarrow x \log 6 - x \log 3 = 2 \log 6$$
$$\Rightarrow x (\log 6 - \log 3) = 2 \log 6$$

- Use $\log_a x - \log_a y = \log_a \left(\frac{x}{y}\right)$ on the bracket:

$$x (\log 2) = 2 \log 6 \Rightarrow x = \frac{2 \log 6}{\log 2} = 5.17 \text{ (3 s.f.)}$$

Tip: Have a look back at page 62 if you can't remember the log laws.

Example 5

Solve the equation $\log_3(2 - 3x) - 2 \log_3 x = 2$.

- First, combine the log terms into one term (you can do this because they both have the same base):

$$\log_3 \frac{2 - 3x}{x^2} = 2$$

Remember that $2\log x = \log x^2$.

- Then take exponentials of both sides using base 3:

$$3^{\log_3 \frac{2 - 3x}{x^2}} = 3^2 \Rightarrow \frac{2 - 3x}{x^2} = 9$$

The exponential ($3^{\cdots}$) and the log ($\log_3 \ldots$) cancel each other.

- Finally, rearrange the equation and solve for x:

$$2 - 3x = 9x^2 \Rightarrow 0 = 9x^2 + 3x - 2$$
$$\Rightarrow 0 = (3x - 1)(3x + 2)$$

- So: $x = \frac{1}{3}$

Tip: Ignore the negative solution $\left(x = -\frac{2}{3}\right)$ because you can't have a log of a negative number — see p.60.

Exponential growth and decay

Logs can be used to solve **real-life** problems involving **exponential growth** and **decay**.

- Exponential **growth** is when the rate of growth **increases** faster and faster as the amount gets bigger.
- Exponential **decay** is just **negative** exponential growth.
- The **rate** of decay gets slower and slower as the amount gets smaller.

For example, if you have money in a bank account that earns interest at a certain percentage per year, the balance will **grow exponentially** over time (if you don't take any money out). So the **more money** you have in the account, the **more interest** you get.

Tip: Don't worry, you haven't accidentally wandered into A-Level Economics. We've just used a real-life example to make this topic fun and engaging. Ahem.

Here is an example of **exponential decay**:

Example

The radioactivity of a substance decays by 20% over a year.
The initial level of radioactivity is 400 Bq (becquerels).
Find the time taken for the radioactivity to fall to 200 Bq (the half-life).

- Write a **formula** for the radioactivity:
$$R = 400 \times 0.8^T$$
where R is the **level of radioactivity** at time T years.

Tip: A becquerel (Bq) is just a unit which measures radioactivity.

- We need $R = 200$, so:
$$200 = 400 \times 0.8^T \quad \Rightarrow \quad 0.8^T = \frac{200}{400} = 0.5$$

Tip: 0.8 is used as every year the radioactivity decreases by 0.2 (or 20%) and so it is 0.8 (or 80%) of what it was the previous year.

- Take logs of both sides, and then use the laws of logs:
$$T \log 0.8 = \log 0.5$$
$$T = \frac{\log 0.5}{\log 0.8} = 3.106 \text{ years (4 s.f.)}$$

Exercise 2.1

Answers to this exercise should be given correct to 3 s.f.

Q1 Solve each of these equations for x:
- a) $2^x = 3$
- b) $4^x = 16$
- c) $7^x = 2$
- d) $1.8^x = 0.4$
- e) $0.7^x = 3$
- f) $0.5^x = 0.2$
- g) $2^{3x-1} = 5$
- h) $10^{3-x} = 8$
- i) $0.4^{5x-4} = 2$

Q2 Find the value of x for each case:
- a) $\log 5x = 3$
- b) $\log_2 (x + 3) = 4$
- c) $\log_3 (5 - 2x) = 2.5$

Q3 Solve each of these equations for x:
- a) $4^{x+1} = 3^{2x}$
- b) $2^{5-x} = 4^{x+3}$
- c) $3^{2x-1} = 6^{3-x}$

Q4 Find the value(s) of x which satisfy each of the following equations:
- a) $\log 2x = \log (x + 1) - 1$
- b) $\log_2 2x = 3 - \log_2 (9 - 2x)$
- c) $\log_6 x = 1 - \log_6 (x + 1)$
- d) $\log_2 (2x + 1) = 3 + 2 \log_2 x$

Q5 Solve the equations $2^{x+y} = 8$ and $\log_2 x - \log_2 y = 1$ simultaneously.

Q6 Solve the equations $9^{x-2} = 3^y$ and $\log_3 2x = 1 + \log_3 y$ simultaneously.

Q7 Find the solutions of each of the following equations:
- a) $2^{2x} - 5(2^x) + 4 = 0$
- b) $4^{2x} - 17(4^x) + 16 = 0$
- c) $3^{2x+2} - 82(3^x) + 9 = 0$
- d) $2^{2x+3} - 9(2^x) + 1 = 0$

Q7 Hint: First rewrite each equation as a quadratic equation.

Q8 Howard bought several cases of wine as an investment. He paid £500 and expects the value to increase by 8% a year. He wants to sell when the wine exceeds £1500 in value. How many full years will this take?

Review Exercise — Chapter 4

Q1 Write the following using log notation:

a) $4^2 = 16$
b) $216^{\frac{1}{3}} = 6$
c) $3^{-4} = \dfrac{1}{81}$

Q2 Write down the values of the following:

a) $\log_3 27$
b) $\log_3 (1 \div 27)$
c) $\log_3 18 - \log_3 2$

Q3 Simplify the following:

a) $\log 3 + 2 \log 5$

b) $\tfrac{1}{2} \log 36 - \log 3$

c) $\log 2 - \tfrac{1}{4} \log 16$

Q4 Simplify $\log_b (x^2 - 1) - \log_b (x - 1)$.

Q5 Find the value of the following, giving your answers to 3 s.f.:

a) $\log_7 12$
b) $\log_5 8$
c) $\log_{16} 125$

Q6 Prove that $\dfrac{2 + \log_a 4}{\log_a 2a} = 2$.

Q7 a) Copy and complete the table for the function $y = 4^x$:

x	−3	−2	−1	0	1	2	3
y							

b) Plot a graph of $y = 4^x$ for $-3 \le x \le 3$.
c) Use the graph to solve the equation $4^x = 20$.
d) Solve the equation $4^x = 20$ algebraically, giving your answer to 3 s.f.

Q8 Solve the following, giving your answer to 3 s.f.:

a) $10^x = 240$
b) $\log_{10} x = 5.3$

c) $10^{2x+1} = 1500$
d) $4^{(x-1)} = 200$

Q9 Find the exact solutions of $2(10^{2x}) - 7(10^x) + 5 = 0$.

Q10 Find the smallest integer P such that $1.5^P > 1\,000\,000$.

Q11 Scientists are monitoring the population of curly-toed spiders at a secret location. It appears to be dropping at the rate of 25% a year. When the population has dropped below 200, the species will be in danger of extinction. At the moment the population is fairly healthy at 2000. How many whole years will it be before the spiders are in danger of extinction?

1 a) Solve the equation

$$2^x = 9$$

giving your answer to 2 decimal places.

(3 marks)

   b) Hence, or otherwise, solve the equation

$$2^{2x} - 13(2^x) + 36 = 0$$

giving each solution to an appropriate degree of accuracy.

(5 marks)

2 a) Solve the equation

$$\log_3 x = -\frac{1}{2}$$

leaving your answer as an exact value.

(3 marks)

   b) Find x, where

$$2 \log_3 x = -4$$

leaving your answer as an exact value.

(2 marks)

3 a) Find x, if

$$6^{(3x + 2)} = 9$$

giving your answer to 3 significant figures.

(3 marks)

   b) Find y, if

$$3^{(y^2 - 4)} = 7^{(y + 2)}$$

giving your answers to 3 significant figures where appropriate.

(5 marks)

4   a)   Write the following expressions in the form $\log_a n$, where n is an integer:

      (i) $\log_a 20 - 2 \log_a 2$

                                                                 (3 marks)

      (ii) $\frac{1}{2} \log_a 16 + \frac{1}{3} \log_a 27$

                                                                 (3 marks)

   b)   Find the value of:

      (i) $\log_2 64$

                                                                   (1 mark)

      (ii) $2 \log_3 9$

                                                                   (2 marks)

   c)   Calculate the value of the following, giving your answer to 4 d.p.

      (i) $\log_6 25$

                                                                   (1 mark)

      (ii) $\log_3 10 + \log_3 2$

                                                                  (2 marks)

5   Solve the equation
$$\log_7 (y + 3) + \log_7 (2y + 1) = 1$$
where $y > 0$.

                                                                   (5 marks)

6   For the positive integers p and q,
$$\log_4 p - \log_4 q = \frac{1}{2}$$

   a)   Show that $p = 2q$.

                                                                   (3 marks)

   b)   The values of p and q are such that $\log_2 p + \log_2 q = 7$.

      Use this information to find the values of p and q.

                                                                   (5 marks)

1. Geometric Sequences and Series

You'll have come across arithmetic sequences and series in C1 — you saw how to find the general term and how to find the sum of a number of terms. In C2 you'll do the same thing for geometric sequences and series.

Geometric sequences

In C1 you came across **arithmetic** sequences, where you get from one term to the next by **adding** a fixed amount each time.

They have a **first term** (*a*), and the amount you add to get from one term to the next is called the **common difference** (*d*).

With **geometric sequences**, rather than adding, you get from one term to the next by **multiplying** by a **constant** called the **common ratio** (*r*).

- This is a **geometric sequence** where you find each term by **multiplying** the previous term by 2:

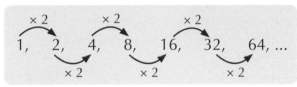

$$\times 2 \qquad \times 2 \qquad \times 2$$
$$1, \quad 2, \quad 4, \quad 8, \quad 16, \quad 32, \quad 64, \ ...$$
$$\times 2 \qquad \times 2 \qquad \times 2$$

- If the common ratio is **negative**, the signs of the sequence will **alternate**. For this geometric sequence the common ratio is –3.

$$2, -6, 18, -54, 162, -486, \ ...$$

- The common ratio might **not** be a **whole number**. Here, it's $\frac{3}{4}$.

$$16, 12, 9, \frac{27}{4}, \frac{81}{16}, \frac{243}{64}, \ ...$$

You get each term by multiplying the first term by the common ratio some number of times. In other words, each term is the **first term** multiplied by **some power** of the **common ratio**.

This is how you describe geometric sequences using **algebra**:

> The **terms** in the sequence are labelled by a letter with a subscript — so the first term is u_1, the second term is u_2, etc.

> The **common ratio** (the number you multiply by) is called '*r*'.

> The **first term** (u_1) is called '*a*'.

$$u_1 = a \qquad\qquad = a$$
$$u_2 = a \times r \qquad = ar$$
$$u_3 = (a \times r) \times r = ar^2$$
$$u_4 = (a \times r \times r) \times r = ar^3$$

So the **formula** that describes **any term** in a geometric sequence is:

$$\boxed{u_n = ar^{n-1}}$$

Learning Objectives:

- Be able to recognise geometric sequences and series.
- Know and be able to use the formula for the general term of a geometric sequence or series.
- Know and be able to use the formula for the sum of the first *n* terms of a geometric sequence or series.
- Be able to recognise convergent geometric series and find their sum to infinity.

Tip: Geometric sequences are also called **geometric progressions**.

Tip: In the first sequence at the top of the page (1, 2, 4, 8, ...) $a = 1$ and $r = 2$.

Tip: The term u_n is called the 'n^{th} term', or '**general term**' of the sequence. This formula for the general term is on the formula sheet.

If you know the values of a and r, you can substitute them into the **general formula** to find an expression that describes the whole sequence:

Example

A chessboard has a 1p piece on the first square, 2p on the second square, 4p on the third, 8p on the fourth and so on until the board is full.

Find a formula for the amount of money on each square.

- This is a **geometric sequence**: $u_1 = 1$, $u_2 = 2$, $u_3 = 4$, $u_4 = 8$...
- You get each term in the sequence by **multiplying** the previous one by **2**.
- So $a = 1$ (because you start with 1p on the first square) and $r = 2$.
- Then the **formula** for the amount of money on each square is:

$$u_n = ar^{n-1} = 1 \times (2^{n-1}) = 2^{n-1}$$

$$\boxed{u_n = 2^{n-1}}$$

Tip: The trick with questions like this is to recognise that you're being asked about a geometric sequence — there's more about this on page 82.

You can also use the formula to find the **first term** a, the **common ratio** r or a **particular term** in the sequence, given other information about the sequence:

Examples

a) Find the 5th term in the geometric sequence 1, 3, 9, ...

- First find the common ratio r. Each term is the previous term multiplied by r, so you find the common ratio by dividing consecutive terms:

$$\text{second term} = \text{first term} \times r \implies r = \frac{\text{second term}}{\text{first term}} = \frac{3}{1} = \boxed{3}$$

- Then find the 5th term. The 3rd term is 9, so:

$$4^{th} \text{ term} = 3^{rd} \text{ term} \times r = 9 \times 3 = 27$$
$$5^{th} \text{ term} = 4^{th} \text{ term} \times r = 27 \times 3 = \boxed{81}$$

b) A geometric sequence has first term 2 and common ratio 1.2. Find the 15th term in the sequence to 3 significant figures.

$a = 2$ and $r = 1.2$, and the formula for the nth term is $u_n = ar^{n-1}$

So $u_n = ar^{n-1} = 2 \times (1.2)^{n-1}$

Then the 15th term is $u_{15} = 2 \times (1.2)^{14} = 2 \times 12.839... = \boxed{25.7}$ to 3 s.f.

c) A geometric sequence has a first term 25 and 10th term 80. Calculate the common ratio. Give your answer to 3 significant figures.

$a = 25$, so the nth term is given by $u_n = ar^{n-1} = 25r^{n-1}$.

The 10th term is $80 = u_{10} = 25r^9$

$$80 = 25r^9 \implies r^9 = \frac{80}{25} \implies r = \sqrt[9]{\frac{80}{25}} = 1.137... = \boxed{1.14} \text{ to 3 s.f.}$$

d) The 8th term of a geometric sequence is 4374 and the common ratio is 3. What is the first term?

The common ratio $r = 3$, so the nth term is $u_n = ar^{n-1} = a(3)^{n-1}$.

Then the 8th term is $4374 = a(3)^7 = 2187a \implies a = \frac{4374}{2187} = \boxed{2}$

Tip: You can choose any two consecutive terms — e.g. dividing the third term by the second will also give you r.

Tip: Working up from the last term you know is a good method for finding a couple more terms in the sequence, but when you're asked for a higher term you're better off using the method in example b).

Tip: Make sure you know how to find the nth root on your calculator — and remember, $\sqrt[n]{x}$ is the same as $x^{\frac{1}{n}}$.

Q1   Find the seventh term in the geometric progression 2, 3, 4.5, 6.75 ...

Q2   The sixth and seventh terms of a geometric sequence are 2187
     and 6561 respectively.
     What is the first term?

Q3   A geometric sequence is 24, 12, 6, ... What is the 9th term?

Q4   The 14th term of a geometric progression is 9216.
     The first term is 1.125.
     Calculate the common ratio.

Q5   The first and second terms of a geometric progression are
     1 and 1.1 respectively.
     How many terms in this sequence are less than 4?

> **Q5 Hint:** Write yourself an equation or inequality and then solve it using logs. Have a flick through Chapter 4 if you need a reminder of how to use logs.

Q6   A geometric progression has a common ratio of 0.6. If the first term
     is 5, what is the difference between the 10th term and the 15th term?
     Give your answer to 5 d.p.

Q7   A geometric sequence has a first term of 25 000 and
     a common ratio of 0.8.
     Which term is the first to be below 1000?

Q8   A geometric sequence is 5, –5, 5, –5, 5, ...
     Give the common ratio.

Q9   The first three terms of a geometric progression are $\frac{1}{4}$, $\frac{3}{16}$ and $\frac{9}{64}$.
     a) Calculate the common ratio.
     b) Find the 8th term. Give your answer as a fraction.

Q10 The 7th term of a geometric sequence is 196.608 and the common
     ratio is 0.8. What is the first term?

Q11 3, –2.4, 1.92,... is a geometric progression.
     a) What is the common ratio?
     b) How many terms are there in the sequence before you reach
         a term with modulus less than 1?

> **Q11 Hint:** The modulus of a number is just its size ignoring the sign, so the modulus of –1 is 1. A number with modulus less than 1 is between –1 and 1.

Geometric series

A **sequence** becomes a **series** when you **add** the terms to find the **sum**.

Geometric series work just like geometric sequences (they have a **first term** and a **common ratio**), but they're written as a **sum of terms** rather than a list:

geometric sequence:	geometric series:
3, 6, 12, 24, 48, ...	$3 + 6 + 12 + 24 + 48 + ...$

Sometimes you'll need to find the **sum** of the **first few terms** of a geometric series:

- The sum of the **first n terms** is called S_n.

- S_n can be written in terms of the first term a and the common ratio r:
$$S_n = u_1 + u_2 + u_3 + u_4 + ... + u_n = a + ar + ar^2 + ar^3 + ... + ar^{n-1}$$

- There's a nice **formula** for finding S_n that doesn't involve loads of adding. You need to know the **proof** of the formula — luckily it's fairly straightforward:

> For any geometric sequence:
> $$S_n = a + ar + ar^2 + ar^3 + ... + ar^{n-2} + ar^{n-1} \quad \text{①}$$
> Multiplying this by r gives:
> $$rS_n = ar + ar^2 + ar^3 + ... + ar^{n-2} + ar^{n-1} + ar^n \quad \text{②}$$
> Subtract equation ② from equation ①: $S_n - rS_n = a - ar^n$
>
> Factorise both sides: $(1 - r)S_n = a(1 - r^n)$
>
> Then divide through by $(1 - r)$: $S_n = \dfrac{a(1 - r^n)}{1 - r}$

So the sum of the first n terms of a geometric series is:

$$S_n = \frac{a(1 - r^n)}{1 - r}$$

This formula is given on the formula sheet (but the proof isn't).

Tip: Geometric series can be infinite sums (i.e. they can go on forever). We're just adding up bits of them for now, but summing a whole series is covered on page 80.

Tip: You could also subtract equation ① from equation ② to get:
$$S_n = \frac{a(r^n - 1)}{r - 1}$$
Both versions are correct.

Examples

a) A geometric series has first term 3.5 and common ratio 5. Find the sum of the first 6 terms.

You're told that $a = 3.5$ and $r = 5$, and you're looking for the sum of the first 6 terms, so just stick these values into the formula for S_6:

$$S_6 = \frac{a(1 - r^6)}{1 - r} = \frac{3.5(1 - 5^6)}{1 - 5} = \boxed{13\ 671}$$

b) The first two terms in a geometric series are 20, 22. To 2 decimal places, the sum of the first k terms of the series is 271.59. Find k.

$a = 20$, $r = \dfrac{\text{second term}}{\text{first term}} = \dfrac{22}{20} = 1.1$, so put these into the sum formula:

$$S_k = \frac{a(1 - r^k)}{1 - r} = \frac{20(1 - (1.1)^k)}{1 - 1.1} = -200(1 - (1.1)^k)$$

So $271.59 = -200(1 - (1.1)^k) \Rightarrow -\dfrac{271.59}{200} - 1 = -(1.1)^k$

$\Rightarrow \quad -2.35795 = -(1.1)^k$

$\Rightarrow \quad 2.35795 = 1.1^k$

$\Rightarrow \quad \log(2.35795) = k\log(1.1)$

$\Rightarrow \quad k = \dfrac{\log(2.35795)}{\log(1.1)} = \boxed{9}$

Tip: You're looking for a number of terms so the answer must be a positive integer.

Sigma notation

You'll have seen in C1 that the **sum** of the first n **terms** of a series (S_n) can also be written using **sigma (Σ) notation**.

For geometric series, sigma notation looks like this:

$$S_n = u_1 + u_2 + u_3 + \ldots + u_n = a + ar + ar^2 + \ldots + ar^{n-1} = \sum_{k=0}^{n-1} ar^k$$

So, using the formula from the previous page, the sum of the first n terms can be written:

$$\boxed{\sum_{k=0}^{n-1} ar^k = \dfrac{a(1 - r^n)}{1 - r}}$$

Tip: Remember, Σ just means sum (it's the Greek letter for S). In this case, it's the sum of ar^k from $k = 0$ to $k = n - 1$. Be careful with the limits — it's $n - 1$ on top of the Σ, but the sum is S_n.

Example

$a + ar + ar^2 + \ldots$ **is a geometric series, and** $\displaystyle\sum_{k=0}^{4} ar^k = 85.2672$.

Given that $r = -1.8$, find the first term a.

You've got the sum of the first 5 terms:

$$85.2672 = \sum_{k=0}^{4} ar^k = S_5 = \dfrac{a(1 - r^5)}{1 - r}$$

So plug the value of r into the formula:

$$85.2672 = \dfrac{a(1 - r^5)}{1 - r} = \dfrac{a(1 - (-1.8)^5)}{1 - (-1.8)} = a\dfrac{19.89568}{2.8} = 7.1056a$$

$$\Rightarrow a = \dfrac{85.2672}{7.1056} = \boxed{12}$$

Exercise 1.2

Q1 The first term of a geometric sequence is 8 and the common ratio is 1.2. Find the sum of the first 15 terms.

Q2 A geometric series has first term $a = 25$ and common ratio $r = 0.7$.
Find $\displaystyle\sum_{k=0}^{9} 25(0.7)^k$.

Q3 The sum of the first n terms of a geometric series is 196 605.
The common ratio of the series is 2 and the first term is 3. Find n.

Q4 A geometric progression starts with 4, 5, 6.25.
The first x terms add up to 103.2 to 4 significant figures. Find x.

Chapter 5 Sequences and Series 77

Q5 The 3rd term of a geometric series is 6 and the 8th term is 192. Find:

    a) the common ratio   b)   the first term

    c) the sum of the first 15 terms

Q6 $k + 10$, k, $2k - 21$, ... is a geometric progression, k is a positive constant.

Q6 a) Hint: The ratio of the first term to the second is the same as the ratio of the second term to the third.

    a) Show that $k^2 - k - 210 = 0$.

    b) Hence show that $k = 15$.

    c) Find the common ratio of this series.

    d) Find the sum of the first 10 terms.

Q7 The first three terms of a geometric series are 1, x, x^2.
The sum of these terms is 3 and each term has a different value.

    a) Find x.   b) Calculate the sum of the first 7 terms.

Q8 a, ar, ar^2, ar^3, ... is a geometric progression.
Given that $a = 7.2$ and $r = 0.38$, find $\sum_{k=0}^{9} ar^k$.

Q9 The sum of the first eight terms of a geometric series is 1.2.
Find the first term of the series, given that the common ratio is $-\frac{1}{3}$.

Q10 a, $-2a$, $4a$, $-8a$, ... is a geometric sequence.
Given that $\sum_{k=0}^{12} a(-2)^k = -5735.1$, find a.

Convergent geometric series

Convergent sequences

Some geometric sequences **tend towards zero** — in other words, they get closer and closer to zero (but they never actually reach it). For example:

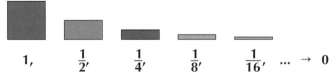

$$1, \quad \frac{1}{2}, \quad \frac{1}{4}, \quad \frac{1}{8}, \quad \frac{1}{16}, \quad ... \rightarrow 0$$

Tip: The arrow here means 'tends to'.

- Sequences like this are called **convergent** — the terms **converge** (get closer and closer) to a **limit** (the number they get close to).
- Geometric sequences either **converge to zero** or **don't converge** at all.
- A sequence that doesn't converge is called **divergent**.

Tip: You ignore the sign of r because you can still have a convergent sequence when r is negative. In that case the terms will alternate between > 0 and < 0, but they'll still be getting closer and closer to zero.

- A geometric sequence a, ar, ar^2, ar^3, ... will converge to **zero** if each term is **closer** to zero than the one before.
- This happens when $-1 < r < 1$. You can write this as $|r| < 1$, where $|r|$ is the modulus of r (the size of r ignoring its sign), so:

$$\boxed{a, ar, ar^2, ar^3, ... \rightarrow 0 \text{ when } |r| < 1}$$

Convergent series

- When you **sum** a sequence that converges to zero you get a **convergent series**.

- Because each term is getting closer and closer to zero, you're **adding smaller and smaller** amounts each time. So the sum gets **closer and closer** to a certain number, but never reaches it — this is the **limit** of the series.

For example, the **sum** of the **convergent sequence** $1, \frac{1}{2}, \frac{1}{4}, \frac{1}{8}, \frac{1}{16}, \ldots$ gets closer and closer to 2:

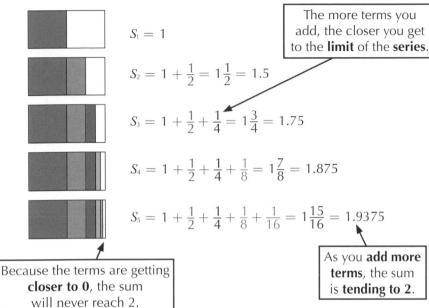

$S_1 = 1$

$S_2 = 1 + \frac{1}{2} = 1\frac{1}{2} = 1.5$

$S_3 = 1 + \frac{1}{2} + \frac{1}{4} = 1\frac{3}{4} = 1.75$

$S_4 = 1 + \frac{1}{2} + \frac{1}{4} + \frac{1}{8} = 1\frac{7}{8} = 1.875$

$S_5 = 1 + \frac{1}{2} + \frac{1}{4} + \frac{1}{8} + \frac{1}{16} = 1\frac{15}{16} = 1.9375$

> The more terms you add, the closer you get to the **limit** of the **series**.

> As you **add more terms**, the sum is **tending to 2**.

> Because the terms are getting **closer to 0**, the sum will never reach 2.

Tip: This is the sum of the sequence from the previous page.

$$1 + \frac{1}{2} + \frac{1}{4} + \frac{1}{8} + \frac{1}{16} + \ldots = 2$$

- So when the **sequence** $a, ar, ar^2, ar^3, \ldots$ **converges** to zero, the **series** $a + ar + ar^2 + ar^3 + \ldots$ **converges** to a **limit**.

- Like sequences, series converge when $|r| < 1$.

Geometric series
$a + ar + ar^2 + ar^3 + \ldots$
with $|r| < 1$ are **convergent**

Geometric series
$a + ar + ar^2 + ar^3 + \ldots$
with $|r| \geq 1$ are **divergent**

Tip: Not all sequences that tend to zero produce a convergent series — this rule is only true for geometric progressions.

For example, the series
$1 + \frac{1}{2} + \frac{1}{3} + \frac{1}{4} + \frac{1}{5} + \ldots$
diverges.

Examples

Determine whether or not each sequence below is convergent.
a) 1, 2, 4, 8, 16, ...   b) 81, –27, 9, –3, 1, ...

a) $r = \frac{2^{nd} \text{ term}}{1^{st} \text{ term}} = \frac{2}{1} = 2$, $|r| = |2| = 2 > 1$, so the series is not convergent

b) $r = \frac{2^{nd} \text{ term}}{1^{st} \text{ term}} = \frac{-27}{81} = -\frac{1}{3}$, $|r| = \left|-\frac{1}{3}\right| = \frac{1}{3} < 1$, the series is convergent

Tip: You can usually spot straight away if a geometric series is convergent — the terms will be getting closer and closer to zero. But you still need to check $|r| < 1$ to prove it.

Summing to infinity

When a series is **convergent** you can find its **sum to infinity**.

- The sum to infinity is called S_∞ — it's the **limit** of S_n as $n \to \infty$.
- This just means that the sum of the first n terms of the series (S_n) gets closer and closer to S_∞ the more terms you add (the bigger n gets).
- In other words, it's the **number** that the **series converges to**.

Example

If $a = 2$ and $r = \frac{1}{2}$, find the sum to infinity of the geometric series.

$u_1 = 2$ $\longrightarrow$ $S_1 = 2$

$u_2 = 2 \times \frac{1}{2} = 1$ $\longrightarrow$ $S_2 = 2 + 1 = 3$

$u_3 = 1 \times \frac{1}{2} = \frac{1}{2}$ $\longrightarrow$ $S_3 = 2 + 1 + \frac{1}{2} = 3\frac{1}{2}$

$u_4 = \frac{1}{2} \times \frac{1}{2} = \frac{1}{4}$ $\longrightarrow$ $S_4 = 2 + 1 + \frac{1}{2} + \frac{1}{4} = 3\frac{3}{4}$

$u_5 = \frac{1}{4} \times \frac{1}{2} = \frac{1}{8}$ $\longrightarrow$ $S_5 = 2 + 1 + \frac{1}{2} + \frac{1}{4} + \frac{1}{8} = 3\frac{7}{8}$

These values are getting closer (**converging**) to 4. So the sum to infinity is **4**.

These values are getting **smaller** each time.

You can show this **graphically**. $\longrightarrow$ The line on the graph is getting **closer and closer** to 4, but it'll never actually get there.

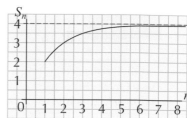

Tip: Luckily you don't have to find a list of sums like this to get the sum to infinity — there's a nifty formula for working it out further down the page. This example is just to show you what's going on when a series converges.

There's a **formula** you use to work out the **sum to infinity** of a geometric series. It comes from the formula for the sum of the first n terms (see p.76):

- The sum of the **first n terms** of a geometric series is $S_n = \dfrac{a(1 - r^n)}{1 - r}$

- If $|r| < 1$ and n is very, very big, then r^n will be very, very **small**, i.e. $r^n \to 0$ as $n \to \infty$.

Tip: Remember, $|r| < 1$ just means $-1 < r < 1$.

- This means $(1 - r^n)$ will get really **close** to 1, so $(1 - r^n) \to 1$ as $n \to \infty$.

- Putting this back into the sum formula gives $S_n \to \dfrac{a \times 1}{1 - r} = \dfrac{a}{1 - r}$ as $n \to \infty$.

- So:

$$S_\infty = \frac{a}{1 - r}$$

This formula is given on the formula sheet.

Examples

Tip: This is the same as the example above, but this time the sum to infinity is worked out using the formula.

a) If $a = 2$ and $r = \frac{1}{2}$, find the sum to infinity of the geometric series.

$|r| = \left|\frac{1}{2}\right| = \frac{1}{2} < 1$, so the series converges and you can find its sum to infinity using the formula for S_∞:

$$S_\infty = \frac{a}{1 - r} = \frac{2}{1 - \frac{1}{2}} = \frac{2}{\frac{1}{2}} = \boxed{4}$$

b) Find the sum to infinity of the series 8 + 2 + 0.5 + 0.125 + ...

First find a and r as before: $a = 8$, $r = \dfrac{2^{nd}\text{ term}}{1^{st}\text{ term}} = \dfrac{2}{8} = 0.25$

Again, $|r| < 1$.

Now find the sum to infinity: $S_\infty = \dfrac{a}{1-r} = \dfrac{8}{1-0.25} = \dfrac{32}{3} = 10\frac{2}{3}$

Divergent series **don't** have a **sum to infinity**.

Because the terms aren't tending to zero, the size of the sum will just keep increasing as you add more terms, so there is **no limit** to the sum.

Exercise 1.3

Q1 State which of these sequences will converge and which will not.
   a) 1, 1.1, 1.21, 1.331, ...   b) 0.8, 0.8^2, 0.8^3, ...
   c) 1, $\dfrac{1}{4}$, $\dfrac{1}{16}$, $\dfrac{1}{64}$, ...   d) 3, $\dfrac{9}{2}$, $\dfrac{27}{4}$, ...
   e) 1, $-\dfrac{1}{2}$, $\dfrac{1}{4}$, $-\dfrac{1}{8}$, $\dfrac{1}{16}$, ...   f) 5, 5, 5, 5, 5, ...

> **Q1 Hint:** If $r = 1$, the sequence is just the same term repeated, so it diverges. If $r = -1$, the sequence alternates between two terms forever.

Q2 A geometric series is 9 + 8.1 + 7.29 +... Calculate the sum to infinity.

Q3 a, ar, ar^2, ... is a geometric sequence. Given that $S_\infty = 2a$, find r.

Q4 The sum to infinity of a geometric progression is 13.5 and the first 3 terms add up to 13.
   a) Find the common ratio r
   b) Find the first term a.

Q5 $a + ar + ar^2 +...$ is a geometric series. $ra = 3$ and $S_\infty = 12$. Find r and a.

Q6 The sum to infinity of a geometric series is 10 and the first term is 6.
   a) Find the common ratio.
   b) What is the 5th term?

Q7 The 2nd term of a geometric progression is –48 and the 5th term is 0.75. Find:
   a) the common ratio
   b) the first term
   c) the sum to infinity

Q8 The sum of the terms after the 10th term of a convergent geometric series is less than 1% of the sum to infinity. The first term is positive. Show that the common ratio $|r| < 0.631$.

> **Q8 Hint:** Find a mathematical statement to express the first sentence.

Real-life problems

Sometimes you'll be given a '**real-life**' situation and you'll have to spot that you're being asked about a **geometric sequence** or **series**.

The trick with these is turning a **wordy** question into the right **maths**.

Example

When a baby is born, £3000 is invested in an account with a fixed interest rate of 4% per year.

a) What will the account be worth at the start of the seventh year?

- The amount of money in the account forms a geometric sequence.

  The first term is: $u_1 = a = 3000$

  The second term is: $u_2 = 3000 + (4\% \text{ of } 3000)$ ← This is the interest.

  $= 3000 + (0.04 \times 3000)$

  $= 3000(1 + 0.04)$

  $= 3000 \times 1.04$

  The third term is: $u_3 = u_2 \times 1.04 = (3000 \times 1.04) \times 1.04$

  $= 3000 \times (1.04)^2$

  And so on — so $r = 1.04$

- So the nth term of the sequence is: $u_n = ar^{n-1} = 3000 \times (1.04)^{n-1}$
- The value of the account at the start of the first year is the 1st term, so the value of the account at the start of the seventh year is the 7th term:

  $u_7 = ar^6 = 3000 \times (1.04)^6 = 3795.957...$

  So it's $\boxed{£3795.96}$ (to the nearest penny)

Tip: You might be able to work out $r = 1.04$ straight away, if you know that a 4% interest rate means you multiply by 1.04.

b) After how many full years will the account have doubled in value?

- You need to know when $u_n > 3000 \times 2 = 6000$
- From part **a)** you know $u_n = 3000 \times (1.04)^{n-1}$

  So $3000 \times (1.04)^{n-1} > 6000 \Rightarrow (1.04)^{n-1} > 2$

- To complete this you need to use logs:

  $\Rightarrow \quad \log(1.04)^{n-1} > \log 2$

  $\Rightarrow \quad (n-1)\log(1.04) > \log 2$ ← This is true because 1.04 > 1 and the log of numbers greater than 1 is positive.

  $\Rightarrow \quad n - 1 > \dfrac{\log 2}{\log 1.04}$

  $\Rightarrow \quad n - 1 > 17.67$

  $\Rightarrow \quad n > 18.67 \quad \text{(to 2 d.p.)}$

- So u_n is more than 6000 when n is more than 18.67. Then u_{19} (the amount at the start of the 19th year) will be more than double the original amount. After $\boxed{18}$ years, the account will have doubled in value.

Tip: It's OK to take logs of both sides of an inequality because logs are increasing functions, so if $x > y$, $\log x > \log y$.

Tip: If you have to give an answer in years, make sure you think through which term belongs to which year carefully — it's easy to give the wrong answer even if you've done all the maths right.

Q1 A collector has 8 china dolls that fit inside each other. The smallest doll is 3 cm high and each doll is 25% taller than the previous one. If he lines them up in height order, how tall is the 8th doll?

Q2 A car depreciates by 15% each year. The value of the car after each year forms a geometric sequence. After 10 years from new the car is valued at £2362. How much was the car when new?

Q2 Hint: 'Depreciates' just means 'gets cheaper'.

Q3 A fishing licence cost £120 in 2005.
The cost rose 3% each year for the next 5 years.
a) How much was a fishing licence in 2006?
b) Nigel bought a fishing licence every year between 2005 and 2010 (including 2005 and 2010). How much in total did he spend?

Q3 Hint: Careful here — part a) is asking about a sequence, but part b) is asking about a series.

Q4 One year, Rob invested in a tangerine farm, and at the end of the year earned £2000 from his investment. He earned money from the investment every year and his annual earnings from the investment increased by 4% each year. Find, to the nearest pound, the total amount he received in the first 8 years of his investment.

Q5 Ron is growing his prize leeks for the Village Show. The bag of compost he's using on his leeks says it will increase their height by 15% every 2 days. After 4 weeks the leeks' height has increased from 5 cm to 25 cm. Has the compost done what it claimed?

Q6 It's predicted that garden gnome value will go up by 2% each year, forming a geometric progression.
Jean-Claude has a garden gnome currently valued at £80 000.
If the predicted rate of inflation is correct:
a) What will Jean-Claude's gnome be worth after 1 year?
b) What is the common ratio of the geometric progression?
c) What will Jean-Claude's gnome be worth after 10 years?
d) It will take k years for the value of Jean-Claude's gnome to exceed £120 000. Show that $k \log 1.02 > \log 1.5$.
e) Find k.

Q7 The thickness of a piece of paper is 0.01 cm. The Moon is 384 000 km from the Earth. The piece of paper is on the Earth. Assuming you can fold the piece of paper as many times as you like, how many times would you have to fold it for it to reach the Moon?

Q7 Hint: Have a go — it really works!*

*Disclaimer: it gets a bit tricky after 7 folds...

Q8 An athlete is preparing for an important event and sets herself a target of running 3% more each day. On day 1 she runs 12 miles.
a) How far does she run on day 10?
b) Her training schedule lasts for 20 days.
To the nearest mile, how far does she run altogether?

Q9 Chardonnay wants to invest her savings for the next 10 years. She wants her investment to double during this time. If interest is added annually, what interest rate does she need?

Q9 Hint: Call the initial value of her savings a.

2. Binomial Expansions

Learning Objectives:

- Be able to use Pascal's triangle to find the coefficients of a binomial expansion.
- Be able to find binomial coefficients using factorials and using the notation $\binom{n}{r}$ or ^nC_r.
- Be able to use the formula to expand binomials of the form $(1 + x)^n$.
- Be able to use the formula to expand binomials of the form $(a + bx)^n$.

Binomials are just polynomials that only have two terms (the 'bi' and 'poly' bits come from the Greek words for 'two' and 'many'). Binomial expansions are all about multiplying out brackets with two terms. This section has a few methods you can use on different types of expansions.

Pascal's triangle

- A **binomial expansion** is what you get when you **multiply out the brackets** of a polynomial with two terms, like $(1 + x)^5$ or $(2 - 3x)^8$.
- It would take ages to multiply out a bracket like this by hand if the power was really big — fortunately binomial expansions **follow a pattern**:

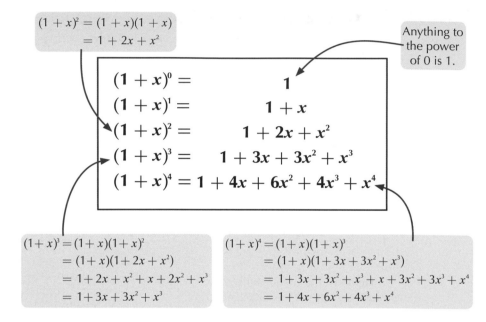

$$(1 + x)^2 = (1 + x)(1 + x)$$
$$= 1 + 2x + x^2$$

Anything to the power of 0 is 1.

$$(1 + x)^0 = 1$$
$$(1 + x)^1 = 1 + x$$
$$(1 + x)^2 = 1 + 2x + x^2$$
$$(1 + x)^3 = 1 + 3x + 3x^2 + x^3$$
$$(1 + x)^4 = 1 + 4x + 6x^2 + 4x^3 + x^4$$

$$(1+x)^3 = (1+x)(1+x)^2$$
$$= (1+x)(1+2x+x^2)$$
$$= 1+2x+x^2+x+2x^2+x^3$$
$$= 1+3x+3x^2+x^3$$

$$(1+x)^4 = (1+x)(1+x)^3$$
$$= (1+x)(1+3x+3x^2+x^3)$$
$$= 1+3x+3x^2+x^3+x+3x^2+3x^3+x^4$$
$$= 1+4x+6x^2+4x^3+x^4$$

Tip: When you expand a binomial you usually write it in increasing powers of x, starting with x^0 and going up to x^n.

Tip: For the moment we're just looking at $(1 + x)^n$. Binomials of the form $(a + bx)^n$, like $(2 - 3x)^8$, are covered on pages 88-92.

A French man called Blaise Pascal spotted the pattern in the **coefficients** and wrote them down in a **triangle**, so it's imaginatively known as '**Pascal's Triangle**'. The pattern's easy — each number is the **sum** of the two above it:

```
              1
           1     1
        1     2     1
     1     3     3  + 1
  1     4     6   = 4     1
```

Tip: The triangle is symmetrical, so once you've got the first half of the coefficients you don't need to work out the rest.

So the next line is: **1   5   10   10   5   1**

Giving: $(1 + x)^5 = 1 + 5x + 10x^2 + 10x^3 + 5x^4 + x^5.$

If you're expanding a binomial with a power that's not too huge, writing out a quick **Pascal's triangle** is a good way to **find the coefficients**.

Example

Find the binomial expansion of $(1 + x)^6$.

Draw **Pascal's triangle** — you're raising the bracket to the **power of 6**, so go down to the **7th row**.

Write the answer out, getting the **coefficients** from the 7th row, and increasing the power of x from left to right:

$$(1 + x)^6 = 1 + 6x + 15x^2 + 20x^3 + 15x^4 + 6x^5 + x^6$$

```
          1
        1   1
      1   2   1
    1   3   3   1
  1   4   6   4   1
1   5  10  10   5   1
1   6  15  20  15   6   1
```

Tip: Make sure you go down to the correct row — you need one more row than the power you're raising the bracket to.

Binomial expansions — $(1 + x)^n$

For expansions with **higher powers** you don't need to write out Pascal's triangle — there's a **formula** you can use instead:

$$(1 + x)^n = 1 + \frac{n}{1}x + \frac{n(n-1)}{1 \times 2}x^2 + \frac{n(n-1)(n-2)}{1 \times 2 \times 3}x^3 + ... + x^n$$

At first glance this looks a bit awful, but each term follows a pattern:

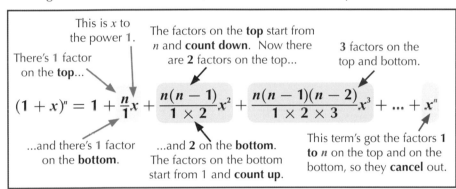

Tip: Try to remember the patterns from term to term — it should sink in if you get plenty of practice using it.

Once you get **halfway** along, the **factors** on the top and bottom start to **cancel**, and the coefficients repeat themselves (they're symmetrical):

$$(1 + x)^5 = 1 + \frac{5}{1}x + \frac{5 \times 4}{1 \times 2}x^2 + \frac{5 \times 4 \times 3}{1 \times 2 \times 3}x^3 + \frac{5 \times 4 \times 3 \times 2}{1 \times 2 \times 3 \times 4}x^4 + \frac{5 \times 4 \times 3 \times 2 \times 1}{1 \times 2 \times 3 \times 4 \times 5}x^5$$

$$= 1 + \frac{5}{1}x + \frac{5 \times 4}{1 \times 2}x^2 + \frac{5 \times 4}{1 \times 2}x^3 + \frac{5}{1}x^4 + x^5$$

$$= 1 + 5x + 10x^2 + 10x^3 + 5x^4 + x^5$$

Example

a) **Expand $(1 + x)^{20}$, giving the first 4 terms only.**

- The binomial formula is:

$$(1 + x)^n = 1 + \frac{n}{1}x + \frac{n(n-1)}{1 \times 2}x^2 + \frac{n(n-1)(n-2)}{1 \times 2 \times 3}x^3 + ... + x^n$$

- You're looking for $(1 + x)^{20}$, so $n = 20$:

$$(1 + x)^{20} = 1 + \frac{20}{1}x + \frac{20 \times 19}{1 \times 2}x^2 + \frac{20 \times 19 \times 18}{1 \times 2 \times 3}x^3 + ... + x^{20}$$

Tip: Make sure you write down your working — if you just bung the numbers straight into your calculator, you won't be able to spot if you've made a mistake.

- The first 4 terms are:

$$(1 + x)^{20} = 1 + \frac{20}{1}x + \frac{\overset{10}{\cancel{20}} \times 19}{1 \times \cancel{2}}x^2 + \frac{\overset{10}{\cancel{20}} \times 19 \times \overset{6}{\cancel{18}}}{1 \times \cancel{2} \times \cancel{3}}x^3 + ...$$
$$= 1 + 20x + (10 \times 19)x^2 + (10 \times 19 \times 6)x^3 + ...$$
$$= \boxed{1 + 20x + 190x^2 + 1140x^3 + ...}$$

b) What is the term in x^7 in this expansion? Give your answer in its simplest form.

- The term in x^7 has 7 factors on the top of the coefficient, and 7 factors on the bottom.
- In this expansion n is 20, so on the top you count down from 20.
- Term in $x^7 = \dfrac{\cancel{20} \times 19 \times \cancel{18} \times 17 \times 16 \times 15 \times \cancel{14}}{1 \times \cancel{2} \times \cancel{3} \times \cancel{4} \times \cancel{5} \times \cancel{6} \times \cancel{7}}x^7$

$$= (19 \times 17 \times 16 \times 15)x^7$$
$$= \boxed{77\,520x^7}$$

There are a few bits of **notation** you need to know that will make writing out the binomial formula a bit easier.

Factorials

- The product on the **bottom** of each binomial coefficient is **$1 \times 2 \times ... \times r$**, where r is the **power x is raised to** in that term.

Tip: $r! = 1 \times 2 \times ... \times r$ is said 'r factorial'.

- This product can be written as a **factorial**: $r! = 1 \times 2 \times ... \times r$. E.g. in the binomial expansion of $(1 + x)^{20}$, the coefficient of the term in x^3 is:

$$\frac{20 \times 19 \times 18}{1 \times 2 \times 3} = \frac{20 \times 19 \times 18}{3!}$$

- In fact, you can write the **whole** coefficient using **factorials**. For example, the coefficient of x^3 above is:

Tip: Here you've multiplied top and bottom by 17! — this is just so you can write the factors on the top as a factorial (you need to multiply all the way down to 1 to do this).

$$\frac{20 \times 19 \times 18}{1 \times 2 \times 3}$$
$$= \frac{20 \times 19 \times 18 \times 17 \times ... \times 2 \times 1}{1 \times 2 \times 3 \times 1 \times 2 \times ... \times 17}$$
$$= \frac{20 \times 19 \times ... \times 1}{(1 \times 2 \times 3)(1 \times 2 \times ... \times 17)}$$
$$= \frac{20!}{3!17!}$$

- For a general binomial expansion of $(1 + x)^n$, the coefficient of x^r is:

$$\frac{n \times (n - 1) \times ... \times (n - (r - 1))}{1 \times 2 \times ... \times r}$$
$$= \frac{n \times (n - 1) \times ... \times (n - (r - 1)) \times (n - r) \times ... \times 2 \times 1}{1 \times 2 \times ... \times r \times 1 \times 2 \times ... \times (n - r)}$$
$$= \frac{n \times (n - 1) \times ... \times 1}{(1 \times 2 \times .... \times r)(1 \times 2 \times ... \times (n - r))}$$
$$= \frac{n!}{r!(n - r)!}$$

Tip: The two numbers on the bottom of the factorial fraction should always add up to the number on the top.

- So the terms in a **binomial expansion** of **$(1 + x)^n$** are of the form:

$$\boxed{\frac{n(n - 1)(n - 2)...(n - (r - 1))}{1 \times 2 \times 3 \times ... \times r}x^r = \frac{n!}{r!(n - r)!}x^r}$$

n is the power you're raising the **bracket** to, r is the power of x in the **term** the coefficient belongs to.

^nC_r notation

There are a couple of even **shorter ways** of writing the **binomial coefficients**:

$$\boxed{\frac{n!}{r!\,(n-r)!}} \;=\; \boxed{\binom{n}{r}} \;=\; \boxed{^nC_r}$$

There's a button on your calculator for finding this.

Tip: This is on the formula sheet.

Tip: The C in ^nC_r stands for 'choose' — you say these coefficients 'n choose r', e.g. $\binom{3}{2}$ is '3 choose 2'.

For example, going back to the coefficient of x^3 in the expansion of $(1+x)^{20}$:

$$\frac{20 \times 19 \times 18}{1 \times 2 \times 3}x^3 = \frac{20!}{3!17!}x^3 = \binom{20}{3}x^3 = {}^{20}C_3 x^3$$

So the **binomial formula** can be written using **any** of these notations, and you need to **be familiar with all of them**:

$$(1+x)^n = 1 + \frac{n!}{1!\,(n-1)!}x + \frac{n!}{2!\,(n-2)!}x^2 + \frac{n!}{3!\,(n-3)!}x^3 + \ldots + x^n$$

$$(1+x)^n = 1 + \binom{n}{1}x + \binom{n}{2}x^2 + \binom{n}{3}x^3 + \ldots + x^n$$

$$(1+x)^n = 1 + {}^nC_1 x + {}^nC_2 x^2 + {}^nC_3 x^3 + \ldots + x^n$$

Tip: These coefficients will always come out as a whole number that's at least 1 — if you get a coefficient that isn't, then you know you've made a mistake somewhere.

- Most **calculators** will have an '**nCr**' button for finding binomial coefficients. To use it you put in **n**, press '**nCr**', then put in **r**.
- This is particularly handy if you're just looking for a **specific term** in a **binomial expansion** and you don't want to write the whole thing out.

Tip: If you get confused with which number's which, remember $n \geq r$.

Example

Find the 6th term of the expansion of $(1+x)^8$.

- You're raising the **bracket** to the power of 8, so **n = 8**.
- The **6th term** is the x^5 term (the first term is $1 = x^0$, the second is the $x = x^1$ term and so on), so **r = 5**.
- So put '**8 nCr 5**' into your calculator: $^8C_5 = $ **56**.
- The question asks for the **whole term** (not just the coefficient), so the answer is: $56x^5$

Exercise 2.1

Q1 Use Pascal's triangle to expand $(1+x)^4$.

Q2 Use your calculator to work out the following:

a) $^6C_2$ 　　b) $\binom{12}{5}$ 　　c) $\dfrac{30!}{4!26!}$ 　　d) $^8C_8$

Q3 Without a calculator work out the following:

a) $\dfrac{9!}{4!5!}$ 　　b) $^{10}C_3$ 　　c) $\dfrac{15!}{11!4!}$ 　　d) $\binom{8}{6}$

Q4　Find the first 4 terms, in ascending powers of x, of the binomial expansion of $(1 + x)^{10}$. Give each term in its simplest form.

Q5　Write down the full expansion of $(1 + x)^6$.

Q6　Find the first 4 terms in the expansion of $(1 + x)^7$.

Binomial expansions — $(1 + ax)^n$

When the **coefficient of x** in your binomial **isn't 1** (e.g. $(1 + 2x)^6$) you have to substitute the **whole x term** (e.g. $2x$) into the **binomial formula**:

$$(1 + ax)^n = 1 + \binom{n}{1}(ax) + \binom{n}{2}(ax)^2 + \binom{n}{3}(ax)^3 + ... + (ax)^n$$

Tip: You're not given the expansion of $(1 + ax)^n$ on the formula sheet, but you can use the one for $(1 + x)^n$ and replace x with (ax).

When a is -1 (i.e. $(1 - x)^n$) the formula looks just like the formula for $(1 + x)^n$, but the **signs** of the terms **alternate**:

$$(1 - x)^n = (1 + (-x))^n$$

$$= 1 + \tfrac{n}{1}(-x) + \frac{n(n-1)}{1 \times 2}(-x)^2 + \frac{n(n-1)(n-2)}{1 \times 2 \times 3}(-x)^3 + ... + (-x)^n$$

$$= 1 - \tfrac{n}{1}x + \frac{n(n-1)}{1 \times 2}x^2 - \frac{n(n-1)(n-2)}{1 \times 2 \times 3}x^3 + ... \pm x^n$$

So for **$(1 - x)^n$** you just use the usual binomial **coefficients**, but with **alternating signs**:

$$(1 - x)^n = 1 - \binom{n}{1}x + \binom{n}{2}x^2 - \binom{n}{3}x^3 + ... \pm x^n$$

Tip: The sign of the last term is **plus** if n is **even** and **minus** if n is **odd**.

Examples

a) **What is the term in x^5 in the expansion of $(1 - 3x)^{12}$?**

The **general** binomial formula for $(1 - 3x)^n$ is:

$(1 - 3x)^n = (1 + (-3x))^n$

$= 1 + \tfrac{n}{1}(-3x) + \frac{n(n-1)}{1 \times 2}(-3x)^2 + \frac{n(n-1)(n-2)}{1 \times 2 \times 3}(-3x)^3 + ... + (-3x)^n$

So for **$n = 12$** this is:

$(1 - 3x)^{12} = 1 + \tfrac{12}{1}(-3x) + \frac{12 \times 11}{1 \times 2}(-3x)^2 + \frac{12 \times 11 \times 10}{1 \times 2 \times 3}(-3x)^3 + ... + (-3x)^{12}$

The **term in x^5** is: $\frac{12!}{5!7!}(-3)^5 x^5 = \binom{12}{5}(-3x)^5 = (792 \times -243)x^5$

$$= -192456x^5$$

b) **Find the coefficient of x^2 in the expansion of $(1 + 6x)^4(1 - 2x)^6$.**

To find the x^2 term in the combined expansion you'll need to find all the **terms up to x^2** in both expansions (because these will form the x^2 term when they're multiplied) and then **multiply** together:

- $(1 + 6x)^4 = 1 + \binom{4}{1}(6x) + \binom{4}{2}(6x)^2 + ... = 1 + 24x + 216x^2 + ...$

- $(1 - 2x)^6 = 1 + \binom{6}{1}(-2x) + \binom{6}{2}(-2x)^2 + ... = 1 - 12x + 60x^2 - ...$

- So: $(1 + 6x)^4(1 - 2x)^6 = (1 + 24x + 216x^2 + ...)(1 - 12x + 60x^2 - ...)$

$$= 1 - 12x + 60x^2 - ... + 24x - 288x^2 + 1440x^3 - ...$$
$$+ 216x^2 - 2592x^3 + 12\,960x^4 - ...$$
$$= 1 + 12x - 12x^2 + (\textit{terms with higher powers}) + ...$$

- So the coefficient of x^2 is: -12

Tip: Most of this line of working isn't needed — you only really need to pick out the terms in the brackets which will multiply to give x^2. These are just $(1 \times 60x^2)$, $(24x \times -12x)$ and $(216x^2 \times 1)$ which give $60x^2$, $-288x^2$ and $216x^2$.

Approximations

- Once you've found a binomial expansion, you can use it to expand **longer expressions** or to find **approximations** of a number raised to a power.
- Approximating usually involves taking a really small value for x so that you can **ignore high powers** of x (because they'll be really, really small).
- For example, if you've expanded $(1 + x)^9$ and you're asked to approximate 1.001^9, then you just stick $x = 0.001$ into your expansion. Because 0.001 is small, 0.001^2 is really small and adding on really small terms won't make much difference, so just the **first few terms** will give a good approximation.

Tip: For approximations like this you're only expected to use the terms of the expansion you've already found the coefficients for (unless you're told otherwise).

Example 1

a) **Find the first 4 terms, in ascending powers of x, of the expansion of $(1 + 2x)^7$.**

$$(1 + 2x)^7 = 1 + {}^7C_1(2x) + {}^7C_2(2x)^2 + {}^7C_3(2x)^3 + ...$$
$$= 1 + 7(2x) + 21(4x^2) + 35(8x^3) + ...$$
$$= 1 + 14x + 84x^2 + 280x^3 + ...$$

b) **When x is small, x^3 and higher powers can be ignored. Hence show that for small x:**
$$(2 - x)(1 + 2x)^7 \approx 1 + 27x + 154x^2$$

Multiply your expansion of $(1 + 2x)^7$ through by $(2 - x)$ — you only need to include the terms up to the one in x^2 as you're told to ignore terms in x^3 and above in the question:

$$(2 - x)(1 + 2x)^7 \approx (2 - x)(1 + 14x + 84x^2)$$
$$= 2 + 28x + 168x^2 - x - 14x^2 - 84x^3$$
$$= 2 + 27x + 154x^2 - 84x^3 \approx 2 + 27x + 154x^2$$

Tip: You can find the binomial coefficients using any of the methods from pages 84-87 — just pick whichever one you prefer.

Example 2

a) **Find the first 3 terms of the expansion of $\left(1 - \frac{x}{4}\right)^9$.**

Use the **formula**, but replace x with $\left(-\frac{x}{4}\right)$:

$$\left(1 - \frac{x}{4}\right)^9 = 1 + \binom{9}{1}\left(-\frac{x}{4}\right) + \binom{9}{2}\left(-\frac{x}{4}\right)^2 + ... = 1 - 9\left(\frac{x}{4}\right) + 36\left(\frac{x^2}{16}\right) - ...$$
$$= 1 - \frac{9}{4}x + \frac{9}{4}x^2 - ...$$

b) **Use your expansion to estimate $(0.998)^9$.**

$(0.998)^9 = (1 - 0.002)^9 = \left(1 - \frac{x}{4}\right)^9$ when $x = 0.008$.

So **substitute** $x = 0.008$ into the expansion you've just found — the first three terms are enough as 0.008^3 and higher powers will be very small:

$$(0.998)^9 = \left(1 - \frac{0.008}{4}\right)^9 \approx 1 - \frac{9}{4}(0.008) + \frac{9}{4}(0.008)^2$$
$$= 1 - 0.018 + 0.000144 = 0.982144$$

Tip: If you do 0.998^9 on your calculator you get $0.98214333...$ so this is a pretty good approximation.

Q1 b) Hint: You'll find some of the terms vanish when you subtract the second expansion.

Q1 Find the full expansions of:
   a) $(1 - x)^6$
   b) $(1 + x)^9 - (1 - x)^9$

Q2 Find the first 3 terms in the expansion of $(1 + x)^3(1 - x)^4$.

Q3 Find the coefficient of x^3y^2 in the expansion of $(1 + x)^5(1 + y)^7$.

Q4 Find the full expansion of $(1 - 2x)^5$.

Q2 Hint: You'll need to go up to x^2 in both expansions, then multiply the expansions together to find all the terms up to x^2 in the combined expansion.

Q5 a) Find the first 4 terms of the binomial expansion $(1 - 3x)^6$.
   b) If x is small, so that x^2 and higher powers can be ignored, show that $(1 + x)(1 - 3x)^6 \approx 1 - 17x$.

Q6 Find the first 4 terms, in ascending powers of x, of the binomial expansion of $(1 + kx)^8$, where k is a non-zero constant.

Q7 a) Find, in their simplest form, the first 5 terms in the expansion of $\left(1 + \frac{x}{2}\right)^{12}$, in ascending powers of x.
   b) Use the expansion to work out the value of 1.005^{12} to 7 d.p.

Binomial expansions — $(a + b)^n$

When your binomial is of the form $(a + b)^n$ (e.g. $(2 + 3x)^7$, where $a = 2$ and $b = 3x$) you can use a slightly **different formula**:

Tip: The powers of a decrease (from n to 0) as the powers of b increase (from 0 to n). The sum of the powers of a and b is always n.

$$(a + b)^n = a^n + \binom{n}{1}a^{n-1}b + \binom{n}{2}a^{n-2}b^2 + ... + \binom{n}{n-1}ab^{n-1} + b^n$$

This formula is on the formula sheet and you don't need to know the proof, but seeing where it comes from might make things a bit clearer.
You can find it from the binomial formula you've already seen:

- First rearrange so the binomial's in a form you can work with:

$$(a + b)^n = \left(a\left(1 + \tfrac{b}{a}\right)\right)^n = a^n\left(1 + \tfrac{b}{a}\right)^n$$

- You expand this by putting '$\frac{b}{a}$' into the **binomial formula** for $(1 + x)^n$, just like in the previous section:

$$= a^n\left(1 + \binom{n}{1}\left(\tfrac{b}{a}\right) + \binom{n}{2}\left(\tfrac{b}{a}\right)^2 + ... + \binom{n}{n-1}\left(\tfrac{b}{a}\right)^{n-1} + \left(\tfrac{b}{a}\right)^n\right)$$

$$= a^n\left(1 + \binom{n}{1}\tfrac{b}{a} + \binom{n}{2}\tfrac{b^2}{a^2} + ... + \binom{n}{n-1}\tfrac{b^{n-1}}{a^{n-1}} + \tfrac{b^n}{a^n}\right)$$

- **Multiply** through by a^n:

$$= a^n + \binom{n}{1}a^{n-1}b + \binom{n}{2}a^{n-2}b^2 + ... + \binom{n}{n-1}ab^{n-1} + b^n$$

This is a general formula that works for any a and b, including 1 and x.
So given **any binomial** you can pop your values for a, b and n into this formula and you'll get the **expansion**.

Example 1

Give the first 3 terms, in ascending powers of x, of the expansion of $(4 - 5x)^7$.

- Use the **formula**:

$$(a + b)^n = a^n + \binom{n}{1}a^{n-1}b + \binom{n}{2}a^{n-2}b^2 + ... + \binom{n}{n-1}ab^{n-1} + b^n$$

- In this case, $a = 4$, $b = -5x$ and $n = 7$.

- $(4 - 5x)^7 = (4 + (-5x))^7 = 4^7 + \left(\binom{7}{1} \times 4^6 \times (-5x)\right) + \left(\binom{7}{2} \times 4^5 \times (-5x)^2\right) + ...$

$= 16\,384 + (7 \times 4096 \times -5x) + (21 \times 1024 \times 25x^2) + ...$

$= \boxed{16\,384 - 143\,360x + 537\,600x^2 + ...}$

Tip: Be careful with b here — there's a minus sign that might catch you out.

Your other option with expansions of $(a + b)^n$ is to **factorise** the binomial so you get $a^n\left(1 + \frac{b}{a}\right)^n$, then plug $\frac{b}{a}$ into the **original binomial formula** (as you did with $(1 + ax)^n$ expansions in the last section).

Example 2

What is the coefficient of x^4 in the expansion of $(2 + 5x)^7$?

- Factorise: $(2 + 5x) = 2\left(1 + \frac{5}{2}x\right)$, so $(2 + 5x)^7 = 2^7\left(1 + \frac{5}{2}x\right)^7$.

- So the expansion $(2 + 5x)^7$ is the same as the expansion of $\left(1 + \frac{5}{2}x\right)^7$ multiplied by 2^7.

- Find the coefficient of x^4 in the expansion of $\left(1 + \frac{5}{2}x\right)^7$.

The term is: $\binom{7}{4} \times \left(\frac{5}{2}x\right)^4 = \frac{7 \times 6 \times 5 \times 4}{1 \times 2 \times 3 \times 4} \times \frac{5^4}{2^4}x^4 = 35 \times \frac{5^4}{2^4}x^4 = \frac{21875}{16}x^4$

So the coefficient is: $\frac{21875}{16}$

Tip: If you do it this way, make sure you don't forget to multiply back through by a^n when you give your final answer.

- Multiply this by 2^7 to get the coefficient of x^4 in the original binomial:

$2^7 \times \frac{21875}{16} = \boxed{175\,000}$

You can find an **unknown** in a binomial expansion if you're given some information about the coefficients:

Example 3

a) The coefficient of x^5 in the binomial expansion of $(4 + kx)^7$ is 81 648. Find k.
- From the $(a + b)^n$ formula, the term in x^5 of this expansion is:
$$^7C_5 4^2(kx)^5 = 21 \times 16 \times k^5 \times x^5 = 336k^5x^5$$

- So the coefficient of x^5 is $336k^5$.

- $336k^5 = 81\,648 \Rightarrow k^5 = 243 \Rightarrow \boxed{k = 3}$

Tip: Just as with the $(1 + x)^n$ formula, you need to be familiar with all the ways of writing the binomial coefficients.

b) In the expansion of $(1 + x)^n$, the coefficient of x^5 is twice the coefficient of x^4. What is the value of n?

- The coefficient of x^5 is $\frac{n!}{5!(n - 5)!}$, the coefficient of x^4 is $\frac{n!}{4!(n - 4)!}$.

- The coefficient of x^5 is twice the coefficient of x^4, so:

$$\frac{n!}{5!(n-5)!} = 2 \times \frac{n!}{4!(n-4)!}$$

$$\Rightarrow \quad \frac{1}{5!(n-5)!} = 2 \times \frac{1}{4!(n-4)!}$$

Cancel the $n!$

$$\Rightarrow \quad \frac{1}{5 \times 4! \times (n-5)!} = 2 \times \frac{1}{4! \times (n-4) \times (n-5)!}$$

$$\Rightarrow \quad \frac{1}{5} = 2 \times \frac{1}{(n-4)}$$

Cancel the 4! and the $(n-5)!$

$$\Rightarrow \quad n - 4 = 10$$

$$\Rightarrow \quad n = 14$$

- To check: $^{14}C_5 = 2002$, $^{14}C_4 = 1001$, i.e. $^{14}C_5 = 2 \times {}^{14}C_4$

Tip: $5! = 5 \times 4!$ and $(n-4)! = (n-4) \times (n-5)!$

Tip: It's really easy to check your answer for questions like this, so make sure you do.

The value of n is 14

Exercise 2.3

Q1 Find the first 4 terms of the binomial expansion of $(3 + x)^6$.

Q2 Find the full expansion of $(2 + x)^4$.

Q3 In the expansion of $(1 + \lambda x)^8$ the coefficient of x^5 is 57344.
   a) Work out the value of λ.
   b) Find the first 3 terms of the expansion.

Q4 a) Find the first 5 terms in the expansion of $(2 + x)^8$.
   b) Use this expansion to find an approximation for 2.01^8 to 5 d.p.

Q4 b) Hint: This is the same method as the approximation example for $(1 + ax)^n$ expansions on page 89.

Q5 Find the first 4 terms in the expansion of $(3 + 5x)^7$.

Q6 a) Find the first 5 terms in the expansion of $(3 + 2x)^6$.
   b) Use this expansion to find the first 5 terms of $(1 + x)(3 + 2x)^6$.

Q7 The term in x^2 for the expansion of $(1 + x)^n$ is $231x^2$.
   a) What is the value of n?   b) What is the term in x^3?

Q8 In the expansion of $(a + 3x)^8$, the coefficient of x^2 is $\frac{32}{27}$ times bigger than the coefficient of x^5. What is the value of a?

Q9 In the expansion of $(1 + 2x)^5(3 - x)^4$, what is the coefficient of x^3?

Q10 In the expansion of $(1 + x)^n$, the coefficient of x^3 is 3 times larger than the coefficient of x^2.
   a) Calculate the value of n.
   b) If the coefficient of x^2 is $a \times$ (the coefficient of x), what is a?

Q11 a) Find the first 3 terms, in ascending powers of x, of the binomial expansion of $(2 + \mu x)^8$, where μ is a constant. Write each term in its simplest form.
   b) If the coefficient of x^2 is 87 808, what are the possible values of μ?

Review Exercise — Chapter 5

Q1 Find the common ratio of the geometric progression 3125, 1875, 1125, 675, 405, ...

Q2 Write an expression for the nth term of the geometric sequence 3, –9, 27, –81, 243, ...

Q3 For the geometric progression 2, –6, 18, ..., find:
a) the 10th term,
b) the sum of the first 10 terms.

Q4 Find the sum of the first 12 terms of the following geometric series:
a) 2 + 8 + 32 + ...
b) 30 + 15 + 7.5 + ...

Q5 A geometric series has first term $a = 7$ and common ratio $r = 0.6$. Find $\sum_{k=0}^{5} 7(0.6)^k$ to 2 d.p.

Q6 Find the common ratio for the following geometric series.
State which ones are convergent and which are divergent.

a) 1 + 2 + 4 + ...

b) 81 + 27 + 9 + ...

c) $1 + \frac{1}{3} + \frac{1}{9} + ...$

d) $4 + 1 + \frac{1}{4} + ...$

Q7 For the geometric progression 24, 12, 6, ..., find:
a) the common ratio,
b) the seventh term,
c) the sum of the first 10 terms,
d) the sum to infinity.

Q8 A geometric progression begins 2, 6, ...
Which term of the geometric progression equals 1458?

Q9 A charity received £20 000 of donations from the public one year (Year 1). The charity predict that the public donations will increase by 8% each year, forming a geometric sequence.
a) Show that their predicted donations from the public the following year are £21 600.
b) Write down the common ratio of the geometric sequence.
c) Write down an expression for the predicted public donations in Year n.
d) Find the total amount of public donations the charity will get
(if their prediction is correct) in the ten years from Year 1 to Year 10.

Q10 Show that the sum of the first n terms of a geometric sequence with first term a
and common ratio r is:
$$S_n = \frac{a(1 - r^n)}{1 - r}$$

Q11 A geometric series has first term $a = 33$, common ratio $r = 0.25$. Find $\displaystyle\sum_{k=0}^{\infty} ar^k$ for this series.

Q12 Give, in their simplest form, the first four terms in the binomial expansion of $(1 + x)^{12}$.

Q13 Find the first 3 terms, in ascending powers of x, of the binomial expansion of $(1 - x)^{20}$.

Q14 What is the term in x^4 in the expansion of $(1 - 2x)^{16}$?

Q15 a) Find the first 4 terms of the expansion of $\left(1 + \frac{x}{3}\right)^9$ in ascending powers of x,
         giving each term in its simplest form.
     b) Use your expansion to estimate the value of $(1.003)^9$,
         giving your answer to 6 decimal places.

Q16 Give the full expansion of the binomial $(1 + 3x)^5$.

Q17 a) Find the first 5 terms, in ascending powers of x, of the expansion of $(1 + ax)^8$,
         where a is a non-zero constant.
     Given that the coefficient of x^2 in this expansion is double the coefficient of x^3,
     b) find the value of a
     c) find the coefficient of x.

Q18 Find the first 3 terms of the binomial expansion of $(4 - 5x)^7$.
     Give each term in its simplest form.

Q19 Find the coefficient of x^2 in the expansion of $(2 + 3x)^5$.

Q20 a) Find the first 5 terms, in ascending powers of x, of the expansion of $\left(3 + \frac{x}{4}\right)^{11}$.
         Give each term in its simplest form.
     b) Use your expansion to find an estimate for the value of $(3.002)^{11}$.
         Give your answer to 3 decimal places.

Q21 a) Find the first 3 terms of the binomial expansion of $(2 + kx)^{13}$,
         where k is a non-zero constant.
     b) Given that the coefficient of x in this expansion is $\frac{1}{6}$ of the coefficient of x^2,
         find the value of k.

1   Find the coefficients of x, x^2, x^3 and x^4 in the binomial expansion of $(4 + 3x)^{10}$.

(4 marks)

2   A geometric series has the first term 12 and is defined by: $u_{n+1} = 12 \times 1.3^n$.

   a)   Is the series convergent or divergent?

(1 mark)

   b)   Find the values of the 3rd and 10th terms.

(2 marks)

3   In a geometric series, $a = 20$ and $r = \frac{3}{4}$.

   Find values for the following, giving your answers to 3 significant figures where necessary:

   a) S_∞

(2 marks)

   b) u_{15}

(2 marks)

   c)   The smallest value of n for which $S_n > 79.76$:

(5 marks)

4   To raise money for charity, Alex, Chris and Heather were sponsored £1 for each kilometre they ran over a 10-day period.
   They receive sponsorship proportionally for partial kilometres completed.
   Alex ran 3 km every day.
   Chris ran 2 km on day 1 and on each subsequent day ran 20% further than the day before.
   Heather ran 1 km on day 1, on each subsequent day she ran 50% further than the previous day.

   a)   How far did Heather run on day 5, to the nearest 10 metres?

(2 marks)

   b)   Show that day 10 is the first day that Chris runs further than 10 km.

(3 marks)

   c)   Find the total amount raised by the end of the 10 days, to the nearest penny.

(4 marks)

5   Two different geometric series have the same second term and sum to infinity:

$$u_2 = 5 \quad \text{and} \quad S_\infty = 36.$$

   a)   Show that $36r^2 - 36r + 5 = 0$, where r represents the two possible ratios.

(4 marks)

   b)   Hence find the values of r, and the corresponding first terms, for both geometric series.

(4 marks)

6   a)   Find the first 3 terms of the expansion of $\left(1 + \frac{x}{3}\right)^8$ in ascending powers of x.
         Give each term in its simplest form.

(3 marks)

   b)   Hence estimate the value of $(1.002)^8$ to 4 decimal places.

(3 marks)

7   a)   Find, to 2 decimal places, the sum of the first 8 terms of the geometric series

$$2 + 5 + 12.5 + 31.25 + ...$$

(3 marks)

     b)   State the condition for an infinite geometric series to be convergent.

(1 mark)

     c)   Find the sum to infinity of the geometric series that has first term 8 and common ratio $-\frac{3}{4}$, giving your answer to 2 decimal places.

(2 marks)

8   a)   Find the first 3 terms, in ascending powers of x, of the binomial expansion of $(3 + kx)^9$, where k is a non-zero constant. Give each term in its simplest form.

(3 marks)

Given that the coefficient of x in this expansion is $\frac{3}{4}$ of the coefficient of x^2

     b)   find the value of the constant k.

(2 marks)

9    $a + ar + ar^2 + ar^3 + ...$ is a geometric series.
The second term of the series is -2 and the sum to infinity of the series is -9.

     a)   Show that $9r^2 - 9r + 2 = 0$.

(3 marks)

     b)   Find the possible values of r.

(2 marks)

     c)   Hence find the possible values of a.

(2 marks)

Given that r takes its smallest possible value,

     d)   find the 7th term in the series to 4 decimal places,

(2 marks)

     e)   find the sum of the first 5 terms to 2 decimal places.

(2 marks)

10   a)   Find, in ascending powers of x, the first three terms of the expansion $(1 + 3x)^6$, giving each term in its simplest form.

(3 marks)

     b)   Assuming x is small so that x^3 and higher powers can be ignored, show that

$$(1 - 2x)(1 + 3x)^6 \approx 1 + 16x + 99x^2$$

(2 marks)

1. Differentiation

In C1 you used differentiation to find the gradient of a curve. In C2, you'll develop this — but we'll start with a quick recap of the rules of differentiation you'll have already covered.

Learning Objective:

- Be able to use all of the rules of differentiation covered in C1.

Differentiation

You came across differentiation in C1 — it's the process of finding the **gradient** of a curve. At any point, the gradient of a curve is the same as the gradient of the **tangent** to the curve at that point.

Tip: Remember, a tangent is a straight line which just touches a curve at a certain point, without going through it.

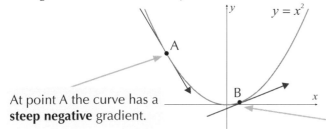

At point A the curve has a **steep negative** gradient.

At point B the curve has a **shallow positive** gradient.

To differentiate a function in the form $f(x) = x^n$, you can use this formula:

$$\text{If } y = x^n \text{ then } \frac{dy}{dx} = nx^{n-1}$$

Tip: In $f(x)$ notation, this just says:
If $f(x) = x^n$
then $f'(x) = nx^{n-1}$.

- If there's a **constant** in front of x^n, it **stays** where it is when you differentiate.

Example 1

Differentiate $y = 4x^3$

$$y = 4x^3 \implies \frac{dy}{dx} = 4(3x^2) = \boxed{12x^2}$$

- Sometimes you have to write n as a **fraction** or **negative** number before you can differentiate.

Tip: Make sure you're really happy with the laws of indices.

Example 2

a) **Differentiate $y = \dfrac{1}{x}$**

$$y = \frac{1}{x} = x^{-1} \implies \frac{dy}{dx} = -1 \times x^{-2} = \boxed{-\frac{1}{x^2}}$$

b) **Find $f'(x)$ if $f(x) = \sqrt{x}$**

$$y = \sqrt{x} = x^{\frac{1}{2}} \implies \frac{dy}{dx} = \frac{1}{2} \times x^{-\frac{1}{2}} = \boxed{\frac{1}{2\sqrt{x}}}$$

- If you have a more complicated function, you might need to **multiply out brackets**, **rearrange** the function and/or **cancel denominators**.

Example 3

a) **Differentiate** $y = \dfrac{x^2 - 4x - 5}{x + 1}$.

$$y = \frac{x^2 - 4x - 5}{x + 1} = \frac{(x + 1)(x - 5)}{x + 1} = x - 5 \;\Rightarrow\; \frac{dy}{dx} = \boxed{1}$$

b) **Find the gradient of the curve** $y = (x + 1)(x + 2)$ **when** $x = 1$.

$$y = (x + 1)(x + 2) = x^2 + 3x + 2 \;\Rightarrow\; \frac{dy}{dx} = 2x + 3$$

When $x = 1$, gradient $= \dfrac{dy}{dx} = 2(1) + 3 = \boxed{5}$

Exercise 1.1

Q1 Find $f'(x)$ for the following functions:

   a) $f(x) = 3x^2 + 7$ b) $f(x) = 5x^4 + x^2 + 1$

   c) $f(x) = (x - 3)^2$ d) $f(x) = (3x + 1)(x - 2)$

Q2 Find $f'(0)$ for the following functions:

Q2 Hint: "Find $f'(0)$" means differentiate the function and then find its value when $x = 0$.

   a) $f(x) = x^6 + 3x^2 + 2x$ b) $f(x) = (x + 3)^2$

   c) $f(x) = (2x + 1)(2x - 1)$ d) $f(x) = 5x^4 + 3x + 7$

Q3 Differentiate the following equations:

   a) $y = \dfrac{2}{x}$ b) $y = x^2 + \dfrac{1}{x}$

   c) $y = \dfrac{2x + 3}{x^3}$ d) $y = x^{-3} + 2x^{-4}$

Q4 Find $\dfrac{dy}{dx}$ for the following equations:

   a) $y = 4\sqrt{x}$ b) $y = x^2 + 7\sqrt{x}$

   c) $y = (x - \sqrt{x})^2$ d) $y = \sqrt{x}(1 + \sqrt{x})$

Q5 Differentiate the following equations:

   a) $y = \sqrt{x}(x + \sqrt{x})^2$ b) $y = (2 + \dfrac{1}{x})(x + 1)$

Q6 a) Given that $f(x) = \sqrt{x}(x^2 + 3)(x^2 - 3)$, show that $f'(x) = \dfrac{9}{2\sqrt{x}}(x^4 - 1)$

   b) Find all the values for which $f'(x) = 0$.

Q7 A function is given by $f(x) = x^3 + kx^2 + kx$. Given that $f'(0) = 3$, find k.

2. Using Differentiation

In this section you'll use differentiation to find a graph's stationary points and determine their nature. You'll also see how to find where a function is increasing or decreasing and, using this information, make accurate sketches of functions.

Stationary points

Stationary points occur when the **gradient** of a graph is **zero**.
There are three types of stationary point:

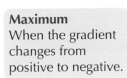
Maximum
When the gradient changes from positive to negative.

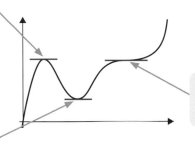

Point of inflexion
When the graph briefly flattens out.

Minimum
When the gradient changes from negative to positive.

Because stationary points occur when the gradient is zero, you can use **differentiation** to find them:

> **1.** Differentiate f(x).
>
> **2.** Set f$'$(x) = 0.
>
> **3.** Solve f$'$(x) = 0 to find the x-values.
>
> **4.** Put the x-values back into the original equation to find the y-values.

Learning Objectives:

- Be able to use differentiation to find all the stationary points on a curve.
- Be able to identify the nature of these stationary points.
- Be able to work out where a function is increasing or decreasing.
- Be able to use this information to make an accurate sketch of the graph of a function.

Tip: Some stationary points are called local maximum or minimum points because the function takes on higher or lower values in other parts of the graph. The maximum and minimum points shown opposite are both local.

Example 1

Find the stationary points on the curve $y = 2x^3 - 3x^2 - 12x + 5$.

- You need to find where $\dfrac{dy}{dx} =$ f$'$(x) = 0, so start by **differentiating** the function:

$$y = 2x^3 - 3x^2 - 12x + 5 \quad \Rightarrow \quad \frac{dy}{dx} = 6x^2 - 6x - 12$$

- Then set the derivative **equal to zero**:

$$6x^2 - 6x - 12 = 0$$

- Now **solve** this equation — it's just a normal quadratic:

$$6x^2 - 6x - 12 = 0 \;\Rightarrow\; x^2 - x - 2 = 0$$
$$\Rightarrow\; (x + 1)(x - 2) = 0$$
$$\Rightarrow\; x = -1 \text{ or } x = 2$$

- You've found the x-values of the **stationary points**. To find the coordinates of the stationary points, just put these x-values into the original equation. This gives the coordinates (–1, 12) and (2, –15).

Tip: Don't forget this last step — once you've found x you need to also find y.

Example 2

Below is a sketch of the graph of $y = x^3(x^2 + x - 3)$. One stationary point occurs at (–1.8, 9.1). Show that the other two occur when $x = 0$ and when $x = 1$, and find their coordinates.

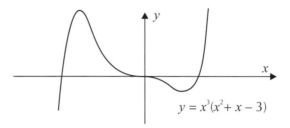

$$y = x^3(x^2 + x - 3)$$

- You need to start by **differentiating** the function, but you can't do that in its current form. So first, **multiply out** the brackets:

$$y = x^3(x^2 + x - 3) = x^5 + x^4 - 3x^3$$

- Then you can just differentiate as normal:

$$y = x^5 + x^4 - 3x^3 \;\Rightarrow\; \frac{dy}{dx} = 5x^4 + 4x^3 - 9x^2$$

- Stationary points occur when the **gradient** is **equal to zero**, so set $\frac{dy}{dx}$ equal to zero and solve for x:

$$5x^4 + 4x^3 - 9x^2 = 0 \;\Rightarrow\; x^2(5x^2 + 4x - 9) = 0$$
$$\Rightarrow\; x^2(5x + 9)(x - 1) = 0$$
$$\Rightarrow\; x = 0, x = -\frac{9}{5} = -1.8 \text{ (given above) and } x = 1.$$

So the other two stationary points occur at $x = 0$ and $x = 1$.

- To find the **coordinates** of these points, just put the x-values into the original equation:

$$y = x^3(x^2 + x - 3) = 0^3(0^2 + 0 - 3) = 0(-3) = 0$$
$$y = x^3(x^2 + x - 3) = 1^3(1^2 + 1 - 3) = 1(-1) = -1$$

- So the coordinates of the stationary points are (0, 0) and (1, –1).

Q1 Without doing any calculations, say how many stationary points the graphs below have in the intervals shown.

a)

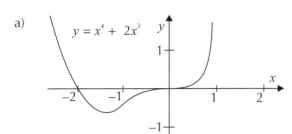

$y = x^4 + 2x^3$

b)

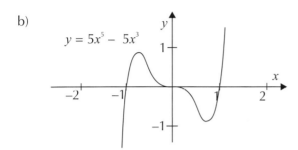

$y = 5x^5 - 5x^3$

Q2 Find the x-coordinates of the stationary points of the curves with the following equations:

a) $y = x^2 + 3x + 2$ 　　　　 b) $y = (3 - x)(4 + 2x)$

Q3 Find the coordinates of the stationary points of the curves with the following equations:

a) $y = 2x^2 - 5x + 2$ 　　　　 b) $y = -x^2 + 3x - 4$

c) $y = 7 - 6x - 3x^2$ 　　　　 d) $y = (x - 1)(2x + 3)$

Q4 Find the coordinates of the stationary points of the curves with the following equations:

a) $y = x^3 - 3x + 2$ 　　　　 b) $y = 4x^3 + 5$

Q5 Show that the graph of the function given by $f(x) = x^5 + 3x + 2$ has no stationary points.

Q5 Hint: If there are no stationary points, there are no values of x for which $f'(x) = 0$.

Q6 a) Differentiate $y = x^3 - 7x^2 - 5x + 2$

b) Hence find the coordinates of the stationary points of the curve with equation $y = x^3 - 7x^2 - 5x + 2$.

Q7 A graph is given by the function $f(x) = x^3 + kx$, where k is a constant. Given that the graph has no stationary points, find the range of possible values for k.

Maximum and minimum points

Tip: That's what a question means when it asks you to "determine the nature of the turning points".

Once you've found where the stationary points are, you might be asked to decide if each one is a **maximum** or **minimum**. Maximum and minimum points are also known as **turning points**.

To decide whether a stationary point is a maximum or minimum, **differentiate again** to find $\frac{d^2y}{dx^2}$ or f''(x) (you did this back in C1).

If $\frac{d^2y}{dx^2} < 0$, it's a maximum

$\frac{dy}{dx} = 0$

$\frac{dy}{dx} > 0$ $\frac{dy}{dx} < 0$

$\frac{dy}{dx}$ is decreasing, so $\frac{d^2y}{dx^2}$ is negative

Tip: If the second derivative is equal to zero, you can't tell what type of stationary point it is.

If $\frac{d^2y}{dx^2} > 0$, it's a minimum

$\frac{dy}{dx} = 0$

$\frac{dy}{dx} < 0$ $\frac{dy}{dx} > 0$

$\frac{dy}{dx}$ is increasing, so $\frac{d^2y}{dx^2}$ is positive

Example

Determine the nature of the stationary points in Example 1 on p.99-100 ($y = 2x^3 - 3x^2 - 12x + 5$).

- The first derivative has been found already: $\frac{dy}{dx} = 6x^2 - 6x - 12$.
  To determine the nature of the stationary points, **differentiate again**:

$$\frac{dy}{dx} = 6x^2 - 6x - 12 \quad \Rightarrow \quad \frac{d^2y}{dx^2} = 12x - 6$$

- Then just put in the x-values of the coordinates of the **stationary points**.

Tip: You found the coordinates of the stationary points for this function on p.100. They are (–1, 12) and (2, –15).

- At $x = -1$, $\frac{d^2y}{dx^2} = -18$, which is **negative** — so (–1, 12) is a maximum.

- And at $x = 2$, $\frac{d^2y}{dx^2} = 18$, which is **positive** — so (2, –15) is a minimum.

- Since you know the **turning points** and the fact that it's a **cubic** with a positive coefficient of x^3, you can now **sketch** the graph (though the points of intersection with the x-axis would be difficult to find accurately).

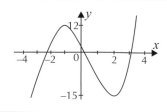

Q1 The diagram below shows a sketch of the graph of $y = f(x)$. For each turning point, say whether $f''(x)$ would be positive or negative.

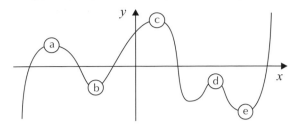

Q2 Find the second derivative of these functions:

a) $y = x^3 + 2x^2 + 5x - 7$
b) $y = 3x^4 - 2$

c) $y = 4x^7 - 3x + 2$
d) $y = x^5 - 10x^3 - 13$

e) $y = 5 - 7x + x^2$
f) $y = 6 - 2x$

g) $y = (x + 2)(3x - 4)$
h) $y = \dfrac{1}{x^2} + 2x^3 + x$

Q3 A function $y = f(x)$ is such that $f(1) = 3$, $f'(1) = 0$ and $f''(1) = 7$.

a) Give the coordinates of one of the turning points of $f(x)$.

b) Determine the nature of this turning point, explaining your answer.

Q4 Find the stationary points on the graphs of the following functions and say whether they're maximum or minimum turning points:

a) $y - 5 - x^2$
b) $y = 2x^3 - 6x + 2$

c) $y = x^3 - 3x^2 - 24x + 15$
d) $y = x^4 + 4x^3 + 4x^2 - 10$

Q5 Find the stationary points on the graphs of the following functions and say whether they're maximum or minimum turning points:

a) $f(x) = 8x^3 + 16x^2 + 8x + 1$
b) $f(x) = \dfrac{27}{x^3} + x$

Q6 a) Given that $f(x) = x^3 - 3x^2 + 4$, find $f'(x)$ and $f''(x)$.

b) Hence find the coordinates of any stationary points on the graph $f(x)$ and say whether they're maximum or minimum turning points.

Q7 Walter makes different sized fishbowls. The volume of each bowl is given by $V = r^2 + \dfrac{2000}{r}$, where r is the radius of the bowl.

a) Find the value of r at which the volume, V, is stationary.

b) Is this a minimum or maximum point?

> **Q7 Hint:** This question's no different to the others in this exercise — just treat V as y and r as x, and carry on as normal.

Q8 The curve given by the function $f(x) = x^3 + ax^2 + bx + c$ has a stationary point with coordinates (3, 10). If $f''(x) = 0$ at (3, 10), find a, b and c.

Q9 a) Given that a curve with the equation $y = x^4 + kx^3 + x^2 + 17$ has only one stationary point, show that $k^2 < \dfrac{32}{9}$.

b) Find the coordinates of the stationary point and say whether it's a maximum or a minimum point.

Increasing and decreasing functions

As differentiation is about finding the gradients of curves, you can use it to find if a function is **increasing** or **decreasing** at a given point. This can help you to sketch the function and determine the nature of turning points.

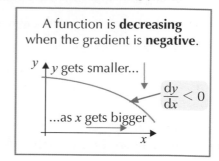

| A function is **increasing** when the gradient is **positive**. | A function is **decreasing** when the gradient is **negative**. |

You can also tell how **quickly** a function is increasing or decreasing by looking at the size of the gradient — the **bigger** the gradient (positive or negative), the **faster** the function is increasing or decreasing.

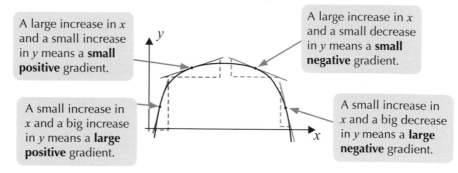

A large increase in x and a small increase in y means a **small positive** gradient.

A large increase in x and a small decrease in y means a **small negative** gradient.

A small increase in x and a big increase in y means a **large positive** gradient.

A small increase in x and a big decrease in y means a **large negative** gradient.

Example

Find the values of x for which the function $y = x^3 - 6x^2 + 9x + 3$, $x > 0$ is increasing.

- You want to know when y is increasing — so **differentiate**.

$$y = x^3 - 6x^2 + 9x + 3 \quad \Rightarrow \quad \frac{dy}{dx} = 3x^2 - 12x + 9$$

- It's an **increasing** function when the derivative is **greater** than zero, so write it down as an inequality and solve it.

$$\frac{dy}{dx} > 0 \quad \Rightarrow \quad 3x^2 - 12x + 9 > 0 \quad \Rightarrow \quad x^2 - 4x + 3 > 0 \quad \Rightarrow \quad (x-3)(x-1) > 0$$

- For the inequality to be true, either **both** brackets must be **positive** or **both** brackets must be **negative**:

$$x - 1 > 0 \text{ and } x - 3 > 0 \quad \Rightarrow \quad x > 1 \text{ and } x > 3 \quad \Rightarrow \quad \boxed{x > 3}$$

$$\text{or } x - 1 < 0 \text{ and } x - 3 < 0 \quad \Rightarrow \quad x < 1 \text{ and } x < 3 \quad \Rightarrow \quad \boxed{x < 1}$$

- $x > 0$, so the function is increasing when $\boxed{0 < x < 1 \text{ and } x > 3.}$

- You could also look at the nature of the **stationary points** — this will tell you where the function goes from increasing to decreasing and vice versa.

$x = 1$ and $x = 3$ at the **stationary points** (as $\frac{dy}{dx} = 0$ at these points).

Tip: You might have come across a different method for solving quadratic inequalities in C1 — use whichever one you prefer.

$$\frac{dy}{dx} = 3x^2 - 12x + 9 \quad \Rightarrow \quad \frac{d^2y}{dx^2} = 6x - 12$$

- When $x = 1$, $\frac{d^2y}{dx^2} = -6$, so it's a **maximum**, which means the function is increasing as it approaches $x = 1$ and starts decreasing after $x = 1$.

- When $x = 3$, $\frac{d^2y}{dx^2} = 6$, so it's a **minimum**, which means the function is decreasing as it approaches $x = 3$ and starts increasing after $x = 3$.

This fits in with what you know already — that the function is increasing when $0 < x < 1$ and $x > 3$.

Exercise 2.3

Q1 For each of these functions, calculate the first derivative and use this to find the range of values for which the function is increasing.

    a) $y = x^2 + 7x + 5$ b) $y = 5x^2 + 3x - 2$ c) $y = 2 - 9x^2$

Q2 For each of these functions, find $f'(x)$ and find the range of values of x for which $f(x)$ is decreasing.

    a) $f(x) = 16 - 3x - 2x^2$ b) $f(x) = (6 - 3x)(6 + 3x)$

    c) $f(x) = (1 - 2x)(7 - 3x)$

Q3 Calculate $\frac{dy}{dx}$ for each of these functions and state the range of values for which the function is increasing.

    a) $y = x^3 - 6x^2 - 15x + 25$ b) $y = x^3 + 6x^2 + 12x + 5$

Q4 Find the first derivative of each function and state the range of values for which the function is decreasing.

    a) $f(x) = x^3 - 3x^2 - 9x + 1$ b) $f(x) = x^3 - 4x^2 + 4x + 7$

Q5 Use differentiation to explain why $f(x) = x^3 + x$ is an increasing function for all real values of x.

> **Q5 Hint:** An increasing function is one where $f'(x) > 0$ for all values of x.

Q6 Is the function $f(x) = 3 - 3x - x^3$ an increasing or decreasing function? Explain your answer.

Q7 Use differentiation to find the range of values of x for which each of these functions is decreasing:

    a) $y = 2x^4 + x$ b) $y = x^4 - 2x^3 - 5x^2 + 6$

Q8 Differentiate these functions and find the range of values for which each function is increasing.

    a) $y = x^2 + \sqrt{x}$, $x > 0$ b) $y = 4x^2 + \frac{1}{x}$, $x \neq 0$

> **Q9, 10 Hint:** See which values of the variable satisfy the conditions of an increasing or decreasing function.

Q9 The function $y = 5 - 3x - ax^5$ is a decreasing function for all $x \in \mathbb{R}$. Find the range of possible values for a.

Q10 The function $y = x^k + x$, where k is a positive whole number, is an increasing function for all $x \in \mathbb{R}$. Find all possible values of k.

Curve sketching

You covered some curve sketching in C1, so you should know the basic shapes of different types of graph. In this topic you'll see how differentiation can be used to find out more about the **shape** of the graph and to work out some **key points** like the turning points. Use the following **step-by-step** method to get all the information you need to draw an accurate graph:

1. **Find where the curve crosses the axes.**
   - To find where it crosses the **y-axis**, just put $x = 0$ into the function and find the value of y.
   - To find where it crosses the **x-axis**, set the function equal to zero and solve for x (you'll probably have to **factorise** and find the **roots**).

2. **Decide on the shape of the graph.**
   - Look at the **highest power** of x and its **coefficient** — this determines the overall **shape** of the graph (have a look back at your C1 notes). The most common ones are **quadratics**, **cubics** and **reciprocals**.
   - A **quadratic** with a **positive** coefficient of x^2 will be **u-shaped**, and if the coefficient is **negative**, it'll be **n-shaped**.
   - A **cubic** will go from **bottom left** to **top right** if the coefficient of x^3 is **positive**, and **top left** to **bottom right** if the coefficient is **negative**. It'll also have a characteristic 'wiggle'.
   - **Reciprocals** (e.g. $\frac{1}{x}$) and other **negative powers** have **two separate curves** in **opposite quadrants**, each with **asymptotes**.

3. **Differentiate to find the stationary points.**
   - Find the **stationary points** (by **differentiating** and setting $f'(x) = 0$).
   - Then **differentiate again** to decide whether these points are **maximums** or **minimums**.

Tip: If you want, you could find where the function is increasing or decreasing as well.

Example 1

Sketch the curve of the equation $y = f(x)$, where $f(x) = x^3 - 4x^2 + 4x$.

- Start by finding where the curve **crosses the axes**. When $x = 0$, $y = 0$, so the curve goes through the origin. Find where it crosses the x-axis by solving the equation $f(x) = 0$:

$$x^3 - 4x^2 + 4x = 0 \quad \Rightarrow \quad x(x^2 - 4x + 4) = 0 \quad \Rightarrow \quad x(x - 2)(x - 2) = 0$$

$$\Rightarrow \boxed{x = 0 \text{ or } x - 2}$$

- Next find the **stationary points** on the graph by finding $f'(x)$ and solving $f'(x) = 0$:

$$f(x) = x^3 - 4x^2 + 4x \quad \Rightarrow \quad f'(x) = 3x^2 - 8x + 4$$

$$f'(x) = 0 \quad \Rightarrow \quad 3x^2 - 8x + 4 = 0 \quad \Rightarrow \quad (3x - 2)(x - 2) = 0$$

$$\Rightarrow \boxed{x = 2 \Rightarrow y = 0} \quad \text{or} \quad \boxed{x = \tfrac{2}{3} \Rightarrow y = \tfrac{32}{27} \ (\approx 1.2)}$$

Tip: If you find it helpful, you can also work out where the graph is increasing and decreasing — it's increasing when $x < \frac{2}{3}$ and when $x > 2$, and decreasing when $\frac{2}{3} < x < 2$.

- Differentiate again to find out if these are **maximums** or **minimums**:
$$f''(x) = 6x - 8$$

At $x = 2$, $f''(x) = 4$, so this is a minimum.

At $x = \frac{2}{3}$, $f''(x) = -4$, so this is a maximum.

- It's a cubic equation with a positive coefficient of x^3, so the graph will go from bottom left to top right.
- Notice that the x-intercept $x = 2$ is also the minimum.
- Now you have all the information you need to sketch the graph:

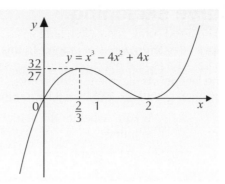

Tip: This is a cubic equation so you should know the general shape from C1.

Example 2

Sketch the graph of $f(x) = -8x^3 + 2x$.

- Again, start by finding where the curve **crosses the axes**. When $x = 0$, $f(x) = 0$ so the curve goes through the origin. Now solve $f(x) = 0$:

$$-8x^3 + 2x = 0 \Rightarrow 2x(-4x^2 + 1) = 0$$

So $x = 0$ or $-4x^2 + 1 = 0 \Rightarrow x^2 = \frac{1}{4} \Rightarrow \boxed{x = \pm\frac{1}{2}}$

- Next **differentiate** the function to find the **stationary point(s)**:

$$f(x) = -8x^3 + 2x \Rightarrow f'(x) = -24x^2 + 2$$

$f'(x) = 0$ so $-24x^2 + 2 = 0 \Rightarrow x^2 = \frac{1}{12} \Rightarrow x = \pm\sqrt{\frac{1}{12}} = \pm\frac{1}{2\sqrt{3}}$

$x = \dfrac{1}{2\sqrt{3}}$ gives $f(x) = \dfrac{2}{3\sqrt{3}}$ and $x = -\dfrac{1}{2\sqrt{3}}$ gives $f(x) = -\dfrac{2}{3\sqrt{3}}$.

- Then **differentiate again** to see if these points are **maximums** or **minimums**:
$f''(x) = -48x$.

At $x = \dfrac{1}{2\sqrt{3}}$, $f''(x) = -\dfrac{24}{\sqrt{3}}$, which is negative so it's a **maximum**.

At $x = -\dfrac{1}{2\sqrt{3}}$, $f''(x) = \dfrac{24}{\sqrt{3}}$, which is positive so it's a **minimum**.

- Finally, think about the overall shape of the graph. The highest power is 3, so it's a cubic and has a negative coefficient, so the graph will go from top left to bottom right.

Tip: Don't forget the characteristic cubic 'wiggle'.

- Now you have all the information you need to sketch the graph:

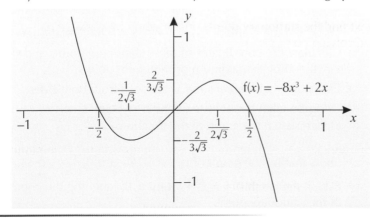

Q1 For the graph $y = x^3 - 2x^2$:

    a) Find the coordinates of the points at which it crosses each axis.

    b) Find $\dfrac{dy}{dx}$ and hence the coordinates of the points where $\dfrac{dy}{dx} = 0$.

    c) Work out whether the stationary points are maximums or minimums.

    d) Sketch the graph of $y = x^3 - 2x^2$.

Q2 a) Solve the equation $x^3 + x^2 = 0$.

    b) Find the stationary points of the graph of $f(x) = x^3 + x^2$ and say whether they're maximum or minimum points.

    c) Use your answers to parts a) and b) to sketch the graph of $f(x) = x^3 + x^2$, labelling the coordinates of the stationary points and places where the curve meets the axes.

Q3 a) Find the first and second derivatives of the function $f(x) = x^4 - x^3$.

    b) Write down the ranges of values of x for which $f(x)$ is increasing and decreasing.

    c) Sketch the graph of $y = f(x)$, labelling the coordinates of all stationary points and the points where the curve crosses the axes.

Q4 Sketch the graphs of the equations below, labelling the coordinates of any stationary points and the points where the curves cross the axes.

    a) $y = 3x^3 + 3x^2$ b) $y = -x^3 + 9x$

    c) $y = x^4 - x^2$ d) $y = x^4 + x^2$

Q5 Given that $x^3 - x^2 - x + 1 = (x + 1)(x - 1)^2$, sketch the graph of $y = x^3 - x^2 - x + 1$, labelling the coordinates of all the stationary points and the points where the curve crosses the axes.

Q6 a) Show that $x^3 - 4x = 0$ when $x = -2$, 0 and 2.

    b) Use first and second derivatives to show that the graph of $y = x^3 - 4x$ has a minimum at $(1.2, -3.1)$ and a maximum at $(-1.2, 3.1)$, where all coordinates are given to 1 d.p.

    c) Use your answers to parts a) and b) to sketch the graph of $y = x^3 - 4x$, labelling the coordinates of the stationary points and the points at which the curve crosses the axes.

Q7 Hint: The function is undefined at $x = 0$, so the graph won't just go straight through the y-axis (it 'jumps' from one side to the other).

Q7 a) Show that the graph of $f(x) = x + \dfrac{1}{x}$, $x \neq 0$ has 2 stationary points.

    b) Calculate the coordinates of these stationary points and say whether they're maximum or minimum points.

    c) Describe what happens to $f(x)$ as $x \to 0$ from both sides.

    d) Describe what happens to $f(x)$ as $x \to \infty$ and $x \to -\infty$.

    e) Hence sketch the graph of the function $x + \dfrac{1}{x}$.

Q8 Hint: Consider what will happen as $x \to 0$ and $x \to \infty$, like you did in Q7.

Q8 a) Show that for the graph of $y = x^4 + \dfrac{8}{\sqrt{x}}$, $x > 0$, $\dfrac{dy}{dx} = 0$ when $x = 1$.

    b) Sketch the graph of $y = x^4 + \dfrac{8}{\sqrt{x}}$, $x > 0$, labelling the coordinates of the stationary point.

3. Real-Life Problems

In real-life contexts you can use differentiation to find an optimum solution to a problem — e.g. finding the maximum size you can get out of a set amount of material, or finding the minimum material you need to use to make an object of given size.

Learning Objective:

- Be able to describe a real-life situation in mathematical terms and use differentiation to find maximum and minimum solutions.

Differentiation in real-life problems

Because differentiation can be used to find the maximum value of a function, it can be used in **real-life problems** to maximise a quantity subject to certain factors, e.g maximising the volume of a box that can be made with a set amount of cardboard.

To find the maximum value of something, all you need is an equation **in terms of only one variable** (e.g. x) — then just **differentiate as normal**.
Often there'll be too many variables in the question, so you've got to know how to manipulate the information to get rid of the unwanted variables.

Example 1

Ceara the farmer wants to build a sheep pen with base of length x m and width y m. She has 20 m of fencing in total, and wants the area inside the pen to be as large as possible. How long should each side of the pen be, and what will the area inside the pen be?

- Start by writing down an expression for the **area** of the pen:

$$\text{Area} = \text{length} \times \text{width} = xy \text{ m}^2$$

- This has **too many variables** for you to be able to work with, so you need to find an expression for y in terms of x. You know how much fencing is available, so find an expression for that in terms of x and y and rearrange it to make y the subject.

$$\text{Perimeter} - 20 \text{ m} = 2x + 2y \quad \rightarrow \quad y = \frac{20 - 2x}{2} = 10 - x.$$

- Now you can substitute this into the expression you wrote down for the area and use **differentiation** to **maximise** it:

$$A = xy = x(10 - x) = 10x - x^2, \quad \text{so} \quad A = 10x - x^2 \Rightarrow \frac{dA}{dx} = 10 - 2x$$

- Now just find when $\frac{dA}{dx} = 0$

$$\frac{dA}{dx} = 0 \quad \Rightarrow \quad 10 - 2x = 0, \quad \text{so} \quad x = 5 \quad \Rightarrow \quad y = 10 - x = 5$$

- To check that this value of x gives a maximum for A, **differentiate again**:

$$\frac{d^2A}{dx^2} = -2, \text{ which is negative, so this will give a maximum for } A.$$

- So both x and y should be 5 m and the total area inside the pen will be $5 \times 5 =$ 25 m².

Example 2

A jewellery box with a lid and dimensions $3x$ cm by x cm by y cm is made using a total of 450 cm² of wood.

Show that the volume of the box can be expressed as: $V = \dfrac{675x - 9x^3}{4}$, and use calculus to find the maximum volume.

Tip: 'Use calculus' here just means differentiate.

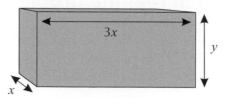

- You know the basic equation for volume:

$$V = \text{length} \times \text{width} \times \text{height} = 3x \times x \times y = \boxed{3x^2y}$$

- This has a y that you want to **get rid of** so look for a way of **replacing** y with an equation in x (like in example 1). You can do this by finding an expression for the surface area of the box, which you know is 450 cm²:

$$\text{Surface area} = 2 \times [(3x \times x) + (3x \times y) + (x \times y)] = 450 \quad \Rightarrow \quad 6x^2 + 8xy = 450$$

$$\Rightarrow \quad y = \frac{450 - 6x^2}{8x} = \frac{225 - 3x^2}{4x}$$

- And now you can substitute this into the expression for the volume of the box:

$$V = 3x^2y = 3x^2\left(\frac{225 - 3x^2}{4x}\right) = \boxed{\frac{675x - 9x^3}{4}}$$

- Now just **differentiate** and find x at the **stationary point(s)**:

$$V = \frac{675x - 9x^3}{4} \quad \Rightarrow \quad \frac{dV}{dx} = \frac{675 - 27x^2}{4}$$

$$\text{When } \frac{dV}{dx} = 0, \ \frac{675 - 27x^2}{4} = 0 \quad \Rightarrow \quad x^2 = \frac{675}{27} = 25 \quad \Rightarrow \quad x = 5$$

Tip: x is a length so it can't have a negative value (–5).

- Check that V is actually a maximum at $x = 5$, then just calculate V with $x = 5$:

$$\frac{d^2V}{dx^2} = -\frac{27x}{2} \quad \text{So when } x = 5, \ \frac{d^2V}{dx^2} = -\frac{135}{2} \ \text{(so } V \text{ will be a maximum)}$$

$$V = \frac{675x - 9x^3}{4} \quad \text{So when } x = 5, \ V = \frac{675(5) - 9(5^3)}{4} = \boxed{562.5 \text{ cm}^3}$$

Differentiation isn't just limited to cuboids — it can be used on **any shape** as long as you can describe its (surface) area or volume with variables (i.e x, y).

Example 3

Ned uses a circular tin to bake his pies in. The tin is t cm high with a d cm diameter. The volume of the pie tin is 1000 cm³.

Prove that the surface area of the tin is given by $A = \frac{\pi}{4}d^2 + \frac{4000}{d}$ and find the minimum surface area.

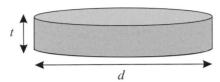

- A = area of tin's base + area of tin's curved face = $\pi(\frac{d}{2})^2 + (\pi d \times t)$.

$$= \frac{\pi d^2}{4} + \pi d t$$

Tip: The pie tin has no lid so you only need to add the surface area of one circle.

- To get rid of the t, find an equation for the volume and **rearrange it** to make t the subject, then sub that into the equation for surface area:

$$V = \pi(\frac{d}{2})^2 t = 1000 \quad \Rightarrow \quad t = \frac{1000}{\pi(\frac{d}{2})^2} = \frac{4000}{\pi d^2}$$

$$\Rightarrow A = \frac{\pi d^2}{4} + \pi d t = \frac{\pi d^2}{4} + (\pi d \times \frac{4000}{\pi d^2}) = \boxed{\frac{\pi d^2}{4} + \frac{4000}{d}}$$

- Next, **differentiate** with respect to d and find the values of d that make $\frac{dA}{dd} = 0$:

$$\frac{dA}{dd} = \frac{\pi d}{2} - \frac{4000}{d^2} \text{ so when } \frac{dA}{dd} = 0, \ \frac{\pi d}{2} - \frac{4000}{d^2} = 0 \Rightarrow d^3 = \frac{2 \times 4000}{\pi}$$

$$\Rightarrow d = \frac{20}{\sqrt[3]{\pi}}$$

- Check to see if this value of d gives a **minimum** for A:

$$\frac{d^2 A}{dd^2} = \frac{\pi}{2} + \frac{8000}{d^3} = \frac{\pi}{2} + \frac{8000}{\left(\frac{8000}{\pi}\right)} = \frac{3\pi}{2}, \text{ so it's a minimum.}$$

- Now calculate the area for that value of d:

$$A = \frac{\pi}{4}\left(\frac{20}{\sqrt[3]{\pi}}\right)^2 + \frac{4000}{\left(\frac{20}{\sqrt[3]{\pi}}\right)} = \boxed{439 \text{ cm}^2} \text{ (to 3 s.f.)}$$

Exercise 3.1

Q1 A rectangular vegetable patch is enclosed by a wall on one side and fencing on three sides as shown in the diagram.

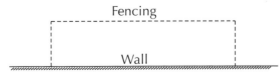

Fencing

Wall

Q1 Hint: It might look like you're not given enough information here, but just call the length x and the width y and you're on your way.

Use calculus to show that the maximum possible area that can be enclosed by 66 m of fencing is 544.5 m².

Q2 Hint: This question is a lot like the example on p.109, but instead of finding the maximum area for a given perimeter, you're finding the minimum perimeter for a given area.

Q2 A farmer wants to enclose a rectangular area of 100 m² with a fence. Find the minimum length of fencing he needs to use.

Q3 A ball is catapulted vertically with an initial speed of 30 m/s. After t seconds the height h of the ball, in m, is given by $h = 30t - 4.9t^2$. Use calculus to find the maximum height the ball reaches.

Q4 A pet food manufacturer designs tins of cat food of capacity 500 cm³ as shown. The radius of the tin is r cm and the height is h cm.

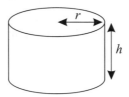

a) Show that the surface area A of the tin is given by $A = 2\pi r^2 + \dfrac{1000}{r}$.

b) Find the value of r which minimises the surface area (to 3 s.f.).

c) Find the minimum possible surface area for the tin (to 3 s.f.).

Q5 A child makes a box by taking a piece of card 40 × 40 cm and cutting squares with side length x cm, as shown in the diagram. The sides are then folded up to make a box.

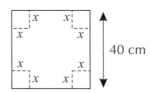

a) Write down a formula for the volume of the box V.

b) Find the maximum possible volume of the box to 3 s.f.

Q6 A chocolate manufacturer designs a new box which is a triangular prism as shown in the diagram. The cross-section of the prism is a right-angled triangle with sides of length x cm, x cm and h cm. The length of the prism is l cm and the volume is 300 cm³.

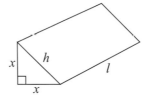

a) Show that the surface area of the prism is given by
$A = x^2 + \dfrac{600(2 + \sqrt{2})}{x}$.

b) Show that the value of x which minimises surfaces area of the box is $\sqrt[3]{600 + 300\sqrt{2}}$.

Review Exercise — Chapter 6

Q1 Find the stationary points of the graph of $y = x^3 - 6x^2 - 63x + 21$.

Q2 a) Find the stationary points of the graph of the function $y = x^3 + \frac{3}{x}$.

     b) Work out whether each stationary point is a maximum or a minimum.

Q3 Find all the stationary points of the graph of $y = 2x^4 - x^2 + 4$ and determine their nature.

Q4 Find when these two functions are increasing and decreasing:

     a) $y = 6(x + 2)(x - 3)$

     b) $y = \frac{1}{x^2}$

Q5 Sketch the graph of $y = 3x^3 - 16x$, clearly showing the coordinates of any turning points.

Q6 Sketch the graph of $y = -3x^3 + 6x^2$, clearly showing the coordinates of any turning points.

Q7 Given that $xy = 20$ and that both x and y are positive find the least possible value of $x^2 + y^2$.

Q8 The height (h m) a firework can reach is related to the mass (m g) of fuel it carries as shown below:

$$h = \frac{m^2}{10} - \frac{m^3}{800}$$

Find the mass of fuel required to achieve the maximum height and state what the maximum height is to 3 s.f.

Q9 The diagram shows a box with dimensions x cm, $2x$ cm and y cm, and volume 200 cm³.

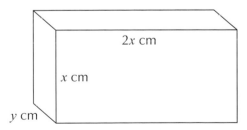

     a) Show that the surface area A of the box is given by $A = 4x^2 + \frac{600}{x}$.

     b) Use calculus to find the value of x that gives the minimum value of A (to 3 s.f.).

     c) Hence find the minimum possible surface area of the box, correct to 3 s.f.

Exam-Style Questions — Chapter 6

1   a)   Find $\dfrac{dy}{dx}$ for the curve with equation $y = 6 + \dfrac{4x^3 - 15x^2 + 12x}{6}$.

                 (3 marks)

     b)   Hence, find the coordinates of the stationary points on the curve.

                 (5 marks)

     c)   Determine the nature of each stationary point.

                 (3 marks)

2   A steam train travels between Haverthwaite and Eskdale at a speed of x miles per hour and burns y units of coal per hour, where $y = 2\sqrt{x} + \dfrac{27}{x}$, for $x > 2$.

     a)   Find the speed that gives the minimum rate of coal consumption.

                 (5 marks)

     b)   Find $\dfrac{d^2y}{dx^2}$, and hence verify that the speed found in part a) gives the minimum rate of coal consumption.

                 (2 marks)

     c)   Calculate the minimum rate of coal consumption.

                 (1 mark)

3   Lotte wants to build an open-top plastic fish tank for her goldfish with a capacity of 40 000 cm³, and with sides of length x, x and y cm, as shown in **Figure 1**.

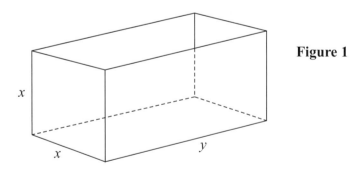

Figure 1

     a)   Show that the surface area of the tank in cm² is given by $A = 2x^2 + \dfrac{120\ 000}{x}$

                 (4 marks)

     b)   Find the value of x at which A is stationary to 3 s.f., and show that this is a minimum value of A.

                 (6 marks)

     c)   Calculate the minimum area of plastic needed to build the tank, to 3 s.f.

                 (2 marks)

4 Ayesha is building a closed-back bookcase. She uses a total of 72 m^2 of wood (not including shelving) to make a bookcase that is x metres high, $\frac{x}{2}$ metres wide and d metres deep, as shown.

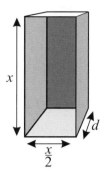

a) Show that the full capacity of the bookcase is given by: $V = 12x - \frac{x^3}{12}$.

(4 marks)

b) Find the value of x for which V is stationary. Leave your answer in surd form.

(4 marks)

c) Show that this gives a maximum value for V and hence calculate the maximum of V.

(4 marks)

5 The function $f(x) = \frac{1}{2}x^4 - 3x$ has a single stationary point.

a) Find the coordinates of the stationary point to 3 d.p.

(3 marks)

b) Determine the nature of the stationary point.

(2 marks)

c) State the range of values of x for which $f(x)$ is:

(i) increasing,

(1 mark)

(ii) decreasing.

(1 mark)

d) Sketch the curve for the function $f(x) = \frac{1}{2}x^4 - 3x$.

(2 marks)

6 a) Determine the coordinates of the stationary points for the curve $y = (x - 1)(3x^2 - 5x - 2)$, giving your answers to 3 s.f.

(4 marks)

b) Find whether each of these points is a maximum or minimum.

(3 marks)

c) Sketch the curve of y, showing where the curve crosses the axes.

(3 marks)

1. Integration

You've already seen indefinite integrals in C1. This chapter will cover definite integrals (integrating between two limits) and how to use them to find areas.

Indefinite integrals

Learning Objectives:

- Be able to find indefinite integrals and evaluate definite integrals.
- Be able to find the area between a curve and the x-axis using definite integration.
- Be able to use integration to calculate the area between a line and a curve.

- Integration is just the **opposite** of differentiation — it gets you from $\frac{dy}{dx}$ back to y.
- The integral of a function f(x) with respect to x is written:

$\int$ means **the integral of.** $\longrightarrow$ $\int f(x)\, dx$ $\longleftarrow$ dx means **with respect to** x.

- **Indefinite** integrals can have lots of different answers, so you need to add a **constant of integration**, C, which represents any number.

- If you need to integrate a function with **lots of terms**, just integrate **each term** separately.

Tip: You should know all this from C1 — this is just a recap.

To integrate a **power of x**, use the following formula:

$$\int x^n\, dx = \frac{x^{n+1}}{n+1} + C$$

(i) **Increase the power** by one — then divide by it.

(ii) Add a **constant**.

- Any **constant** multiplying the **power of x** stays as it is — you can take it **outside** the integral if you want.

Examples

Tip: You can check your answer is right by differentiating it — you should end up with what you started with.

a) Find $\int 2x^5\, dx$.

Increase the power to 6...

$$\int 2x^5\, dx = \frac{2x^6}{6} + C = \frac{x^6}{3} + C$$

...divide by 6....

...and add a constant of integration.

b) Find $\int \frac{1}{\sqrt[6]{x}}\, dx$.

Write the function as a **power of x** before trying to integrate.

Add 1 to the power...

$$\int \frac{1}{\sqrt[6]{x}}\, dx = \int \frac{1}{x^{\frac{1}{6}}}\, dx = \int x^{-\frac{1}{6}}\, dx = \frac{x^{\frac{5}{6}}}{\left(\frac{5}{6}\right)} + C = \frac{6}{5}\sqrt[6]{x^5} + C$$

...then divide by $\frac{5}{6}$...

...and add a constant of integration.

- Integrating the expression for the **gradient** of a curve gives you the **equation** of the curve.
- But because of the constant of integration, C, it actually gives a **family of curves** which all have the **same gradient**.
- If you're given the coordinates of a **point** on a curve as well as its gradient, you can **find C** and use it to find the **equation** of that curve.

Example

The curve $y = f(x)$ passes through the point (1, 5) and $\frac{dy}{dx} = 18x^5 - 12x^2 + 5x$. Find the equation of the curve.

- Integrate each term separately to find y:

$$y = \int \frac{dy}{dx}\,dx = \int (18x^5 - 12x^2 + 5x)\,dx$$

$$= \frac{18x^6}{6} - \frac{12x^3}{3} + \frac{5x^2}{2} + C = 3x^6 - 4x^3 + \frac{5}{2}x^2 + C$$

Tip: Each integration will give a constant of integration, but you can just group them into one constant C.

- To find the full equation, you now need to find C — you do this by using the fact that it goes through the point (1, 5).

$$y = 3x^6 - 4x^3 + \frac{5}{2}x^2 + C$$

Putting $x = 1$ and $y = 5$ in the above equation gives...

$$5 = 3(1)^6 - 4(1)^3 + \frac{5}{2}(1)^2 + C$$

$$\Rightarrow C = \frac{7}{2}$$

- So the equation of the curve is: $\quad y = 3x^6 - 4x^3 + \frac{5}{2}x^2 + \frac{7}{2}$

Exercise 1.1

Q1 Integrate the following functions with respect to x:

a) $9x^8$

b) $3x^2$

c) $8x^7 + 2x$

d) $-8x^3 + 5x^2 - 3$

e) $6x^2 - 2x^3$

f) $x^4 - x^2 - 3$

g) $\frac{3}{4}x^2 + 7x$

h) $\frac{1}{3}x^8 - \frac{1}{3}x^2 + 6$

i) $-\frac{3}{x^4}$

j) $\frac{1}{3}x^{-\frac{2}{3}}$

k) $\sqrt{x}$

l) $\frac{2}{x^3} + 2x$

m) $(\sqrt{x})^3$

n) $(2x + 3)^2$

o) $\frac{x^5 + 7}{x^2}$

p) $\frac{(3 - x)(2 + 3x)}{\sqrt{x}}$

Q2 A curve with equation $y = f(x)$ passes through the point (−1, 1) and $\frac{dy}{dx} = 5x^4 - 4x^3 + 5$. Find $f(x)$.

Q3 a) A curve with gradient $\frac{dy}{dx} = 4\sqrt{x}$ passes through the origin. Find the equation of the curve.

b) A curve with gradient $\frac{dy}{dx} = \frac{4}{\sqrt[3]{x}}$ passes through the point (1, 2). Find the equation of the curve.

Definite integrals

Tip: If you're integrating with respect to a different variable, say t, then the limits tell you the range of t-values instead.

Definite integrals have **limits** (little numbers) next to the integral sign. The limits just tell you the **range of x-values** to integrate the function between.

> The definite integral of **f(x)** with respect to x between the limits $x = a$ and $x = b$ is written:
>
> The upper limit goes here.
>
> $$\int_a^b f(x)\,dx$$
>
> The lower limit goes here.

Tip: You might be asked to 'evaluate' a definite integral — it just means to find the value.

Finding a definite integral isn't really any harder than an indefinite one — there's just an **extra stage** you have to do.

- Integrate the function as normal but **don't** add a **constant of integration**.
- Once you've integrated the function, work out the **value** of the definite integral by **putting in the limits**:

> If you know that the integral of f(x) is $\int f(x)\,dx = g(x) + C$ then:
>
> $$\int_a^b f(x)\,dx = [g(x)]_a^b = g(b) - g(a)$$
>
> **Subtract** the value of g at the **lower** limit from the value of g at the **upper** limit.

Tip: The proper way to write out definite integrals is to use square brackets with the limits to the right like this.

Example

Evaluate $\int_1^3 (x^2 + 2)\,dx$.

- Find the integral in the normal way — but put the integrated function in **square brackets** and rewrite the **limits** on the right-hand side.

$$\int_1^3 (x^2 + 2)\,dx = \left[\frac{x^3}{3} + 2x\right]_1^3$$

Notice that there's no constant of integration.

- Put in the limits:

Put the upper limit into the integral...

...then subtract the value of the integral at the lower limit.

$$\left[\frac{x^3}{3} + 2x\right]_1^3 = \left(\frac{3^3}{3} + 6\right) - \left(\frac{1^3}{3} + 2\right)$$

$$= 15 - \frac{7}{3} = \frac{38}{3}$$

Tip: A definite integral always comes out as a number.

The area under a curve

The value of a **definite integral** represents the **area under** the graph of the function you're integrating between the two limits.

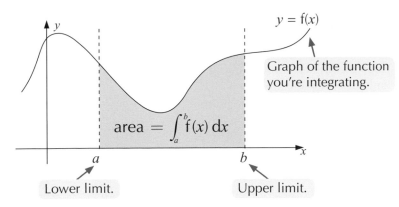

area $= \int_a^b f(x)\,dx$

Graph of the function you're integrating.

Lower limit.

Upper limit.

Example

Find the area between the graph of $y = x^2$, the x-axis and the lines $x = -1$ and $x = 2$.

You just need to integrate the function $f(x) = x^2$ between -1 and 2 with respect to x.

The limits of integration are -1 and 2. $f(x) = x^2$

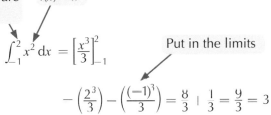

$$\int_{-1}^{2} x^2\,dx = \left[\frac{x^3}{3}\right]_{-1}^{2}$$

Put in the limits

$$-\left(\frac{2^3}{3}\right) - \left(\frac{(-1)^3}{3}\right) = \frac{8}{3} + \frac{1}{3} = \frac{9}{3} = 3$$

So the area is 3.

If you integrate a function to find an area that lies **below** the x-axis, it'll give a **negative** value.

If you need to find an area like this, you'll need to make your answer **positive** at the end as you can't have **negative** area.

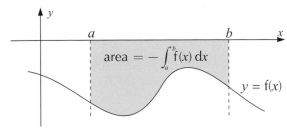

area $= -\int_a^b f(x)\,dx$

$y = f(x)$

Tip: It's important to note that you're actually finding the area between the curve and the **x-axis**, not the area under the curve (the area below a curve that lies under the x-axis will be infinite).

Example

Find the area between the graph of
$y = 4x - 3x^2 - x^3$ and the x-axis
between $x = -4$ and $x = 0$.

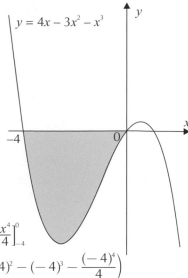

- You can see from the sketch of the graph that the area you're trying to find lies **below** the x-axis.

- So all you have to do is integrate the curve between the given limits and then make the area **positive** at the end.

$$\int_{-4}^{0}(4x - 3x^2 - x^3)\,dx = \left[2x^2 - x^3 - \frac{x^4}{4}\right]_{-4}^{0}$$
$$= (0) - \left(2(-4)^2 - (-4)^3 - \frac{(-4)^4}{4}\right)$$
$$= 0 - (32 + 64 - 64)$$
$$= -32$$

- So the area between the curve and the x-axis between $x = -4$ and $x = 0$ is 32.

If you need to find the area for a portion of a curve which lies both **above** and **below** the x-axis, you'll need to find the areas above and below **separately** and add them up at the end so that the negative and positive integrals don't **cancel each other out**.

Examples

a) **Evaluate $\int_{-2}^{2} x^3\,dx$.**

$$\int_{-2}^{2} x^3\,dx = \left[\frac{x^4}{4}\right]_{-2}^{2} = \left(\frac{2^4}{4}\right) - \left(\frac{(-2)^4}{4}\right) = \frac{16}{4} - \frac{16}{4} = 0$$

b) **Find the area between the graph of $y = x^3$, the x-axis**
and the lines $x = -2$ and $x = 2$.

- You'd usually just integrate the function between the limits, but this gives 0 in part a).

- But you can see from the diagram on the next page that the area is not 0.

- The 'negative area' below the axis has **cancelled out** the positive area.

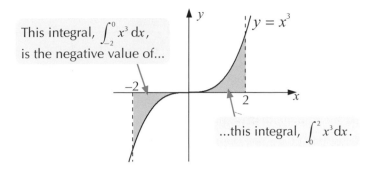

This integral, $\int_{-2}^{0} x^3 \, dx$, is the negative value of...

...this integral, $\int_{0}^{2} x^3 \, dx$.

- So to work out the total area, you need to work out the positive and negative areas separately and then **add** them together.

- The area **above** the x-axis is between 0 and 2, so integrate the function between these limits:

$$\int_{0}^{2} x^3 = \left[\frac{x^4}{4}\right]_{0}^{2} = \frac{1}{4}\left[x^4\right]_{0}^{2} = \frac{1}{4}(2^4 - 0^4) = \frac{16}{4} = 4$$

Tip: Sometimes it's easier to take a common factor outside the square brackets — here, I've taken out ¼.

- The area **below** the x-axis lies between −2 and 0, so integrate the function between these limits:

$$\int_{-2}^{0} x^3 = \left[\frac{x^4}{4}\right]_{-2}^{0} = \frac{1}{4}\left[x^4\right]_{-2}^{0} = \frac{1}{4}(0^4 - (-2)^4) = -\frac{16}{4} = -4$$

The area is just the positive value of this integral, so the area is 4.

- Finally, **add** together the areas to get the total area:

$$\text{Area} = 4 + 4 = \boxed{8}$$

Integrating to Infinity

- You can find **definite integrals** where one of the limits is ∞ or $-\infty$ — it just means that you keep integrating forever in the x-direction.

- Most integrals to infinity (or minus infinity) will give you an **infinite** answer.

- However, if a curve **tends to 0** as $x \to \infty$ (or $x \to -\infty$) then the integral to infinity (or minus infinity) can sometimes be **finite**.

$y = f(x)$

f(x) tends to 0 as x tends to infinity.

This area is $\int_{a}^{\infty} f(x) \, dx$.

It goes on forever but gets closer and closer to a **finite** value.

Tip: Remember, the notation $x \to \infty$ means 'x tends to infinity' and just means x gets larger and larger.

- In these cases, you can find the **area under the curve** for $x \geq a$ (integrate from a to ∞) or $x \leq a$ (integrate from $-\infty$ to a).

- An integral to infinity is written in the same way as a definite integral except one of the limits is **replaced** with ∞ or $-\infty$.

To work out the value of an integral to infinity:

$$\int_a^\infty f(x)\,dx = \int_a^n f(x)\,dx \text{ as } n \to \infty \qquad \int_{-\infty}^a f(x)\,dx = \int_n^a f(x)\,dx \text{ as } n \to -\infty$$

- Replace the ∞ (or $-\infty$) with an n.
- Integrate the function and put in the limits as normal — you'll end up with an expression containing n's.
- Work out what happens to the expression as $n \to \infty$ (or $n \to -\infty$).

Tip: You'll often find that terms in the expression which contain n will tend to 0 and disappear.

Example

Find the area under the curve $y = \dfrac{15}{x^2} - \dfrac{30}{x^3}$ **for** $x \geq 2$.

- To find this area, A, you need to integrate from $x = 2$ up to infinity.

$$A = \int_2^\infty \left(\frac{15}{x^2} - \frac{30}{x^3}\right) dx$$

$y = \frac{15}{x^2} - \frac{30}{x^3}$

Curve continues → forever in x direction

- Replace ∞ with n and integrate:

$$\int_2^n \left(\frac{15}{x^2} - \frac{30}{x^3}\right) dx = 15 \int_2^n (x^{-2} - 2x^{-3})\,dx$$

$$= 15\left[\frac{x^{-1}}{-1} - \frac{2x^{-2}}{(-2)}\right]_2^n = 15\left[-\frac{1}{x} + \frac{1}{x^2}\right]_2^n$$

$$= 15\left[\left(-\frac{1}{n} + \frac{1}{n^2}\right) - \left(-\frac{1}{2} + \frac{1}{4}\right)\right] = 15\left(-\frac{1}{n} + \frac{1}{n^2} + \frac{1}{4}\right)$$

- As $n \to \infty$, $\dfrac{1}{n} \to 0$ and $\dfrac{1}{n^2} \to 0$.

So $15\left(-\dfrac{1}{n} + \dfrac{1}{n^2} + \dfrac{1}{4}\right) \to 15\left(-0 + 0 + \dfrac{1}{4}\right) = \dfrac{15}{4}$ as $n \to \infty$.

- So the area under the curve for $x \geq 2 = \int_2^\infty \left(\dfrac{15}{x^2} - \dfrac{30}{x^3}\right) dx = \boxed{\dfrac{15}{4}}$

Exercise 1.2

Q1 Hint: Don't let minus signs catch you out. Remember you're subtracting the whole bracket so if there's a minus sign in the bracket, it'll become positive.

Q1 Find the value of the following, giving exact answers:

a) $\displaystyle\int_1^3 3x^2\,dx$ b) $\displaystyle\int_{-2}^0 (4x^3 + 2x)\,dx$ c) $\displaystyle\int_{-2}^5 (x^3 + x)\,dx$

d) $\displaystyle\int_{-5}^{-2} (x + 1)^2\,dx$ e) $\displaystyle\int_1^4 x^{-2}\,dx$ f) $\displaystyle\int_2^7 (x^{-3} + x)\,dx$

g) $\displaystyle\int_3^4 (6x^{-4} + x^{-2})\,dx$ h) $\displaystyle\int_1^2 \left(x^2 + \frac{1}{x^2}\right) dx$

Q2 Given that $\displaystyle\int_0^a x^3\,dx = 64$, find a, where $a > 0$.

Q3 Evaluate the following, giving exact answers:

a) $\int_0^1 \sqrt{x}\, dx$

b) $\int_8^{27} \sqrt[3]{x}\, dx$

c) $\int_0^9 (x^2 + \sqrt{x})\, dx$

d) $\int_1^4 (3x^{-4} + \sqrt{x})\, dx$

e) $\int_0^1 (2x + 3)(x + 2)\, dx$

f) $\int_1^4 \frac{1}{\sqrt{x}}\, dx$

Q4 Find the exact value of the following definite integrals:

a) $\int_1^4 \frac{x^2 + 2}{\sqrt{x}}\, dx$

b) $\int_0^1 (\sqrt{x} + 1)^2\, dx$

c) $\int_4^9 \left(\frac{1}{x} + \sqrt{x}\right)^2 dx$

Q5 Calculate the exact shaded area in the following diagrams:

a)  $y = x^3 + x$

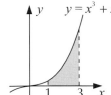

b) $y = x + \sqrt{x}$

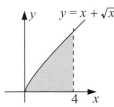

c) $y = 4 - x^2$

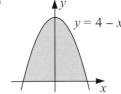

d) $y = x(x - 1)(x - 3)$

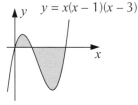

Q6 Find the area between the graph of $y = x^2 + x$, the x-axis and the lines $x = 1$ and $x = 3$.

Q7 Find the area enclosed by the graph of $y = 5x^3$, the x-axis and the lines $x = -5$ and $x = -3$.

Q8 Find the area enclosed by the curve with equation $y = (x - 1)(3x + 9)$, the x-axis and the lines $x = -2$ and $x = 2$.

Q9 Find the area enclosed by the graph of $y = \frac{20}{x^5}$, the x-axis and the lines $x = 1$ and $x = 2$.

Q6-9 Hint: Think about whether the curve lies above or below the x-axis in the interval (it could do both). It may help to sketch the graph.

Q8 Hint: It'll help to work out where the graph crosses the x-axis.

Q10 Find the shaded area in each diagram below:

a) $y = \frac{1}{x^2}$

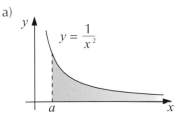

b) $y = 18\left(\frac{2}{x^3} - \frac{1}{x^2}\right)$

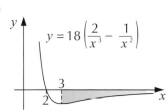

Q11 Find the value of the following integrals:

a) $\int_1^\infty \frac{3}{x^2}\, dx$

b) $\int_3^\infty \left(\frac{2}{x^2} + \frac{7}{x^3}\right) dx$

c) $\int_{-\infty}^{-2} \left(\frac{3}{x^4} + \frac{5}{x^3}\right) dx$

Finding the area between a curve and a line

You've seen how to find the area between a curve and the *x*-axis using integration, but you can also use integration to find the area between a **curve** and a **line** (or even **two curves**).

You'll either have to **add** or **subtract** integrals to find the area you're after — it's always best to **draw a diagram** of the area.

Example 1

Find the area enclosed by the curve $y = x^2$, the line $y = 2 - x$ and the *x*-axis.

- Draw a diagram of the curve and the line.

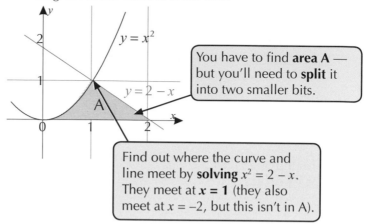

You have to find **area A** — but you'll need to **split** it into two smaller bits.

Find out where the curve and line meet by **solving** $x^2 = 2 - x$. They meet at **x = 1** (they also meet at $x = -2$, but this isn't in A).

Tip: You need to find the *x*-coordinate of the point of intersection between the line and the curve so that you know the limits to integrate between.

- The area A is the area under the **red** curve between 0 and 1 **added to** the area under the **blue** line between 1 and 2.

A_1 is the area under the curve $y = x^2$ between 0 and 1, so integrate between these limits to find the area:

$$\int_0^1 x^2\, dx = \left[\frac{x^3}{3}\right]_0^1 = \frac{1}{3} - 0 = \frac{1}{3}$$

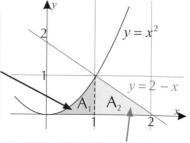

A_2 is the area under the line $y = 2 - x$ between 1 and 2, so integrate between these limits to find the area:

$$\int_1^2 (2 - x)\, dx = \left[2x - \frac{x^2}{2}\right]_1^2$$

$$= \left(2(2) - \frac{2^2}{2}\right) - \left(2(1) - \frac{1^2}{2}\right)$$

$$= 2 - \frac{3}{2} = \frac{1}{2}$$

Tip: A_2 is just a triangle with base 1 and height 1, so you could also calculate its area using the formula for the area of a triangle.

- **Add** the areas together to find the area A:

$$A = A_1 + A_2 = \frac{1}{3} + \frac{1}{2} = \boxed{\frac{5}{6}}$$

Sometimes you'll need to find the area **enclosed** by the graphs of two functions — this usually means **subtracting** one area from another. Here is an example with two curves:

Example 2

Find the area enclosed by the curves $y = x^2 + 1$ and $y = 9 - x^2$.

- Draw a diagram of the two curves.

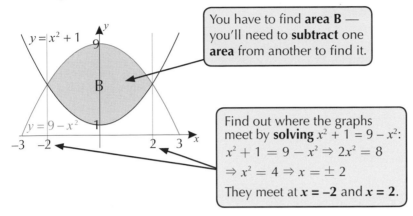

You have to find **area B** — you'll need to **subtract** one **area** from another to find it.

Find out where the graphs meet by **solving** $x^2 + 1 = 9 - x^2$:
$x^2 + 1 = 9 - x^2 \Rightarrow 2x^2 = 8$
$\Rightarrow x^2 = 4 \Rightarrow x = \pm 2$
They meet at **$x = -2$ and $x = 2$**.

- The area B is the area under the **blue** curve between -2 and 2 **minus** the area under the **red** curve between -2 and 2.

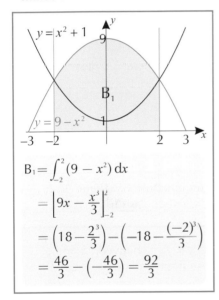

$B_1 = \int_{-2}^{2} (9 - x^2)\, dx$

$= \left[9x - \dfrac{x^3}{3} \right]_{-2}^{2}$

$= \left(18 - \dfrac{2^3}{3} \right) - \left(-18 - \dfrac{(-2)^3}{3} \right)$

$= \dfrac{46}{3} - \left(-\dfrac{46}{3} \right) = \dfrac{92}{3}$

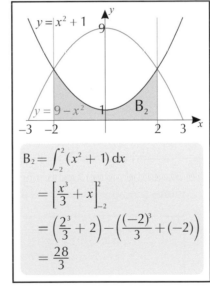

$B_2 = \int_{-2}^{2} (x^2 + 1)\, dx$

$= \left[\dfrac{x^3}{3} + x \right]_{-2}^{2}$

$= \left(\dfrac{2^3}{3} + 2 \right) - \left(\dfrac{(-2)^3}{3} + (-2) \right)$

$= \dfrac{28}{3}$

Tip: Instead of integrating each bit separately and then subtracting, you could try 'subtracting the curves' and then integrating.

Just subtract the equation of one curve from the equation of the other and then integrate the resulting expression over the limits.

So A can be expressed:
$\int_{-2}^{2} ((9 - x^2) - (x^2 + 1))\, dx$

- **Subtract** the areas to find the area B:

$$B = B_1 - B_2 = \dfrac{92}{3} - \dfrac{28}{3} = \boxed{\dfrac{64}{3}}$$

Sometimes you might need to add **and** subtract integrals to find the right area. You'll often need to do this when the curve goes **below** the x-axis. The integrations you need to do should be obvious if you draw a **picture**.

The next page has an example of subtracting integrals and then adding them.

Example 3

Find the area enclosed by the curve $y = x^2 - 2x$ and the line $y = 2x$.

- Draw a diagram of the curve and the line.

Tip: The curve $y = x^2 - 2x$ crosses the x-axis at $x = 0$ and $x = 2$.

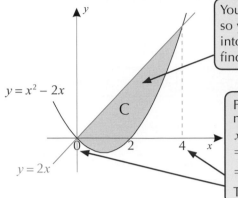

You have to find **area C** — so you need to split it up into areas which you can find by integration.

Find out where the graphs meet by **solving** $x^2 - 2x = 2x$:
$$x^2 - 2x = 2x \Rightarrow x^2 - 4x = 0$$
$$\Rightarrow x(x - 4) = 0$$
$$\Rightarrow x = 0 \text{ or } x = 4$$
They meet at **$x = 0$** and **$x = 4$**.

- Split up the area C:

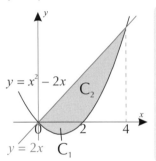

- The area C_1 is easy — it's the integral of $y = x^2 - 2x$ between 0 and 2:
$$\int_0^2 (x^2 - 2x)\, dx = \left[\frac{x^3}{3} - x^2\right]_0^2$$
$$= \left(\frac{2^3}{3} - 2^2\right) - \left(\frac{0^3}{3} - 0^2\right)$$
$$= \frac{8}{3} - 4 = -\frac{4}{3}$$
Area is positive so the area C_1 is $\frac{4}{3}$.

- C_2 is more difficult — it can't be found by a single integration. You'll have to **subtract** integrals like you did in Example 2.

  The area C_2 is the area under the **blue** line between 0 and 4 **minus** the area under the **red** curve between 2 and 4.

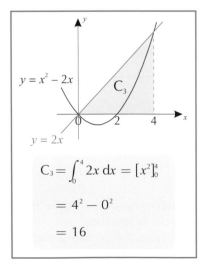

$$C_3 = \int_0^4 2x\, dx = \left[x^2\right]_0^4$$
$$= 4^2 - 0^2$$
$$= 16$$

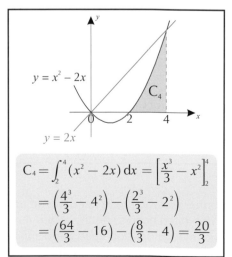

$$C_4 = \int_2^4 (x^2 - 2x)\, dx = \left[\frac{x^3}{3} - x^2\right]_2^4$$
$$= \left(\frac{4^3}{3} - 4^2\right) - \left(\frac{2^3}{3} - 2^2\right)$$
$$= \left(\frac{64}{3} - 16\right) - \left(\frac{8}{3} - 4\right) = \frac{20}{3}$$

So the area $C_2 = C_3 - C_4 = 16 - \frac{20}{3} = \frac{28}{3}$

- Now the **total area** is given by **adding** the two areas:

$$C = C_1 + C_2 = \frac{4}{3} + \frac{28}{3} = \boxed{\frac{32}{3}}$$

Exercise 1.3

Q1   Find the shaded area in the following diagrams:

a)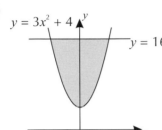
$y = 3x^2 + 4$
$y = 16$

b)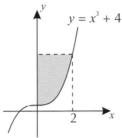
$y = x^3 + 4$
2

Q1 Hint: If it doesn't look like you've been given enough information, look for any hints on the graph that may tell you what line to use.

c)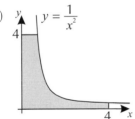
$y = \frac{1}{x^2}$
4
4

d)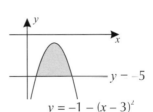
$y = -5$
$y = -1 - (x - 3)^2$

e)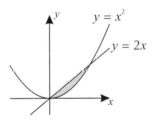
$y = x^2$
$y = 2x$

f)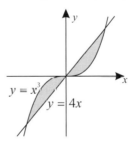
$y = x^3$
$y = 4x$

Q2   For each part, find the area enclosed by the curve and line:

a) $y = x^2 + 4$ and $y = x + 4$

b) $y = 3x^2 + 11x + 6$ and $y = 9x + 6$

c) $y = x^2 + 2x - 3$ and $y = 4x$

Q2 c) Hint: This one's a bit tricky — the best way to do it is to consider the bits above and below the x-axis separately.

Q3   The area between the graphs of $y = x^2$ and $y = ax$ is 36, where a is a constant and $a > 0$. Find a.

2. The Trapezium Rule

Learning Objective:

- Be able to use the trapezium rule to approximate the value of definite integrals.

It's not always possible to integrate a function using the methods you learn at A-level, and some functions can't be integrated at all. When this happens, all is not lost — you can approximate the integral using the trapezium rule.

The trapezium rule

When you find yourself with a function which is too difficult to integrate, you can **approximate** the area under the curve using lots of **trapeziums**, which gives an approximate value of the integral.

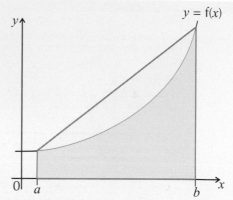

- The **area** under this curve between **a** and **b** can be approximated by the green **trapezium** shown.

- It has height $(b - a)$ and parallel sides of length f(a) and f(b).

- The area of the trapezium is an **approximation** of the integral $\int_a^b f(x)\, dx$.

- It's not a very good approximation, but if you split the area up into **more** trapeziums, the approximation will get more and more **accurate** because the **difference** between the trapeziums and the curve will get **smaller**.

Tip: The area of a trapezium is given by the formula $\frac{h}{2}(a + b)$.

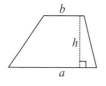

The **trapezium rule** for approximating $\int_a^b f(x)\, dx$ works like this:

- n is the **number** of strips i.e. trapeziums.

- h is the **width** of each strip — it's equal to $\frac{(b - a)}{n}$.

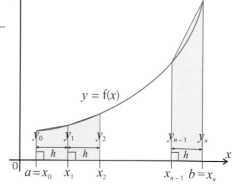

- The **x-values** go up in steps of **h**, starting with $x_0 = a$.

- The **y-values** are found by putting the x-values into the equation of the curve — so $y_1 = f(x_1)$. They give the **heights** of the sides of the trapeziums.

- The **area** of each trapezium is $A = \frac{h}{2}(y_r + y_{r+1})$.

Then an **approximation** for $\int_a^b f(x)\, dx$ is found by **adding** the **areas** of all the trapeziums:

$$\int_a^b f(x)\, dx \approx \frac{h}{2}(y_0 + y_1) + \frac{h}{2}(y_1 + y_2) + \dots + \frac{h}{2}(y_{n-1} + y_n)$$

$$= \frac{h}{2}[y_0 + 2(y_1 + y_2 + \dots + y_{n-1}) + y_n]$$

Tip: If the number of strips is n, the number of y-values is $n + 1$. The y-values are sometimes called **ordinates**.

Tip: This just says 'Add the first and last heights ($y_0 + y_n$) and add this to twice all the other heights added up — then multiply by $\frac{h}{2}$.'

So the **trapezium rule** says:

$$\int_a^b f(x)\,dx \approx \frac{h}{2}\left[y_0 + 2(y_1 + y_2 + \ldots + y_{n-1}) + y_n\right]$$

Tip: The trapezium rule is in the formula booklet — always look it up. It's actually given as $\int_a^b y\,dx$, but this is the same since $y = f(x)$.

This may seem like a lot of information, but it's simple if you follow this **step by step** method:

To approximate the integral $\int_a^b f(x)\,dx$:

- **Split** the interval up into a number of equal sized strips, n. You'll always be told what n is (it could be 4, 5 or even 6).

- Work out the **width** of each strip: $h = \dfrac{(b-a)}{n}$

- Make a **table** of x and y values:

x	$x_0 = a$	$x_1 = a + h$	$x_2 = a + 2h$	...	$x_n = b$
y	$y_0 = f(x_0)$	$y_1 = f(x_1)$	$y_2 = f(x_2)$	...	$y_n = f(x_n)$

- Put all the values into the **trapezium rule**:
$$\int_a^b f(x)\,dx \approx \frac{h}{2}\left[y_0 + 2(y_1 + y_2 + \ldots + y_{n-1}) + y_n\right]$$

Let's have a look at an example:

Example 1

Find an approximate value for $\int_0^2 \sqrt{4 - x^2}\,dx$ using 4 strips. Give your answer to 4 s.f.

- You're told that you need **4 strips** so $n = 4$.

- Work out the **width** of each strip: $h = \dfrac{(b-a)}{n} = \dfrac{(2-0)}{4} = 0.5$

- Set up a **table** and work out the y-values or heights.

Work out the y-values from the x-values, using the function in the integral.

The x-values increase in steps of h.

x	$y = \sqrt{4 - x^2}$
$x_0 = 0$	$y_0 = \sqrt{4 - 0^2} = 2$
$x_1 = 0.5$	$y_1 = \sqrt{4 - 0.5^2} = \sqrt{3.75} = 1.936491673$
$x_2 = 1.0$	$y_2 = \sqrt{4 - 1.0^2} = \sqrt{3} = 1.732050808$
$x_3 = 1.5$	$y_3 = \sqrt{4 - 1.5^2} = \sqrt{1.75} = 1.322875656$
$x_4 = 2.0$	$y_4 = \sqrt{4 - 2.0^2} = 0$

Tip: The x-values should go up in nice jumps — make sure that you use the right value for x when calculating y. If $x_2 = 1$ make sure you use 1 instead of 2.

- Now put all the y-values into the formula with h and n:
$$\int_0^2 \sqrt{4 - x^2}\,dx \approx \frac{0.5}{2}[2 + 2(1.9365 + 1.7321 + 1.3229) + 0]$$
$$= 0.25[2 + (2 \times 4.9915)]$$
$$= 2.996 \ (4 \text{ s.f.})$$

Tip: If you want a more accurate approximation you just need to use more strips.

Tip: Ordinates are just *y*-values so don't make the mistake of writing *n* = 7 — if there are 7 *y*-values there are only 6 strips.

Example 2

Use the trapezium rule with 7 ordinates to find an approximation to $\int_1^{2.2} 2\log_{10}x \, dx$, giving your answer to 3 d.p.

- Remember, **7 ordinates** means **6 strips** — so *n* = 6.

- Calculate the **width** of the strips: $h = \dfrac{(b-a)}{n} = \dfrac{(2.2-1)}{6} = 0.2$

- Set up a **table** and work out the *y*-values using $y = 2\log_{10} x$:

x	$y = 2\log_{10} x$ (5 d.p.)
$x_0 = 1$	$y_0 = 2\log_{10} 1 = 0$
$x_1 = 1.2$	$y_1 = 2\log_{10} 1.2 = 0.15836$
$x_2 = 1.4$	$y_2 = 0.29226$
$x_3 = 1.6$	$y_3 = 0.40824$
$x_4 = 1.8$	$y_4 = 0.51055$
$x_5 = 2.0$	$y_5 = 0.60206$
$x_6 = 2.2$	$y_6 = 0.68485$

- Putting all these values in the **formula** gives:

$$\int_1^{2.2} 2\log_{10}x \, dx \approx \frac{0.2}{2}[0 + 2(0.15836 + 0.29226 + 0.40824$$
$$+ 0.51055 + 0.60206) + 0.68485]$$
$$= 0.1 \times [0.68485 + (2 \times 1.97147)]$$
$$= 0.462779$$
$$= 0.463 \ (3 \text{ d.p.})$$

The **approximation** that the trapezium rule gives will either be an **overestimate** (too big) or an **underestimate** (too small).

This will depend on the **shape** of the graph — draw the graph and see whether the tops of the trapeziums lie **above** the curve or stay **below** it.

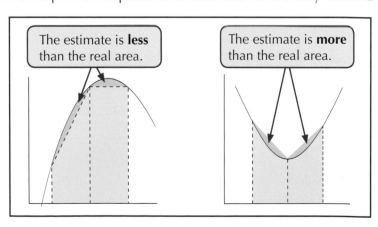

The estimate is **less** than the real area.

The estimate is **more** than the real area.

Example 3

Use the trapezium rule with 8 intervals to find an approximation to $\int_0^\pi \sin x \, dx$, and say whether it is an overestimate or underestimate.

- There are **8 intervals**, so $n = 8$.

- Calculate the **width** of the strips: $h = \dfrac{(b-a)}{n} = \dfrac{(\pi - 0)}{8} = \dfrac{\pi}{8}$

- Set up a **table** and work out the y-values. Make sure you keep your x-values in terms of π — it'll make it a lot easier:

x	$y = \sin x \ (5 \text{ d.p.})$
$x_0 = 0$	$y_0 = \sin 0 = 0$
$x_1 = \dfrac{\pi}{8}$	$y_1 = 0.38268$
$x_2 = \dfrac{\pi}{4}$	$y_2 = 0.70711$
$x_3 = \dfrac{3\pi}{8}$	$y_3 = 0.92388$
$x_4 = \dfrac{\pi}{2}$	$y_4 = 1$
$x_5 = \dfrac{5\pi}{8}$	$y_5 = 0.92388$
$x_6 = \dfrac{3\pi}{4}$	$y_6 = 0.70711$
$x_7 = \dfrac{7\pi}{8}$	$y_7 = 0.38268$
$x_8 = \pi$	$y_8 = 0$

- So, putting all this in the **formula** gives:

$$\int_0^\pi \sin x \, dx \approx \frac{1}{2} \times \frac{\pi}{8}[0 + 2(0.38268$$
$$+ 0.70711 + 0.92388$$
$$+ 1 + 0.92388 + 0.70711$$
$$+ 0.38268) + 0]$$
$$= \frac{\pi}{16} \times [2 \times 5.02734]$$
$$= 1.97 \ (3 \text{ s.f.})$$

- If you sketch the graph of $y = \sin x$, you'll be able to work out if this estimate is an **overestimate** or an **underestimate**.

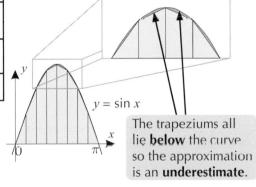

$y = \sin x$

The trapeziums all lie **below** the curve so the approximation is an **underestimate**.

Tip: Whenever you get a question using trig functions, you have to use radians (see p.23). You'll probably be given a limit with π in, which is a good reminder.

Tip: The values of y in the table are quicker to work out if you know that the graph is symmetrical — the values will repeat.

Tip: In C3, you'll see how to integrate trig functions — this integral has the exact value 2.

Exercise 2.1

Q1 Use the trapezium rule with 4 strips to estimate the following, giving your answer to 2 d.p. Give all values in your table to 4 d.p.

a) $\displaystyle\int_1^3 \frac{1}{x} \, dx$

b) $\displaystyle\int_0^2 \sqrt{1 + 3^x} \, dx$

c) $\displaystyle\int_1^6 \ln x \, dx$

d) $\displaystyle\int_0^3 2^x \, dx$

Q1 c) Hint: Use the 'ln' button on your calculator for this one.

Q2 a) Complete the table for the function $y = \sin x$, giving exact answers.

x	0	$\frac{\pi}{6}$	$\frac{\pi}{3}$	$\frac{\pi}{2}$
y				

b) Use the trapezium rule with 3 strips and the values in the table of part a) to estimate $\int_0^{\frac{\pi}{2}} \sin x \, dx$. Give your answer correct to 2 d.p.

Q3 Hint: See p.118-122 for more on finding definite integrals.

Q3 a) Evaluate $I = \int_{-4}^{0}(x^2 + 2x + 3)\,dx$ using integration.
b) Use the trapezium rule with 5 ordinates to estimate the value of I.
c) Comment on your answers to part a) and part b).

Q4 Hint: 'Continuous' just means there are no breaks or jumps in the curve. You need a continuous curve if you're using the trapezium rule — but you'll always be given continuous curves so you don't need to worry.

Q4 The curve $y = f(x)$ is continuous and gives the table:

x	−2	−1	0	1	2
y	10	8	7	6.5	3

Use the trapezium rule with 4 strips to estimate $\int_{-2}^{2} f(x)\,dx$.

Q5 Use the trapezium rule to estimate the following correct to 3 d.p. :
a) $\int_0^5 2^{-x}\,dx$ with 4 strips,   b) $\int_0^5 2^{-x}\,dx$ with 5 strips.
c) Explain which is likely to be the more accurate estimate.

Q6 a) Use the trapezium rule to estimate $\int_1^3 (3 - (2 - x)^2)\,dx$ with 9 ordinates.
b) Sketch the graph of $y = 3 - (2 - x)^2$ for $0 \le x \le 4$.
c) Explain whether your approximation in part a) is an overestimate or an underestimate.

Q7 Hint: See p.31 for the cos values of some common angles.

Q7 a) Complete the table giving exact values for $y = \cos x$.

x	$-\frac{\pi}{2}$	$-\frac{\pi}{3}$	$-\frac{\pi}{6}$	0	$\frac{\pi}{6}$	$\frac{\pi}{3}$	$\frac{\pi}{2}$
y							

b) Use your answers to part a) to show that an approximation for $\int_{-\frac{\pi}{2}}^{\frac{\pi}{2}} \cos x \, dx$ is $\frac{\pi(2 + \sqrt{3})}{6}$.

c) Say whether the approximation in part b) is an overestimate or an underestimate. Explain your answer.

Review Exercise — Chapter 7

Q1 Find y in terms of x:

a) $\dfrac{dy}{dx} = \dfrac{5}{7}x^4 + \dfrac{2}{3}x + \dfrac{1}{4}$

b) $\dfrac{dy}{dx} = \dfrac{1}{\sqrt{x}} + \sqrt{x}$

c) $\dfrac{dy}{dx} = \dfrac{3}{x^2} + \dfrac{3}{\sqrt[3]{x}}$

Q1 Hint:
Remember that
$\int \dfrac{dy}{dx}\, dx = y$

Q2 Evaluate the following definite integrals:

a) $\displaystyle\int_0^1 (4x^3 + 3x^2 + 2x + 1)\, dx$

b) $\displaystyle\int_1^2 \left(\dfrac{8}{x^5} + \dfrac{3}{\sqrt{x}}\right) dx$

c) $\displaystyle\int_1^6 \dfrac{3}{x^2}\, dx$

Q3 Evaluate:

a) (i) $\displaystyle\int_{-3}^3 (9 - x^2)\, dx$

(ii) $\displaystyle\int_1^\infty \dfrac{3}{x^2}\, dx$.

b) Sketch the areas represented by these integrals.

Q4 Find the shaded area in the diagrams below:

a)

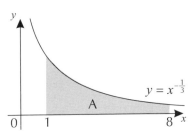

b)

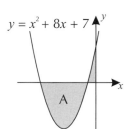

$y = x^2 + 8x + 7$

c)

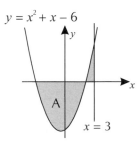

$y = x^2 + x - 6$

Q5 Use integration to find the shaded area in each of these graphs:

a)

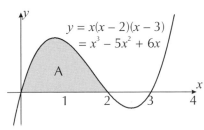

$y = x(x-2)(x-3)$
$= x^3 - 5x^2 + 6x$

b)

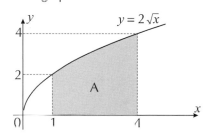

$y = 2\sqrt{x}$

c)

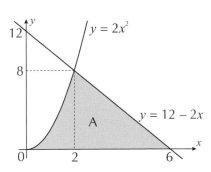

$y = 2x^2$
$y = 12 - 2x$

d)

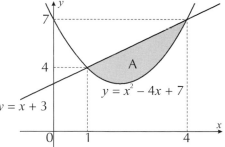

$y = x^2 - 4x + 7$
$y = x + 3$

Q6 Use the trapezium rule with n intervals to estimate the following to 3 s.f. Give y values to 5 d.p. where appropriate:

a) $\displaystyle\int_0^3 (9 - x^2)^{\frac{1}{2}}\, dx$ with $n = 3$

b) $\displaystyle\int_{0.2}^{1.2} x^{x^2}\, dx$ with $n = 5$

c) $\displaystyle\int_1^3 2^{x^2}\, dx$ with $n = 4$

d) $\displaystyle\int_1^3 2^{x^2}\, dx$ with $n = 5$

1   Find the value of $\int_{2}^{7} (2x - 6x^2 + \sqrt{x})\,dx$. Give your answer to 4 d.p.

(5 marks)

2   a)   Using the trapezium rule with n intervals, estimate the values of:

    (i) $\int_{2}^{8} \left(\sqrt{3x^3} + \dfrac{2}{\sqrt{x}} \right) dx, \quad n = 3$

(4 marks)

    (ii) $\int_{1}^{5} \left(\dfrac{x^3 - 2}{4} \right) dx, \quad n = 4$

(4 marks)

  b)   How could you change your application of the trapezium rule
to get better approximations?

(1 mark)

3   Curve C, $y = (x - 3)^2 (x + 1)$, is sketched on the diagram below:

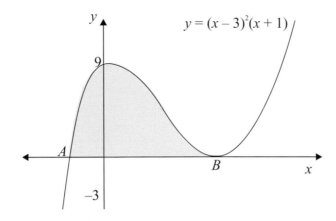

Calculate the shaded area under curve C between point A, where C intersects the
x-axis, and point B, where C touches the x-axis.

(8 marks)

4 Complete the table and hence use the trapezium rule with 6 ordinates to estimate $\int_{1.5}^{4} y \, dx$.
Give your answers to 3 s.f. where appropriate.

x	$x_0 = 1.5$	$x_1 =$	$x_2 =$	$x_3 =$	$x_4 = 3.5$	$x_5 = 4$
$y = 3x - \sqrt{2^x}$	$y_0 =$	$y_1 = 4$	$y_2 = 5.1216$	$y_3 =$	$y_4 =$	$y_5 = 8$

(7 marks)

5 a) Complete the table for the function $y = \sqrt{1 + x}$,
giving your answers to 4 d.p.

x	$x_0 = 0$	$x_1 = 0.5$	$x_2 = 1$	$x_3 = 1.5$	$x_4 = 2$
y	$y_0 =$	$y_1 =$	$y_2 =$	$y_3 =$	$y_4 =$

(3 marks)

b) Use the trapezium rule with all the values calculated in the table to approximate
the value of $\int_{0}^{2} \sqrt{1 + x} \, dx$. Give your answer correct to 2 d.p.
(3 marks)

6 The diagram below shows the curve $y = (x + 1)(x - 5)$.
Points $J(-1, 0)$ and $K(4, -5)$ lie on the curve.

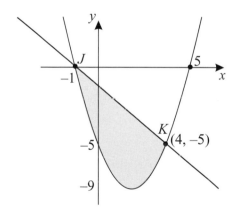

a) Find the equation of the straight line joining J and K in the form $y = mx + c$.
(2 marks)

b) Calculate $\int_{-1}^{4} (x + 1)(x - 5) \, dx$.
(5 marks)

c) Find the area of the shaded region.
(4 marks)

Answers

Chapter 1: Algebra and Functions

1. Algebraic Division

Exercise 1.1 — Algebraic division

Q1 a)
$$\begin{array}{r} x^2 + x - 1 \\ x-3\overline{)x^3 - 2x^2 - 4x + 8} \end{array}$$
$$-\ \underline{x^3 - 3x^2}$$
$$x^2 - 4x$$
$$-\ \underline{x^2 - 3x}$$
$$-x + 8$$
$$-\ \underline{-x + 3}$$
$$5$$

Quotient: $x^2 + x - 1$
Remainder: 5

b)
$$\begin{array}{r} x^2 - 3x - 5 \\ x+2\overline{)x^3 - x^2 - 11x - 10} \end{array}$$
$$-\ \underline{x^3 + 2x^2}$$
$$-3x^2 - 11x$$
$$-\ \underline{-3x^2 - 6x}$$
$$-5x - 10$$
$$-\ \underline{-5x - 10}$$
$$0$$

Quotient: $x^2 - 3x - 5$
Remainder: 0

c)
$$\begin{array}{r} x^2 + x - 1 \\ x-2\overline{)x^3 - x^2 - 3x + 3} \end{array}$$
$$-\ \underline{x^3 - 2x^2}$$
$$x^2 - 3x$$
$$-\ \underline{x^2 - 2x}$$
$$-x + 3$$
$$-\ \underline{-x + 2}$$
$$1$$

Quotient: $x^2 + x - 1$
Remainder: 1

d)
$$\begin{array}{r} x^2 - 5x + 10 \\ x+3\overline{)x^3 - 2x^2 - 5x + 6} \end{array}$$
$$-\ \underline{x^3 + 3x^2}$$
$$-5x^2 - 5x$$
$$-\ \underline{-5x^2 - 15x}$$
$$10x + 6$$
$$-\ \underline{10x + 30}$$
$$-24$$

Quotient: $x^2 - 5x + 10$
Remainder: -24

e)
$$\begin{array}{r} x^2 - 11x + 29 \\ x+2\overline{)x^3 - 9x^2 + 7x + 33} \end{array}$$
$$-\ \underline{x^3 + 2x^2}$$
$$-11x^2 + 7x$$
$$-\ \underline{-11x^2 - 22x}$$
$$29x + 33$$
$$-\ \underline{29x + 58}$$
$$-25$$

Quotient: $x^2 - 11x + 29$
Remainder: -25

Q2
$$\begin{array}{r} x^2 + x - 4 \\ x-1\overline{)x^3 + 0x^2 - 5x + 4} \end{array}$$
$$-\ \underline{x^3 - x^2}$$
$$x^2 - 5x$$
$$-\ \underline{x^2 - x}$$
$$-4x + 4$$
$$-\ \underline{-4x + 4}$$
$$0$$

So $(x^3 - 5x + 4) \div (x - 1) = x^2 + x - 4$ remainder 0.

Q3
$$\begin{array}{r} x^2 + 2x + 6 \\ x-3\overline{)x^3 - x^2 + 0x - 11} \end{array}$$
$$-\ \underline{x^3 - 3x^2}$$
$$2x^2 + 0x$$
$$-\ \underline{2x^2 - 6x}$$
$$6x - 11$$
$$-\ \underline{6x - 18}$$
$$7$$

So $(x^3 - x^2 - 11) \div (x - 3) = x^2 + 2x + 6$ remainder 7.

Q4 a)
$$\begin{array}{r} 2x^2 + x - 12 \\ x+1\overline{)2x^3 + 3x^2 - 11x - 6} \end{array}$$
$$-\ \underline{2x^3 + 2x^2}$$
$$x^2 - 11x$$
$$-\ \underline{x^2 + x}$$
$$-12x - 6$$
$$-\ \underline{-12x - 12}$$
$$6$$

So $(2x^3 + 3x^2 - 11x - 6) \div (x + 1) = 2x^2 + x - 12$ remainder 6.

b)
$$\begin{array}{r} 2x^2 - x - 9 \\ x+2\overline{)2x^3 + 3x^2 - 11x - 6} \end{array}$$
$$-\ \underline{2x^3 + 4x^2}$$
$$-x^2 - 11x$$
$$-\ \underline{-x^2 - 2x}$$
$$-9x - 6$$
$$-\ \underline{-9x - 18}$$
$$12$$

So $(2x^3 + 3x^2 - 11x - 6) \div (x + 2) = 2x^2 - x - 9$ remainder 12.

c)

$$\begin{array}{r} 2x^2 + 5x - 6 \\ x-1\overline{\smash{\big)}\,2x^3 + 3x^2 - 11x - 6} \\ \underline{-\ 2x^3 - 2x^2} \\ 5x^2 - 11x \\ \underline{-\ 5x^2 - 5x} \\ -6x - 6 \\ \underline{-\ -6x + 6} \\ -12 \end{array}$$

So $(2x^3 + 3x^2 - 11x - 6) \div (x - 1) = 2x^2 + 5x - 6$ remainder -12.

Q5

$$\begin{array}{r} x^2 + 4x + 1 \\ x-2\overline{\smash{\big)}\,x^3 + 2x^2 - 7x - 2} \\ \underline{-\ x^3 - 2x^2} \\ 4x^2 - 7x \\ \underline{-\ 4x^2 - 8x} \\ x - 2 \\ \underline{-\ x - 2} \\ 0 \end{array}$$

So $f(x) = (x - 2)(x^2 + 4x + 1)$.

2. The Remainder and Factor Theorems

Exercise 2.1 — The Remainder Theorem

Q1 **a)** $a = 1$, $f(1) = 2(1)^3 - 3(1)^2 - 39(1) + 20 = -20$

b) $a = -1$, $f(-1) = (-1)^3 - 3(-1)^2 + 2(-1) = -6$

c) $a = -1$, $f(-1) = 6(-1)^3 + (-1)^2 - 5(-1) - 2 = -2$

d) $a = -3$, $f(-3) = (-3)^3 + 2(-3)^2 - 7(-3) - 2 = 10$

e) $a = 2$ and $b = -1$

$$f\left(-\tfrac{1}{2}\right) = 4\left(-\tfrac{1}{2}\right)^3 - 6\left(-\tfrac{1}{2}\right)^2 - 12\left(-\tfrac{1}{2}\right) - 6 = -2$$

f) $a = 2$ and $b = 1$

$$f\left(\tfrac{1}{2}\right) = \left(\tfrac{1}{2}\right)^3 - 3\left(\tfrac{1}{2}\right)^2 - 6\left(\tfrac{1}{2}\right) + 8 = \tfrac{35}{8} \text{ or } 4\tfrac{3}{8}$$

Q2 $a = -2$, $f(-2) = (-2)^3 + p(-2)^2 - 10(-2) - 19 = 5$
$4p - 7 = 5$
$4p = 12 \Rightarrow p = 3$
So $f(x) = x^3 + 3x^2 - 10x - 19$

Q3 $a = -2$, $f(-2) = (-2)^3 - d(-2)^2 + d(-2) + 1 = -25$
$-6d - 7 = -25$
$-6d = -18 \Rightarrow d = 3$
So $f(x) = x^3 - 3x^2 + 3x + 1$

Q4 $a = -1$, $f(-1) = (-1)^3 - 2(-1)^2 + 7(-1) + k = -8$
$-10 + k = -8$
$k = 2$
So $f(x) = x^3 - 2x^2 + 7x + 2$

Q5 $a = 2$, $f(2) = (2)^4 + 5(2)^3 + 2p + 156 = 2p + 212$
$a = -1$, $f(-1) = (-1)^4 + 5(-1)^3 - p + 156 = 152 - p$
$\Rightarrow 2p + 212 = 152 - p$
$3p = -60$
$p = -20$
So, $f(x) = x^4 + 5x^3 - 20x + 156$

Exercise 2.2 — The Factor Theorem

Q1 **a)** $a = 1$, find $f(a)$ and show the result is 0:
$f(1) = (1)^3 - (1)^2 - 3(1) + 3 = 1 - 1 - 3 + 3 = 0$
So by the Factor Theorem, $(x - 1)$ is a factor.
The question asked you to use the Factor Theorem, but you could also show it was a factor by adding the coefficients (1, −1, −3, 3) to get 0. (If the coefficients in a polynomial add up to 0, then (x − 1) is a factor.)

b) $a = -1$, find $f(a)$ and show the result is 0:
$f(-1) = (-1)^3 + 2(-1)^2 + 3(-1) + 2$
$= -1 + 2 - 3 + 2 = 0$
So by the Factor Theorem, $(x + 1)$ is a factor.

c) $a = -2$, find $f(a)$ and show the result is 0:
$f(-2) = (-2)^3 + 3(-2)^2 - 10(-2) - 24$
$= -8 + 12 + 20 - 24 = 0$
So by the Factor Theorem, $(x + 2)$ is a factor.

Q2 **a)** Substitute $x = \tfrac{1}{2}$ and show the result is 0:
$$f\left(\tfrac{1}{2}\right) = 2\left(\tfrac{1}{2}\right)^3 - \left(\tfrac{1}{2}\right)^2 - 8\left(\tfrac{1}{2}\right) + 4$$
$$= \tfrac{2}{8} - \tfrac{1}{4} - 4 + 4 = 0$$
So by the Factor Theorem, $(2x - 1)$ is a factor.

b) Substitute $x = \tfrac{2}{3}$ and show the result is 0:
$$f\left(\tfrac{2}{3}\right) = 3\left(\tfrac{2}{3}\right)^3 - 5\left(\tfrac{2}{3}\right)^2 - 16\left(\tfrac{2}{3}\right) + 12$$
$$= \tfrac{8}{9} - \tfrac{20}{9} - \tfrac{32}{3} + 12$$
$$\tfrac{8}{9} - \tfrac{20}{9} - \tfrac{96}{9} + 12 = 0$$
So by the Factor Theorem, $(3x - 2)$ is a factor.

Q3 **a)** $f(3) = (3)^3 - 2(3)^2 - 5(3) + 6 = 27 - 18 - 15 + 6 = 0$
So by the Factor Theorem, $(x - 3)$ is a factor.

b) $1 - 2 - 5 + 6 = 0$
So by the Factor Theorem, $(x - 1)$ is a factor.

Q4 **a)** $f(-4) = 3(-4)^3 - 5(-4)^2 - 58(-4) + 40$
$= -192 - 80 + 232 + 40 = 0$
So by the Factor Theorem, $(x + 4)$ is a factor.

b) $f\left(\tfrac{2}{3}\right) = 3\left(\tfrac{2}{3}\right)^3 - 5\left(\tfrac{2}{3}\right)^2 - 58\left(\tfrac{2}{3}\right) + 40$
$= \tfrac{8}{9} - \tfrac{20}{9} - \tfrac{116}{3} + 40 = -\tfrac{120}{3} + 40 = 0$
So by the Factor Theorem, $(3x - 2)$ is a factor.

Q5 $(x - 3)$ is a factor of $f(x) = qx^3 - 4x^2 - 7qx + 12$, so $f(3) = 0$. Using the Factor Theorem in reverse:
$f(3) = q(3)^3 - 4(3)^2 - 7q(3) + 12$
$= 27q - 36 - 21q + 12 = 6q - 24$
$\Rightarrow 6q - 24 = 0$
$6q = 24$
$q = 4$
So, $f(x) = 4x^3 - 4x^2 - 28x + 12$

Q6 $(x - 1)$ and $(x - 2)$ are factors, so using the Factor Theorem in reverse, $f(1) = 0$ and $f(2) = 0$.

$f(1) = (1)^3 + c(1)^2 + d(1) - 2 = 1 + c + d - 2 = 0$

$c + d = 1$ (equation 1)

$f(2) = (2)^3 + c(2)^2 + d(2) - 2 = 8 + 4c + 2d - 2 = 0$

$4c + 2d = -6$ (equation 2)

Rearrange (1) to get $d = 1 - c$, and sub into (2):

$\Rightarrow 4c + 2(1 - c) = -6$

$4c + 2 - 2c = -6$

$2c = -8$

$c = -4$

Sub c into rearranged (1):

$d = 1 - c = 1 + 4 = 5$

So, $f(x) = x^3 - 4x^2 + 5x - 2$

Exercise 2.3 — Factorising a cubic

Q1 a) x is a common factor, so you get:

$x(x^2 - 3x + 2) \Rightarrow x(x - 1)(x - 2)$

b) Adding the coefficients gives you –6, so $(x - 1)$ is not a factor. Using trial and error, $f(2) = 0$, so $(x - 2)$ is a factor. Factorise to get:

$(x - 2)(2x^2 + 7x + 3) \Rightarrow (x - 2)(x + 3)(2x + 1)$

c) Add the coefficients $(1 - 3 + 3 - 1)$ to get 0, so $(x - 1)$ is a factor. Factorise to get:

$(x - 1)(x^2 - 2x + 1) \Rightarrow (x - 1)^3$

d) Adding the coefficients gives you 2, so $(x - 1)$ is not a factor. Using trial and error, $f(2) = 0$, so $(x - 2)$ is a factor. Factorise to get:

$(x - 2)(x^2 - x - 2) \Rightarrow (x - 2)(x - 2)(x + 1)$

e) Adding the coefficients gives you 0, so $(x - 1)$ is a factor. Factorise to get:

$(x - 1)(x^2 - 2x - 35) \Rightarrow (x - 1)(x + 5)(x - 7)$

f) Adding the coefficients gives you 21, $(x - 1)$ is not a factor. Using trial and error, $f(2) = 0$, so $(x - 2)$ is a factor. Factorise to get:

$(x - 2)(x^2 + 2x + 48) \Rightarrow (x - 2)(x - 4)(x + 6)$

Q2 a) Adding the coefficients gives you 3, so $(x - 1)$ is not a factor. Using trial and error, $f(-2) = 0$, so $(x + 2)$ is a factor. Factorise to get:

$(x + 2)(x^2 - 4x + 4) \Rightarrow (x - 2)^2(x + 2)$

b) So the solutions are $x = 2$ and -2.

Q3 Add the coefficients $(1 - 1 - 3 + 3)$ to get 0, so $(x - 1)$ is a factor. Factorise to get:

$(x - 1)(x^2 - 3)$. So the roots are $x = 1$ and $\pm \sqrt{3}$.

Q4 a) 5 is a factor, so $f(5) = 0$

$f(5) = (5)^3 - p(5)^2 + 17(5) - 10$

$= 125 - 25p + 85 - 10 = 200 - 25p$

$\Rightarrow 200 - 25p = 0$

$200 = 25p$

$p = 8$

b) $(x - 5)(x^2 - 3x + 2) \Rightarrow (x - 5)(x - 1)(x - 2)$

c) So the roots are $x = 5$, 1 and 2.

Review Exercise — Chapter 1

Q1 a)

$$\begin{array}{r} x^2 - 4x + 9 \\ x + 3\overline{)x^3 - x^2 - 3x + 3} \\ -\underline{x^3 + 3x^2} \\ -4x^2 - 3x \\ -\underline{-4x^2 - 12x} \\ 9x + 3 \\ -\underline{9x + 27} \\ -24 \end{array}$$

Quotient: $x^2 - 4x + 9$, remainder: –24

b)

$$\begin{array}{r} x^2 - x - 7 \\ x - 2\overline{)x^3 - 3x^2 - 5x + 6} \\ -\underline{x^3 - 2x^2} \\ -x^2 - 5x \\ -\underline{-x^2 + 2x} \\ -7x + 6 \\ -\underline{-7x + 14} \\ -8 \end{array}$$

Quotient: $x^2 - x - 7$, remainder: –8

c)

$$\begin{array}{r} x^2 + 0 + 3 \\ x + 2\overline{)x^3 + 2x^2 + 3x + 2} \\ -\underline{x^3 + 2x^2} \\ 0 + 3x \\ -\underline{0 + 0} \\ 3x + 2 \\ -\underline{3x + 6} \\ -4 \end{array}$$

Quotient: $x^2 + 3$, remainder: –4

Q2 a)

$$\begin{array}{r} 3x^2 - 10x + 15 \\ x + 2\overline{)3x^3 - 4x^2 - 5x - 6} \\ -\underline{3x^3 + 6x^2} \\ -10x^2 - 5x \\ -\underline{-10x^2 - 20x} \\ 15x - 6 \\ -\underline{15x + 30} \\ -36 \end{array}$$

So, $f(x) = (x + 2)(3x^2 - 10x + 15) - 36$

b)

$$\begin{array}{r} x^2 - 0x - 3 \\ x + 2\overline{)x^3 + 2x^2 - 3x + 4} \\ -\underline{x^3 + 2x^2} \\ 0 - 3x \\ -\underline{0 - 0} \\ -3x + 4 \\ -\underline{-3x - 6} \\ 10 \end{array}$$

So, $f(x) = (x + 2)(x^2 - 3) + 10$

c)

$$\begin{array}{r} 2x^2 - 4x + 14 \\ x + 2\overline{)2x^3 + 0x^2 + 6x - 3} \\ -\underline{2x^3 + 4x^2} \\ -4x^2 + 6x \\ -\underline{-4x^2 - 8x} \\ 14x - 3 \\ -\underline{14x + 28} \\ -31 \end{array}$$

So, $f(x) = (x + 2)(2x^2 - 4x + 14) - 31$

Q3 a) (i) You just need to find f(–1).
This is $-6 - 1 + 3 - 12 = -16$.

  (ii) Now find f(1). This is $6 - 1 - 3 - 12 = -10$.

  (iii) Now find f(2). This is $48 - 4 - 6 - 12 = 26$.

b) (i)   f(–1) = –1

  (ii)   f(1) = 9

  (iii)   f(2) = 38

c) (i)   f(–1) = –2

  (ii)   f(1) = 0

  (iii)   f(2) = 37

Q4 a) You need to find f(–2). This is
$(-2)^4 - 3(-2)^3 + 7(-2)^2 - 12(-2) + 14$
$= 16 + 24 + 28 + 24 + 14 = 106$.

b) You need to find f(–4/2) = f(–2). You found this in part a, so remainder = 106. You might also have noticed that $2x + 4$ is a multiple of $x + 2$ (from part a), so the remainder must be the same.

c) You need to find f(3). This is
$(3)^4 - 3(3)^3 + 7(3)^2 - 12(3) + 14$
$= 81 - 81 + 63 - 36 + 14 = 41$.

d) You need to find f(6/2) = f(3). You found this in part c), so remainder = 41. You might also have noticed that $2x - 6$ is a multiple of $x - 3$, so the remainder must be the same.

Q5 f(–3) = –16,
$f(-3) = (-3)^3 + c(-3)^2 + 17(-3) - 10$
$= -27 + 9c - 51 - 10 = -88 + 9c$
$\Rightarrow -88 + 9c = -16$
$9c = 72$
$c = 8$
So $f(x) = x^3 + 8x^2 + 17x - 10$

Q6 a) You need to find f(1) — if f(1) = 0, then $(x - 1)$ is a factor:
$f(1) = 1 - 4 + 3 + 2 - 2 = 0$, so $(x - 1)$ is a factor.

b) You need to find f(–1) — if f(–1) = 0, then $(x + 1)$ is a factor:
$f(-1) = -1 - 4 - 3 + 2 - 2 = -8$,
so $(x + 1)$ is not a factor.

c) You need to find f(2) — if f(2) = 0, then $(x - 2)$ is a factor:
$f(2) = 32 - (4 \times 16) + (3 \times 8) + (2 \times 4) - 2$
$= 32 - 64 + 24 + 8 - 2 = -2$, so $(x - 2)$ is not a factor.

d) The remainder when you divide by $(2x - 2)$ is the same as the remainder when you divide by $x - 1$. $(x - 1)$ is a factor (i.e. remainder = 0), so $(2x - 2)$ is also a factor.

Q7 Sub in $x = -2$, if $(x + 2)$ is a factor, then f(–2) = 0.
$f(-2) = (-2 + 5)(-2 - 2)(-2 - 1) + k$
$= (3)(-4)(-3) + k = 36 + k$
$\Rightarrow 36 + k = 0$
$k = -36$

Q8 If $f(x) = 2x^4 + 3x^3 + 5x^2 + cx + d$, then to make sure f(x) is exactly divisible by $(x - 2)(x + 3)$, you have to make sure f(2) = f(–3) = 0.
$f(2) = 32 + 24 + 20 + 2c + d = 0$, i.e. $\underline{2c + d = -76}$.
$f(-3) = 162 - 81 + 45 - 3c + d = 0$, i.e. $\underline{3c - d = 126}$.
Add the two underlined equations to get: $5c = 50$, and so $c = 10$. Then $d = -96$.

Q9 Get $(x - 3)(x^2 - 6x - 11)$ by either dividing the cubic by $(x - 3)$ or using the factorising cubics method. To solve this quadratic, you will need to use the quadratic formula, to find the roots 3 and $3 \pm 2\sqrt{5}$.

Q10 Use trial and error to find $(x + 1)$ as a factor. Then get divide the cubic by $(x + 1)$ or use the factorising quadratics method to get:
$f(x) = (x + 1)(x^2 + 5x + 6) = (x + 1)(x + 2)(x + 3)$
So f(x) has roots of –1, –2 and –3.

Exam-Style Questions — Chapter 1

Q1 a) (i) Remainder = $f(1) = 2(1)^3 - 5(1)^2 - 4(1) + 3$
[1 mark]
$= -4$ *[1 mark]*.

  (ii) Remainder = $f\left(-\frac{1}{2}\right)$
$f\left(-\frac{1}{2}\right) = 2\left(-\frac{1}{8}\right) - 5\left(\frac{1}{4}\right) - 4\left(-\frac{1}{2}\right) + 3$
[1 mark]
$= \frac{7}{2}$ *[1 mark]*.

b) If f(–1) = 0 then $(x + 1)$ is a factor.
$f(-1) = 2(-1)^3 - 5(-1)^2 - 4(-1) + 3$ *[1 mark]*
$= -2 - 5 + 4 + 3 = 0$, so $(x + 1)$ is a factor of f(x).
[1 mark]

c) $(x + 1)$ is a factor, so find the other factors by taking out $(x + 1)$:
$2x^3 - 5x^2 - 4x + 3 = (x + 1)(2x^2 \qquad)$
$= (x + 1)(2x^2 \qquad + 3)$
$= (x + 1)(2x^2 - 7x + 3)$
$= (x + 1)(2x - 1)(x - 3)$
So $f(x) = (2x - 1)(x - 3)(x + 1)$.
[4 marks available — 1 mark for taking out $(x + 1)$ to form a quadratic, 1 mark for correct quadratic factor, 1 mark for attempt to factorise quadratic, 1 mark for correct factorisation of quadratic.]

Q2 a) $f(p) = (4p^2 + 3p + 1)(p - p) + 5$
$= (4p^2 + 3p + 1) \times 0 + 5$
$= 5$ *[1 mark]*.

b) f(–1) = –1.
$f(-1) = (4(-1)^2 + 3(-1) + 1)((-1) - p) + 5$
$= (4 - 3 + 1)(-1 - p) + 5$
$= 2(-1 - p) + 5 = 3 - 2p$ *[1 mark]*
So: $3 - 2p = -1$, $p = 2$ *[1 mark]*.

c) $f(x) = (4x^2 + 3x + 1)(x - 2) + 5$
$f(1) = (4 + 3 + 1)(1 - 2) + 5 = -3$ *[1 mark]*.

Q3 **a)** If $f(-2) = 0$ then $(x + 2)$ is a factor.
$f(-2) = 3(-2)^3 + 8(-2)^2 + 3(-2) - 2$ *[1 mark]*
$= -24 + 32 - 6 - 2 = 0$, so $(x + 2)$ is a factor
of $f(x)$. *[1 mark]*

b) $(x + 2)$ is a factor, so find the other factors by
taking out $(x + 2)$:
$3x^3 + 8x^2 + 3x - 2 = (x + 2)(3x^2 \qquad)$
$\qquad\qquad\qquad\qquad = (x + 2)(3x^2 \qquad - 1)$
$\qquad\qquad\qquad\qquad = (x + 2)(3x^2 + 2x - 1)$
$\qquad\qquad\qquad\qquad = (x + 2)(3x - 1)(x + 1)$
So $f(x) = (x + 2)(3x - 1)(x + 1)$.
[4 marks available — 1 mark for taking out
(x + 2) to form a quadratic, 1 mark for correct
quadratic factor, 1 mark for attempt to factorise
quadratic, 1 mark for correct factorisation of
quadratic.]

c) The solutions to the equation are -2, -1 and $\frac{1}{3}$.
[1 mark]

Chapter 2: Circles

1. Equation of a Circle

Exercise 1.1 — Equation of a circle with centre (*a*, *b*)

Q1 $x^2 + y^2 = 25$

Q2 $x^2 + y^2 = 49$

Q3 **a)** $a = 2$, $b = 5$, $r = 3$
$(x - 2)^2 + (y - 5)^2 = 9$

b) $a = -3$, $b = 2$, $r = 5$
$(x + 3)^2 + (y - 2)^2 = 25$

c) $a = -2$, $b = -3$, $r = 7$
$(x + 2)^2 + (y + 3)^2 = 49$

d) $a = 3$, $b = 0$, $r = 4$
$(x - 3)^2 + y^2 = 16$

Q4 **a)** $a = 1$, $b = 5$, $r = \sqrt{4}$
So the centre is $(1, 5)$ and the radius is 2.

b) $a = 3$, $b = 5$, $r = \sqrt{64}$
So the centre is $(3, 5)$ and the radius is 8.

c) $a = 3$, $b = -2$, $r = \sqrt{25}$
So the centre is $(3, -2)$ and the radius is 5.

Q5 $a = 5$, $b = 3$, $r = 8$
$(x - 5)^2 + (y - 3)^2 = 64$

Q6 $a = 3$, $b = 1$, $r = \sqrt{31}$
$(x - 3)^2 + (y - 1)^2 = 31$

Q7 **a)** $a = 6$, $b = 4$, so the centre is $(6, 4)$

b) Radius $= \sqrt{20} = \sqrt{4 \times 5} = 2\sqrt{5}$

Q8 $a = -3$, $b = -2$, $r = \sqrt{5}$
$(x + 3)^2 + (y + 2)^2 = 5$

Exercise 1.2 — Rearranging circle equations

Q1 **a)** Complete the square for the x's and y's:
$x^2 + y^2 + 2x - 6y - 6 = 0$
$x^2 + 2x + y^2 - 6y - 6 = 0$
$(x + 1)^2 - 1 + (y - 3)^2 - 9 - 6 = 0$
$(x + 1)^2 + (y - 3)^2 = 16$
Radius = 4, centre is $(-1, 3)$.

b) Complete the square for the x's and y's:
$x^2 + y^2 - 2y - 4 = 0$
$x^2 + (y - 1)^2 - 1 - 4 = 0$
$x^2 + (y - 1)^2 = 5$
Radius $= \sqrt{5}$, centre is $(0, 1)$

c) Complete the square for the x's and y's:
$x^2 + y^2 - 6x - 4y = 12$
$x^2 - 6x + y^2 - 4y = 12$
$(x - 3)^2 - 9 + (y - 2)^2 - 4 = 12$
$(x - 3)^2 + (y - 2)^2 = 25$
Radius = 5, centre is $(3, 2)$

d) Complete the square for the x's and y's:
$x^2 + y^2 - 10x + 6y + 13 = 0$
$x^2 - 10x + y^2 + 6y + 13 = 0$
$(x - 5)^2 - 25 + (y + 3)^2 - 9 + 13 = 0$
$(x - 5)^2 + (y + 3)^2 = 21$
Radius $= \sqrt{21}$, centre is $(5, -3)$

Q2 a) Complete the square for the x's and y's:
$x^2 + y^2 + 2x - 4y - 3 = 0$
$x^2 + 2x + y^2 - 4y - 3 = 0$
$(x + 1)^2 - 1 + (y - 2)^2 - 4 - 3 = 0$
$(x + 1)^2 + (y - 2)^2 = 8$
Centre is $(-1, 2)$.

b) Radius $= \sqrt{8} = \sqrt{2 \times 4} = 2\sqrt{2}$

Q3 a) Complete the square for the x's and y's:
$x^2 + y^2 - 3x + 1 = 0$
$x^2 - 3x + y^2 + 1 = 0$
$(x - \frac{3}{2})^2 - \frac{9}{4} + y^2 + 1 = 0$
$(x - \frac{3}{2})^2 + y^2 = \frac{5}{4}$
Centre is $(\frac{3}{2}, 0)$

b) Radius $= \sqrt{\frac{5}{4}} = \frac{\sqrt{5}}{2}$

2. Circle Properties

Exercise 2.1 — Using circle properties

Q1 a) Centre is $(3, 1)$

b) Gradient of radius $= \frac{1 - 4}{3 - 4} = \frac{-3}{-1} = 3$

c) Gradient of the tangent is $-\frac{1}{3}$,
use $y - y_1 = m(x - x_1)$ to find equation of tangent:
$y - 4 = -\frac{1}{3}(x - 4)$
$3y - 12 = -x + 4$
$x + 3y = 16$
You're asked for the equation in a particular form, so don't forget to rearrange it.

Q2 Centre of the circle is $(-1, 2)$
Gradient of radius $= \frac{2 - (-1)}{-1 - (-3)} = \frac{3}{2}$

Gradient of the tangent $= -\frac{2}{3}$
Use $y - y_1 = m(x - x_1)$ to find equation of tangent:
$y - (-1) = -\frac{2}{3}(x - (-3))$
$3y + 3 = -2(x + 3)$
$3y + 3 = -2x - 6$
$2x + 3y + 9 = 0$

Q3 Centre of the circle is $(3, 4)$,
Gradient of radius $= \frac{4 - 1}{3 - 7} = -\frac{3}{4}$

Gradient of the tangent $= \frac{4}{3}$
Use $y - y_1 = m(x - x_1)$ to find equation of tangent:
$y - 1 = \frac{4}{3}(x - 7)$
$3y - 3 = 4x - 28$
$4x - 3y = 25$

Q4 Rearrange $x^2 + y^2 + 2x - 7 = 0$ and complete the square for the x terms to get:
$(x + 1)^2 + y^2 = 8$
Centre of the circle is $(-1, 0)$

Gradient of radius $= \frac{0 - 2}{-1 - (-3)} = \frac{-2}{2} = -1$
Gradient of the tangent $= 1$

Use $y - y_1 = m(x - x_1)$ to find equation of tangent:
$y - 2 = 1(x - (-3))$
$y - 2 = x + 3$
$y = x + 5$

Q5 Rearrange $x^2 + y^2 + 2x + 4y = 5$ and complete the square for the x and y terms to get:
$(x + 1)^2 + (y + 2)^2 = 10$
Centre of the circle is $(-1, -2)$
Gradient of the radius $= \frac{-2 - (-5)}{-1 - 0} = \frac{3}{-1} = -3$

Gradient of the tangent $= \frac{1}{3}$
Use $y - y_1 = m(x - x_1)$ to find equation of tangent:
$y - (-5) = \frac{1}{3}(x - 0)$
$3y + 15 = x$
$x - 3y = 15$

Q6 a) The line l is perpendicular to the chord AB.
So find the gradient of AB:

Gradient of $AB = \frac{1 - 7}{-1 - (-3)} = -\frac{6}{2} = -3$

So the gradient of l is $\frac{1}{3}$.
Then sub the gradient of l and point M $(-1, 1)$ into $y - y_1 = m(x - x_1)$ to find the equation:
$y - 1 = \frac{1}{3}(x - (-1))$
$3y - 3 = x + 1$
$x - 3y + 4 = 0$

b) The centre is $(2, 2)$, so $a = 2$ and $b = 2$ in the equation $(x - a)^2 + (y - b)^2 = r^2$.
The radius is the length CA, which can be found using Pythagoras:
$r^2 = CA^2 = (-3 - 2)^2 + (7 - 2)^2 = (-5^2) + 5^2 = 50$
So the equation of the circle is:
$(x - 2)^2 + (y - 2)^2 = 50$

Q7 If AC is the diameter, the angle ABC will be 90°.
So find out if AB and BC are perpendicular.

Gradient of AB $= \frac{14 - 12}{4 - (-2)} = \frac{2}{6} = \frac{1}{3}$

Gradient of BC $= \frac{2 - 14}{8 - 4} = \frac{-12}{4} = -3$

Use the gradient rule to check:
$m_1 \times m_2 = -1$
$\frac{1}{3} \times -3 = -1$
As AB and BC are perpendicular, the angle ABC must be 90° and so AC must be the diameter of the circle.

3. Arcs and Sectors

Exercise 3.1 — Radians

Q1 a) π **b)** $\dfrac{3\pi}{4}$ **c)** $\dfrac{3\pi}{2}$

    d) $\dfrac{7\pi}{18}$ **e)** $\dfrac{5\pi}{6}$ **f)** $\dfrac{5\pi}{12}$

Q2 a) $45°$ **b)** $90°$ **c)** $60°$

    d) $450°$ **e)** $135°$ **f)** $420°$

Exercise 3.2 — Arc length and sector area

Q1 $s = r\theta = 6 \times 2 = 12\,\text{cm}$

$A = \dfrac{1}{2}r^2\theta = \dfrac{1}{2} \times 6^2 \times 2 = 36\,\text{cm}^2$

Q2 Get the angle in radians:

$46° = \dfrac{46 \times \pi}{180} = 0.802...\,\text{radians}$

$s = r\theta = 8 \times 0.802... = 6.4\,\text{cm}$

$A = \dfrac{1}{2}r^2\theta = \dfrac{1}{2} \times 8^2 \times 0.802... = 25.7\,\text{cm}^2\,(1\text{d.p.})$

Q3 a) $s = r\theta = 5 \times 1.2 = 6\,\text{cm}$

    $A = \dfrac{1}{2}r^2\theta = \dfrac{1}{2} \times 5^2 \times 1.2 = 15\,\text{cm}^2$

b) $s = r\theta = 4 \times 0.6 = 2.4\,\text{cm}$

    $A = \dfrac{1}{2}r^2\theta = \dfrac{1}{2} \times 4^2 \times 0.6 = 4.8\,\text{cm}^2$

c) Get the angle in radians:

    $80° = \dfrac{80 \times \pi}{180} = \dfrac{4\pi}{9}\,\text{radians}$

    $s = 9 \times \dfrac{4\pi}{9} = 4\pi\,\text{cm} = 12.6\,\text{cm}$

    $A = \dfrac{1}{2} \times 9^2 \times \dfrac{4\pi}{9} = 18\pi\,\text{cm}^2 = 56.5\,\text{cm}^2$

d) $s = 4 \times \dfrac{5\pi}{12} = \dfrac{5\pi}{3}\,\text{cm} = 5.24\,\text{cm}$

    $A = \dfrac{1}{2} \times 4^2 \times \dfrac{5\pi}{12} = \dfrac{10\pi}{3}\,\text{cm}^2 = 10.5\,\text{cm}^2$

Q4 Find the radius, r:

$A = \dfrac{1}{2}r^2\theta = \dfrac{1}{2} \times r^2 \times 0.9 = 16.2\,\text{cm}^2$

$16.2 = 0.45r^2 \Rightarrow 36 = r^2 \Rightarrow r = 6$

$s = r\theta = 6 \times 0.9 = 5.4\,\text{cm}$

Q5 Get the angle in radians:

$20° = \dfrac{20 \times \pi}{180} = \dfrac{\pi}{9}\,\text{radians}$

$s = r\theta = 3 \times \dfrac{\pi}{9} = \dfrac{\pi}{3}\,\text{cm}$

$A = \dfrac{1}{2}r^2\theta = \dfrac{1}{2} \times 3^2 \times \dfrac{\pi}{9} = \dfrac{\pi}{2}\,\text{cm}^2$

Q6 Find the radius, r:

$s = r\theta \Rightarrow r = \dfrac{S}{\theta} = \dfrac{7}{1.4} = 5\,\text{cm}$

$A = \dfrac{1}{2}r^2\theta = \dfrac{1}{2} \times 5^2 \times 1.4 = 17.5\,\text{cm}^2$

Review Exercise — Chapter 2

Q1 a) Radius = 3, centre = (0, 0)

b) Radius = 2, centre = (2, –4)

c) Complete the square for the x's and y's:

$x(x + 6) = y(8 - y)$

$x^2 + 6x = 8y - y^2$

$x^2 + 6x + y^2 - 8y = 0$

$(x + 3)^2 - 9 + (y - 4)^2 - 16 = 0$

$(x + 3)^2 + (y - 4)^2 = 25$

Radius = 5, centre = (–3, 4)

Q2 a) $(x - 3)^2 + (y - 2)^2 = 36$

b) $(x + 4)^2 + (y + 8)^2 = 64$

c) $x^2 + (y + 3)^2 = 14$

Q3 Complete the square for the x's and y's:

$x^2 + y^2 - 4x + 6y - 68 = 0$

$x^2 - 4x + y^2 + 6y - 68 = 0$

$(x - 2)^2 - 4 + (y + 3)^2 - 9 - 68 = 0$

$(x - 2)^2 + (y + 3)^2 = 81$

Centre is (2, –3), radius = 9

Q4 Centre is (2, 1)

Gradient of the radius = $\dfrac{1 - 7}{2 - 10} = \dfrac{-6}{-8} = \dfrac{3}{4}$

Gradient of the tangent $= -\dfrac{4}{3}$

Use $y - y_1 = m(x - x_1)$ to find equation of tangent:

$y - 7 = -\dfrac{4}{3}(x - 10)$

$3y - 21 = -4x + 40$

$3y = -4x + 61 \Rightarrow 4x + 3y - 61 = 0$

Q5 Rearrange $x^2 + y^2 - 12x + 2y + 11 = 0$, to get:

$(x - 6)^2 + (y + 1)^2 = 26$

Centre is (6, –1)

Gradient of the radius = $\dfrac{-1 - (-2)}{6 - 1} = \dfrac{1}{5}$

Gradient of the tangent = –5

Use $y - y_1 = m(x - x_1)$ to find equation of tangent:

$y - (-2) = -5(x - 1)$

$y = 3 - 5x$

Q6 $s = r\theta = 10 \times 0.7 = 7\,\text{cm}$

Area $= \dfrac{1}{2}r^2\theta = \dfrac{1}{2} \times 10^2 \times 0.7 = 35\,\text{cm}^2$

Q7 Get the angle in radians:

$50° = \dfrac{50 \times \pi}{180} = \dfrac{5\pi}{18}\,\text{radians}$

Find the radius, r, first:

Area $= \dfrac{1}{2}r^2\theta = \dfrac{1}{2} \times r^2 \times \dfrac{5\pi}{18} = 20\pi\,\text{cm}^2$

$20\pi = \dfrac{5\pi}{36}r^2 \Rightarrow 144 = r^2 \Rightarrow r = 12$

BC $= s = r\theta = 12 \times \dfrac{5\pi}{18} = \dfrac{10\pi}{3}\,\text{cm}$

Exam-Style Questions — Chapter 2

Q1 a) Rearrange equation and complete the square:
$x^2 - 2x + y^2 - 10y + 21 = 0$ *[1 mark]*
$(x - 1)^2 - 1 + (y - 5)^2 - 25 + 21 = 0$ *[1 mark]*
$(x - 1)^2 + (y - 5)^2 = 5$ *[1 mark]*
Compare with $(x - a)^2 + (y - b)^2 = r^2$:
centre = (1, 5) *[1 mark]*,
radius = $\sqrt{5}$ = 2.24 (to 3 s.f.) *[1 mark]*.

b) The point (3, 6) and centre (1, 5) both lie on the diameter.
Gradient of the diameter = $\frac{6 - 5}{3 - 1}$ = 0.5.
Q (q, 4) also lies on the diameter, so $\frac{4 - 6}{q - 3}$ = 0.5.
$-2 = 0.5q - 1.5$
So $q = (-2 + 1.5) \div 0.5 = -1$.
[3 marks available — 1 mark for finding the gradient of the diameter, 1 mark for linking this with the point Q, and 1 mark for correct calculation of q.]

c) Tangent at Q is perpendicular to the diameter at Q, so gradient $m = -\frac{1}{0.5} = -2$
$y - y_1 = m(x - x_1)$, and (−1, 4) is a point on the line, so:
$y - 4 = -2(x + 1)$
$y - 4 = -2x - 2$
$2x + y - 2 = 0$ is the equation of the tangent.
[5 marks available — 1 mark for gradient = −1 ÷ gradient of diameter, 1 mark for correct value for gradient, 1 mark for substituting Q in straight-line equation, 2 marks for correct substitution of values in the correct form, or 1 mark if not in the form ax + by + c = 0.]

Q2 a) Centre of C = (−4, 4), so $a = -4$ and $b = 4$ *[1 mark]*.
Radius r is the length of MJ.
Using Pythagoras, $r^2 = (0 - -4)^2 + (7 - 4)^2 = 25$ *[1 mark]*
So substituting into the equation gives:
$(x + 4)^2 + (y - 4)^2 = 25$ *[1 mark]*

b) Angle MJH between the radius and tangent is a right angle, so use trig ratios to calculate angle JMH (or θ).
E.g. Length $MJ = r = \sqrt{25} = 5$.
Using Pythagoras,
Length $JH = \sqrt{(6 - 0)^2 + (-1 - 7)^2} = 10$.
$\tan\theta = \frac{10}{5} = 2$
$\theta = \tan^{-1}2 = 1.1071$ rad
[4 marks available — 1 mark for identifying that MJH is a right angle and that trig ratios can be used, 1 mark for calculation of two lengths of the triangle, 1 mark for correct substitution into a trig ratio, and 1 mark for correct answer in radians.]

c) Arc length S = $r\theta$ *[1 mark]*
Angle of sector = θ = 1.1071 rad and r = 5 (both from (b)) so S = 5 × 1.1071 = 5.54 to 3 s.f.
[1 mark].

Q3 a) Area of cross-section = $\frac{1}{2}r^2\theta$
$= \frac{1}{2} \times 20^2 \times \frac{\pi}{4} = 50\pi$ cm².
Volume = area of cross-section × height,
so V = $50\pi \times 10 = 500\pi$ cm³.
[3 marks available — 1 mark for correct use of area formula, 1 mark for 50π, and 1 mark for correct final answer.]

b) Surface area is made up of:
2 × cross-sectional area + 2 × side rectangles + 1 curved end rectangle.
Cross-sectional area = 50π (from part (a))
Area of each side rectangle = 10 × 20 = 200
Area of end rectangle = 10 × arc length
$= 10 \times (20 \times \frac{\pi}{4})$
$= 50\pi$.
S = $(2 \times 50\pi) + (2 \times 200) + 50\pi$
$= (150\pi + 400)$ cm².
[5 marks available — 1 mark for each correct shape area, 1 mark for correct combination, and 1 mark for correct final answer.]

Q4 a) The line through P is a diameter, and as such is perpendicular to the chord AB at the midpoint M.
Gradient of AB = Gradient of AM = $\frac{(7 - 10)}{(11 - 9)}$
$= -\frac{3}{2}$.
$\Rightarrow$ Gradient of $PM = \frac{2}{3}$.
Gradient of $PM = \frac{(7 - 3)}{(11 - p)} = \frac{2}{3}$
$\Rightarrow 3(7 - 3) = 2(11 - p)$
$\Rightarrow 12 = 22 - 2p \Rightarrow p = 5$.
[5 marks available — 1 mark for identifying that PM and AM are perpendicular, 1 mark for correct gradient of AM, 1 mark for correct gradient of PM, 1 mark for substitution of the y-value of P into the gradient or equation of the line PM, and 1 mark for correct final answer.]

b) The centre of C is P(5, 3), so $a = 5$ *[1 mark]* and $b = 3$ *[1 mark]* in the equation
$(x - a)^2 + (y - b)^2 = r^2$.
The radius is the length of AP, which can be found using Pythagoras as follows:
$r^2 = AP^2 = (9 - 5)^2 + (10 - 3)^2 = 65$ *[1 mark]*
So the equation of C is:
$(x - 5)^2 + (y - 3)^2 = 65$ *[1 mark]*.

Chapter 3 — Trigonometry

1. The Sine and Cosine Rules

Exercise 1.1 — The sine and cosine rules

Q1 $a^2 = b^2 + c^2 - 2bc \cos A$
$QR^2 = 9^2 + 10^2 - (2 \times 9 \times 10 \times \cos 42°)$
$= 47.2...$
$QR = 6.87 \text{ cm} \,(3\,\text{s.f.})$

Q2 $\dfrac{a}{\sin A} = \dfrac{b}{\sin B}$
$\Rightarrow TW = \dfrac{FW \times \sin F}{\sin T}$
$= \dfrac{15 \times \sin 39°}{\sin 82°} = 9.53 \text{ cm} \,(3\,\text{s.f.})$

Q3 Angle $C = 180° - 48° - 65° = 67°$
Using the sine rule:
$\dfrac{b}{\sin B} = \dfrac{c}{\sin C}$
$\Rightarrow AC = \dfrac{AB \times \sin B}{\sin C}$
$= \dfrac{11 \times \sin 65°}{\sin 67°} = 10.8 \text{ cm} \,(3\,\text{s.f.})$

Q4 Using the cosine rule:
$a^2 = b^2 + c^2 - 2bc \cos A$
$\Rightarrow \cos A = \dfrac{b^2 + c^2 - a^2}{2bc}$
$\Rightarrow D = \cos^{-1}\!\left(\dfrac{6^2 + 9^2 - 8^2}{2 \times 6 \times 9}\right) = 60.6° \,(3\,\text{s.f.})$

Q5 Using the cosine rule:
$a^2 = b^2 + c^2 - 2bc \cos A$
$\Rightarrow (JK)^2 = 24^2 + 29^2 - (2 \times 24 \times 29 \times \cos 62°)$
$\Rightarrow JK = \sqrt{763.4...} = 27.6 \text{ cm} \,(3\,\text{s.f.})$

Q6 Using the cosine rule:
$a^2 = b^2 + c^2 - 2bc \cos A$
$\Rightarrow (GI)^2 = 6.4^2 + 8.3^2 - (2 \times 6.4 \times 8.3 \times \cos 2.3)$
$\Rightarrow GI = \sqrt{180.6...} = 13.4 \text{ cm} \,(3\,\text{s.f.})$

Q7 Using the cosine rule:
$a^2 = b^2 + c^2 - 2bc \cos A$
$\Rightarrow \cos A = \dfrac{b^2 + c^2 - a^2}{2bc}$
$\Rightarrow C = \cos^{-1}(\dfrac{14^2 + 11^2 - 23^2}{2 \times 14 \times 11}) = 133.5° \,(1\,\text{d.p.})$

Q8 Angle $Q = \pi - 0.66 - 0.75 = 1.731...$ rad
There are π radians in a triangle (180°).
Using the sine rule:
$\dfrac{a}{\sin A} = \dfrac{b}{\sin B}$
$\Rightarrow PQ = \dfrac{PR \times \sin R}{\sin Q}$
$= \dfrac{48 \times \sin 0.75}{\sin 1.731...} = 33.1 \text{ m} \,(3\,\text{s.f.})$

Q9 The smallest angle is between the two biggest sides, so angle F is the smallest angle.
To be safe you could just work out all 3 angles and then see which is smallest.

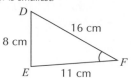

Using the cosine rule:
$a^2 = b^2 + c^2 - 2bc \cos A$
$\Rightarrow \cos F = \dfrac{11^2 + 16^2 - 8^2}{2 \times 11 \times 16}$
$\Rightarrow F = \cos^{-1}\!\left(\dfrac{11^2 + 16^2 - 8^2}{2 \times 11 \times 16}\right) = 27.2° \,(3\,\text{s.f.})$

Q10 Area $= \dfrac{1}{2}ab \sin C$
$= \dfrac{1}{2} \times 9 \times 11 \times \sin 68°$
$= 45.9 \text{ cm}^2 \,(3\,\text{s.f.})$

Q11 Area $= \dfrac{1}{2}ab \sin C$
$= \dfrac{1}{2} \times 12 \times 10.5 \times \sin 53°$
$= 50.3 \text{ cm}^2 \,(3\,\text{s.f.})$

Q12 Start by finding any angle (M here).
Using the cosine rule:
$a^2 = b^2 + c^2 - 2bc \cos A$
$\Rightarrow \cos M = \dfrac{5^2 + 7^2 - 4.2^2}{2 \times 5 \times 7}$
$\Rightarrow M = \cos^{-1}\!\left(\dfrac{5^2 + 7^2 - 4.2^2}{2 \times 5 \times 7}\right) = 36.3...°$

Now you can find the area.
Area $= \dfrac{1}{2}ab \sin C$
$= \dfrac{1}{2} \times 5 \times 7 \times \sin 36.3...°$
$= 10.38 \text{ cm}^2 \,(2\,\text{d.p.})$

You could have found any angle to start off then used the corresponding sides.

Q13 Start by sketching the triangle:

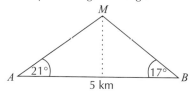

a) Angle $M = 180 - 21 - 17 = 142°$.
Using the sine rule:
$\dfrac{a}{\sin A} = \dfrac{m}{\sin M}$
$\Rightarrow a = \dfrac{5 \times \sin 21°}{\sin 142°} = 2.91 \text{km} \,(3\,\text{s.f.})$

Here a is the distance BM and m is the distance AB.

b) To find the height, draw a line through the triangle from M at a right angle to AB (the dotted line shown in the diagram in part a)).

Height = $\sin 17° \times 2.91 = 0.8509$ km = 851 m (to the nearest m).

The final step just uses SOHCAHTOA — height is the opposite side and 2.91 is the hypotenuse.

Q14 a)

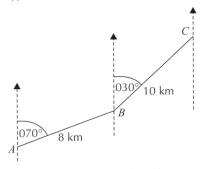

b) The angle anticlockwise from the vertical at B to A is $180° - 70° = 110°$ (parallel lines), so the angle ABC is $110° + 30° = 140°$. Now you can use the cosine rule to find the distance AC.

$$a^2 = b^2 + c^2 - 2bc\cos A$$
$$\Rightarrow AC = \sqrt{(BC)^2 + (AB)^2 - 2(BC)(AB)\cos B}$$
$$\Rightarrow AC = \sqrt{100 + 64 - 160\cos 140°}$$
$$= 16.928... = 16.9\,\text{km (3 s.f.)}$$

c) Start by finding the angle ACB using the cosine rule:

$$a^2 = b^2 + c^2 - 2bc\cos A$$
$$\Rightarrow \cos ACB = \frac{10^2 + 16.928...^2 - 8^2}{2 \times 10 \times 16.928...}$$
$$\Rightarrow ACB = \cos^{-1}\left(\frac{10^2 + 16.928...^2 - 8^2}{2 \times 10 \times 16.928...}\right)$$
$$= 17.68...°$$

The bearing required is therefore $180° + 17.68° + 30°$ (parallel lines) $= 227.68...° = 228°$ (3 s.f.).

Q15 a) Using the cosine rule:

$$a^2 = b^2 + c^2 - 2bc\cos A$$
$$\Rightarrow AC = \sqrt{9^2 + 11^2 - (2 \times 9 \times 11 \times \cos 2.58)}$$
$$= 19.224... = 19.2\,\text{m (3 s.f.)}$$

b) Find the area of each triangle individually. For the top triangle:

$$\text{Area} = \frac{1}{2}ab\sin C$$
$$= \frac{1}{2} \times 9 \times 11 \times \sin 2.58$$
$$= 26.36...\,\text{m}^2 \text{ (3 s.f.)}$$

For the bottom triangle:

$$\text{Area} = \frac{1}{2}ab\sin C$$
$$= \frac{1}{2} \times 8 \times 19.2... \times \sin 1.38$$
$$= 75.50...\,\text{m}^2$$

So the area of the quadrilateral is $26.36... + 75.50... = 101.86... = 102\,\text{m}^2$ (3 s.f.).

Q16 Use the cosine rule to find one of the angles. Here a is 12 cm, b is 15 cm and c is 18 cm.

$$a^2 = b^2 + c^2 - 2bc\cos A$$
$$\Rightarrow A = \cos^{-1}\left(\frac{b^2 + c^2 - a^2}{2bc}\right)$$
$$\Rightarrow A = \cos^{-1}\left(\frac{15^2 + 18^2 - 12^2}{2 \times 15 \times 18}\right) = 41.40...°$$

Now find the area of one face:

$$\text{Area} = \frac{1}{2}ab\sin C$$
$$= \frac{1}{2} \times 15 \times 18 \times \sin(41.40...°)$$
$$= 89.29...\,\text{cm}^2$$

So the total area needed is $4 \times 89.29... = 357.17... = 357\,\text{cm}^2$ (3 s.f.).

2. Trig Identities
Exercise 2.1 — Trig identities

Q1 Use $\tan\theta \equiv \frac{\sin\theta}{\cos\theta}$:
$$\frac{\sin\theta}{\tan\theta} - \cos\theta \equiv \frac{\sin\theta}{\left(\frac{\sin\theta}{\cos\theta}\right)} - \cos\theta$$
$$\equiv \cos\theta - \cos\theta \equiv 0$$

Q2 Use $\sin^2\theta + \cos^2\theta \equiv 1$:
$$\cos^2\theta \equiv 1 - \sin^2\theta \equiv (1 - \sin\theta)(1 + \sin\theta)$$

Q3 Use $\sin^2 x + \cos^2 x \equiv 1$:
$$\cos^2 x \equiv 1 - \sin^2 x$$
$$\Rightarrow \cos x = \sqrt{1 - \sin^2 x} = \sqrt{1 - \left(\frac{1}{2}\right)^2} = \sqrt{\frac{3}{4}} = \frac{\sqrt{3}}{2}$$

Q4 Use $\sin^2 x + \cos^2 x \equiv 1$:
$$4\sin^2 x - 3\cos x + 1 \equiv 4(1 - \cos^2 x) - 3\cos x + 1$$
$$\equiv 4 - 4\cos^2 x - 3\cos x + 1$$
$$\equiv 5 - 3\cos x - 4\cos^2 x$$

Q5 $\cos^2 x \equiv 1 - \sin^2 x$, $\tan x \equiv \frac{\sin x}{\cos x}$
$$\Rightarrow \tan x \equiv \frac{\sqrt{\sin^2 x}}{\sqrt{1 - \sin^2 x}} \equiv \frac{\frac{\sqrt{3}}{2}}{\frac{1}{2}} = \sqrt{3}$$

Q6 Use $\tan x \equiv \frac{\sin x}{\cos x}$ and $\sin^2 x + \cos^2 x \equiv 1$:
$$(\tan x + 1)(\tan x - 1) \equiv \tan^2 x - 1$$
$$\equiv \frac{\sin^2 x}{\cos^2 x} - 1$$
$$\equiv \frac{1 - \cos^2 x}{\cos^2 x} - 1$$
$$\equiv \frac{1}{\cos^2 x} - \frac{\cos^2 x}{\cos^2 x} - 1$$
$$\equiv \frac{1}{\cos^2 x} - 2$$

Q7 Use $\sin^2\theta + \cos^2\theta \equiv 1$:

$(\sin\theta + \cos\theta)^2 + (\sin\theta - \cos\theta)^2$

$\equiv \sin^2\theta + 2\sin\theta\cos\theta + \cos^2\theta$
$\quad + \sin^2\theta - 2\sin\theta\cos\theta + \cos^2\theta$

$\equiv 2(\sin^2\theta + \cos^2\theta)$

$\equiv 2$

Don't make the mistake of writing $(\sin\theta + \cos\theta)^2$
$= \sin^2\theta + \cos^2\theta$

Q8 Use $\sin^2 x + \cos^2 x \equiv 1$ and $\tan x \equiv \dfrac{\sin x}{\cos x}$:

$\tan x + \dfrac{1}{\tan x} \equiv \dfrac{\sin x}{\cos x} + \dfrac{\cos x}{\sin x}$

$\equiv \dfrac{\sin^2 x + \cos^2 x}{\sin x \cos x}$

$\equiv \dfrac{1}{\sin x \cos x}$

Q9 Use $\sin^2 x + \cos^2 x \equiv 1$:

$4 + \sin x - 6\cos^2 x \equiv 4 + \sin x - 6(1 - \sin^2 x)$

$\equiv -2 + \sin x + 6\sin^2 x$

$\equiv (2\sin x - 1)(3\sin x + 2)$

If you're struggling to factorise, let $y = \sin x$, then it
becomes $-2 + y + 6y^2$.

Q10 Use $\sin^2 x + \cos^2 x \equiv 1$:

$\sin^2 x \cos^2 y - \cos^2 x \sin^2 y$

$\equiv (1 - \cos^2 x)\cos^2 y - \cos^2 x(1 - \cos^2 y)$

$\equiv \cos^2 y - \cos^2 x \cos^2 y - \cos^2 x + \cos^2 x \cos^2 y$

$\equiv \cos^2 y - \cos^2 x$

3. Trig Functions

Exercise 3.1 — Graphs of trig functions

Q1 $y = \cos(x + 90°)$

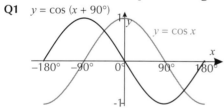

Q2

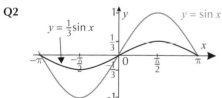

Q3

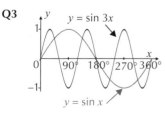

Q4

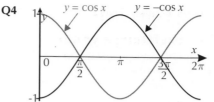

Q5 **a)**

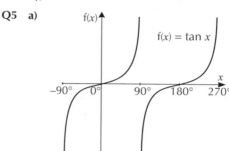

b)

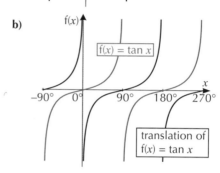

c) $f(x) = \tan(x + 90°)$

Because the graph of tan x repeats every 180°, the
transformation could also be $f(x) = \tan(x - 90°)$.

Q6 **a)**

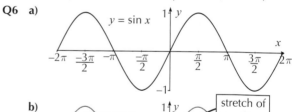

b)

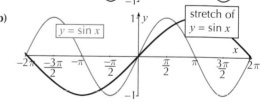

c) $y = \sin \frac{1}{2}x$.

Q7 **a)** The graph has been translated to the left by $\frac{\pi}{2}$.

   b) $y = \sin\left(x + \frac{\pi}{2}\right)$.

You might have noticed that this graph is exactly the
same as the graph of $y = \cos x$.

Q8 **a)** The graph has been stretched vertically by a
   factor of 2.

   b) $y = 2\cos x$.

4. Solving Trig Equations
Exercise 4.1 — Sketching a graph

Q1 a) Find the first solution using a calculator:
$\sin x = 0.75 \Rightarrow x = 48.6°$ (1 d.p.).
Then sketch a graph:

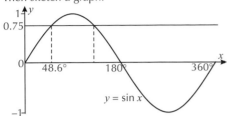

You can see from the graph that there are 2 solutions in the given interval.
Using the symmetry of the graph, if one solution is at 48.6°, the other will be at $180° - 48.6° = 131.4°$ (1 d.p.).

b) Find the first solution: $\cos x = 0.31 \Rightarrow x = 71.9°$ (1 d.p.). Then sketch a graph:

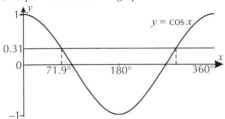

Using the symmetry of the graph to find the second solution: if one solution is at 71.9°, the other will be at $360° - 71.9° = 288.1°$ (1 d.p.).

c) Find the first solution: $\tan x = -1.5 \Rightarrow x = -56.3°$ (1 d.p.). This is outside the given interval, so add on 180° to find the first solution: $-56.3° + 180° = 123.7°$ (1 d.p.). Then sketch a graph:

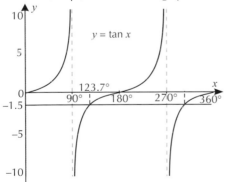

You can see from the graph that the next solution will be at $180° + 123.7° = 303.7°$ (1 d.p.) as tan x repeats every 180°.

d) Find the first solution: $\sin x = -0.42 \Rightarrow x = -24.8°$ (1 d.p.). This is outside the given interval, so add on 360° to find the first solution: $-24.8° + 360° = 335.2°$ (1 d.p.). Then sketch a graph:

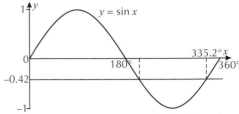

Using the symmetry of the graph, the first solution is $360° - 335.2° = 24.8°$ away from 360°, so the other solution will be 24.8° away from 180°, i.e. at $180° + 24.8° = 204.8°$ (1 d.p.).

e) Find the first solution: $\cos x = -0.56 \Rightarrow x = 124.1°$ (1 d.p.). Then sketch a graph:

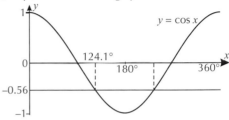

Using the symmetry of the graph to find the second solution: if one solution is at 124.1°, the other will be at $360° - 124.1° = 235.9°$ (1 d.p.).

f) Find the first solution: $\tan x = -0.67 \Rightarrow x = -33.8°$ (1 d.p.). This is outside the given interval, so add on 180° to find the first solution: $-33.8° + 180° = 146.2°$ (1 d.p.). Then sketch a graph:

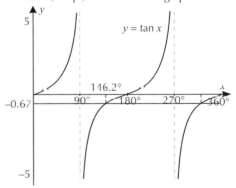

You can see from the graph that the next solution will be at $180° + 146.2° = 326.2°$ (1 d.p.) as tan x repeats every 180°.

Q2 a) Using your knowledge of common angles, the first solution is at $x = \frac{\pi}{4}$. Then sketch a graph:

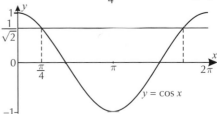

Using the symmetry of the graph, the second solution is at $2\pi - \frac{\pi}{4} = \frac{7\pi}{4}$.

You're asked for exact values, so leave your answers in terms of π.

b) Using common angles, the first solution is at $x = \frac{\pi}{3}$. Then sketch a graph:

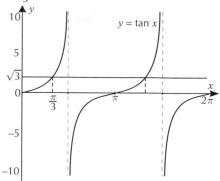

By the symmetry of the graph, the second solution is at $\pi + \frac{\pi}{3} = \frac{4\pi}{3}$.

c) Using common angles, the first solution is at $x = \frac{\pi}{6}$. Then sketch a graph:

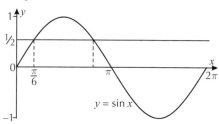

By the symmetry of the graph, the second solution is at $\pi - \frac{\pi}{6} = \frac{5\pi}{6}$.

d) Using common angles, the first solution is at $x = \frac{\pi}{6}$. Then sketch a graph:

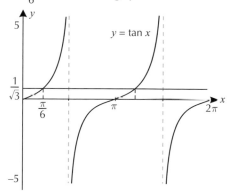

By the symmetry of the graph, the second solution is at $\pi + \frac{\pi}{6} = \frac{7\pi}{6}$.

e) Using common angles, the first solution is at $x = \frac{\pi}{4}$. Then sketch a graph:

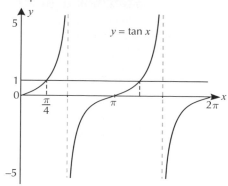

By the symmetry of the graph, the second solution is at $\pi + \frac{\pi}{4} = \frac{5\pi}{4}$.

f) Using your knowledge of common angles, the first solution is at $x = \frac{\pi}{6}$. Then sketch a graph:

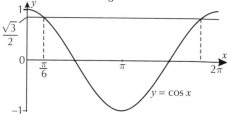

Using the symmetry of the graph, the second solution is at $2\pi - \frac{\pi}{6} = \frac{11\pi}{6}$.

Q3 You're told that there is a solution at 143.1°, and from the graph you can see that there is another solution in the given interval. The first solution is $180° - 143.1° = 36.9°$ away from 180°, so the other solution will be at $180° + 36.9° = 216.9°$ (1 d.p.).

You could also have worked this one out by doing $360° - 143.1°$.

Q4 Use a calculator to find the first solution: $\tan x = 2.5 \Rightarrow x = 68.2°$ (1 d.p.). Then sketch a graph — this time the interval is bigger, so you'll need more repetitions of the tan shape:

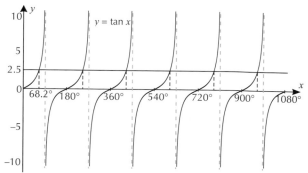

You can see from the graph that there are 6 solutions in the given interval — so just keep adding lots of 180° onto the first solution: $x = 68.2°$, 248.2°, 428.2°, 608.2°, 788.2°, 968.2° (all to 1 d.p.).

You don't have to draw out the whole graph if you don't want — just sketch the first part to find the first solution, then keep adding on lots of 180° until the solutions are bigger than 1080°.

Q5 Find the first solution: $\sin x = 0.81 \Rightarrow x = 0.944$ (3 s.f.). Then sketch a graph — this time for the interval $-2\pi \leq x \leq 2\pi$:

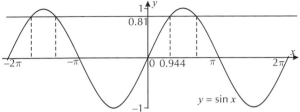

You can see from the graph that there are 4 solutions. Using the symmetry of the graph, there's another solution at $\pi - 0.944 = 2.20$ (3 s.f.). To find the other 2 solutions, subtract 2π from the values you've just found: $0.944 - 2\pi = -5.34$ (3 s.f.) and $2.20 - 2\pi = -4.09$ (3 s.f.).

Exercise 4.2 — Using a CAST diagram

Q1 0.45 is positive, so look at the quadrants where $\sin x$ is positive:

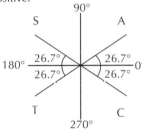

The second quadrant is the other one, so the only other solution is $180° - 26.7° = 153.3°$.

Q2 a) Use a calculator to find the first solution: $\cos x = 0.8 \Rightarrow x = 36.9°$ (1 d.p.). 0.8 is positive, so look at the other quadrants where cos is positive:

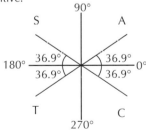

Cos is positive in the 4th quadrant, so the other solution is at $360° - 36.9° = 323.1°$ (1 d.p.).

b) Use a calculator to find the first solution: $\tan x = 2.7 \Rightarrow x = 69.7°$ (1 d.p.). 2.7 is positive, so look at the other quadrants where tan is positive:

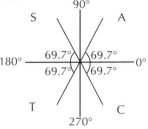

Tan is positive in the 3rd quadrant, so the other solution is at $180° + 69.7° = 249.7°$ (1 d.p.).

c) Use a calculator to find the first solution: $\sin x = -0.15 \Rightarrow x = -8.6°$ (1 d.p.). −0.15 is negative, so look at the quadrants where sin is negative:

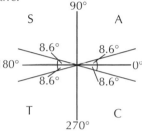

Sin is negative in the 3rd and 4th quadrants, so the solutions are at $180° + 8.6° = 188.6°$ and $360° - 8.6° = 351.4°$ (both to 1 d.p.).

d) Use a calculator to find the first solution: $\tan x = 0.3 \Rightarrow x = 16.7°$ (1 d.p.). 0.3 is positive, so look at the other quadrants where tan is positive:

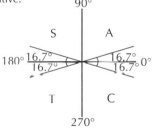

Tan is positive in the 3rd quadrant, so the other solution is at $180° + 16.7° = 196.7°$ (1 d.p.).

e) Use a calculator to find the first solution: $\tan x = -0.6 \Rightarrow x = -31.0°$ (1 d.p.). −0.6 is negative, so look at the quadrants where tan is negative:

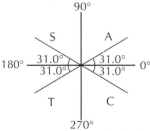

Tan is negative in the 2nd and 4th quadrants. So the solutions are at $180° - 31.0° = 149.0°$ and $360° - 31.0° = 329.0°$ (both to 1 d.p.).

f) Use a calculator to find the first solution: $\sin x = -0.29 \Rightarrow x = -16.9°$ (1 d.p.). -0.29 is negative, so look at the quadrants where sin is negative:

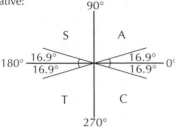

Sin is negative in the 3rd and 4th quadrants, so the solutions are at $180° + 16.9° = 196.9°$ and $360° - 16.9° = 343.1°$ (both to 1 d.p.).

g) Rearranging $4\sin x - 1 = 0$ to make $\sin x$ the subject gives $\sin x = 0.25$. Use a calculator to find the first solution: $\sin x = 0.25 \Rightarrow x = 14.5°$ (1 d.p.). 0.25 is positive, so look at the other quadrants where sin is positive:

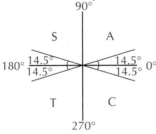

Sin is positive in the 2nd quadrant, so the other solution is at $180° - 14.5° = 165.5°$ (1 d.p.).

h) Rearranging $4\cos x - 3 = 0$ to make $\cos x$ the subject gives $\cos x = 0.75$. Use a calculator to find the first solution: $\cos x = 0.75 \Rightarrow x = 41.4°$ (1 d.p.). 0.75 is positive, so look at the other quadrants where cos is positive:

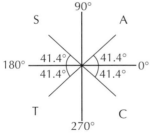

Cos is positive in the 4th quadrant, so the other solution is at $360° - 41.4° = 318.6°$ (1 d.p.).

i) Rearranging $5\tan x + 7 = 0$ to make $\tan x$ the subject gives $\tan x = -1.4$. Use a calculator to find the first solution: $\tan x = -1.4 \Rightarrow x = -54.5°$ (1 d.p.). -1.4 is negative, so look at the quadrants where tan is negative:

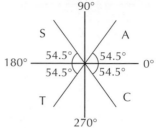

Tan is negative in the 2nd and 4th quadrants, so the solutions are at $180° - 54.5° = 125.5°$ and $360° - 54.5° = 305.5°$ (both to 1 d.p.).

Q3 Use a calculator to find the first solution: $\tan x = -8.4 \Rightarrow x = -1.45$ (3 s.f.). -8.4 is negative, so look at the quadrants where tan is negative:

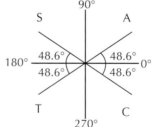

Tan is negative in the 2nd and 4th quadrants, so the solutions are at $\pi - 1.45 = 1.69$ and $2\pi - 1.45 = 4.83$ (both to 3 s.f.).

Q4 The first solution is $x = 48.6°$ (1 d.p.).

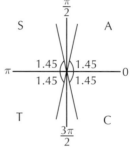

Using a CAST diagram, the solutions of sin x are also positive in the second quadrant. So the next solution is $180° - 48.6° = 131.4°$ (1 d.p.). To find the other solutions in the given interval, add on $360°$ to the solutions already found: $48.6° + 360° = 408.6°$ (1 d.p.) and $131.4° + 360° = 491.4°$ (1 d.p.).

Q5 The first solution is $x = 71.9°$ (1 d.p.).

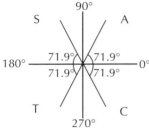

Using a CAST diagram, the solutions of cos x are also positive in the 4th quadrant. So the next solution is $360° - 71.9° = 288.1°$, but this is outside the given interval for x. To find the other solutions in the given interval, subtract $360°$ from the solutions already found: $71.9° - 360° = -288.1°$ (outside interval) and $288.1° - 360° = -71.9°$. So the solutions to 1 d.p. are $x = 71.9°$ and $-71.9°$.

You could have found the negative solutions more directly by reading the CAST diagram in the negative (i.e. clockwise) direction. Reading clockwise from O°, the angle in the 4th quadrant is $-71.9°$.

Q6 The first solution is $x = 0.961$ (3 s.f.).

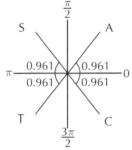

Using a CAST diagram, the solutions of sin x are also positive in the second quadrant. So the next solution is $\pi - 0.961 = 2.18$ (3 s.f.). To find the other solutions in the given interval, add on 2π to the solutions already found: $0.961 + 2\pi = 7.24$ (3 s.f.) and $2.18 + 2\pi = 8.46$ (3 s.f.).

Exercise 4.3 — Changing the interval

For all the questions in this exercise, you can either sketch a graph or use a CAST diagram.

Q1 $\sin 2x = 0.6$, so look for solutions in the interval $0° \leq 2x \leq 720°$. The first solution is $2x = 36.87°$ (2 d.p.). Using a CAST diagram, sin is also positive in the 2nd quadrant, so there's another solution at $2x = 180° - 36.87° = 143.13°$ (2 d.p.). The sin graph repeats every $360°$, so add $360°$ onto the answers already found: $2x = 396.87°$, $503.13°$ (2 d.p.). These are solutions for $2x$, so divide them all by 2: $x = 18.4°$, $71.6°$, $198.4°$, $251.6°$ (1 d.p.).

Q2 $\tan 4x = 4.6$, so look for solutions in the interval $0° \leq 4x \leq 1440°$. The first solution is $4x = 77.74°$ (2 d.p.). From the pattern of the graph, there will be another solution at $4x = 77.74° + 180° = 257.74°$ (2 d.p.). Then just keep adding on $180°$ to find the rest of the solutions within the interval: $4x = 437.74°$, $617.74°$, $797.74°$, $977.74°$, $1157.74°$, $1337.74°$

(2 d.p.). These are solutions for $4x$, so divide them all by 4: $x = 19.4°$, $64.4°$, $109.4°$, $154.4°$, $199.4°$, $244.4°$, $289.4°$, $334.4°$ (1 d.p.).

Q3 $\cos 3x = -0.24$, so look for solutions in the interval $0° \leq 3x \leq 1080°$. The first solution is $3x = 103.89°$ (2 d.p.). Using the symmetry of the graph, there's another solution at $3x = 360° - 103.89° = 256.11°$ (2 d.p.). To find the other solutions within the interval, add on multiples of $360°$: $3x = 463.89°$, $616.11°$, $823.89°$, $976.11°$ (2 d.p.). These are solutions for $3x$, so divide them all by 3: $x = 34.6°$, $85.4°$, $154.6°$, $205.4°$, $274.6°$, $325.4°$ (1 d.p.).

Q4 $\cos 2x = 0.72$, so look for solutions in the interval $0 \leq 2x \leq 4\pi$. The first solution is $2x = 0.7670$ (4 s.f.). Looking at the symmetry of the graph of cos x, the other solutions are $2\pi - 0.7670 = 5.516$ (4 s.f.), $2\pi + 0.7670 = 7.050$ and $4\pi - 0.7670 = 11.80$ (4 s.f.). These are solutions for $2x$, so divide them all by 2: $x = 0.383$, 2.76, 3.53, 5.90 (3 s.f.).

Q5 $\sin 3x = -0.91$, so look for solutions in the interval $0 \leq 3x \leq 6\pi$. The first solution is $3x = -1.143$ (4 s.f.). This is outside the interval, but putting 1.143 into a CAST diagram and looking at the quadrants where sin is negative gives $3x = \pi + 1.143 = 4.285$ and $3x = 2\pi - 1.143 = 5.140$ (4 s.f.). Add on multiples of 2π to find the other solutions in the interval: $3x = 10.57$, 11.42, 16.85, 17.71 (4 s.f.). These are solutions for $3x$, so divide them all by 3: $x = 1.43$, 1.71, 3.52, 3.81, 5.62, 5.90 (3 s.f.).

Q6 $\tan \frac{x}{2} = 2.1$, so look for solutions in the interval $0° \leq \frac{x}{2} \leq 180°$. The first solution is $\frac{x}{2} = 64.54°$ (2 d.p.). This is the only solution in the interval, as tan doesn't repeat any values between $0°$ and $180°$ (looking at its graph). Multiply by 2 to get the value of x: $x = 129.1°$ (1 d.p.).

Q7 $\cos (x - 27°) = 0.64$, so look for solutions in the interval $-27° \leq x - 27° \leq 333°$. The first solution is $x - 27° = 50.2°$ (1 d.p.). Using the symmetry of the graph, there's another solution at $x - 27° = 360° - 50.2° = 309.8°$ (1 d.p.). So the solutions are $x = 77.2°$ and $336.8°$ (1 d.p.).

Q8 $\tan (x - 140°) = -0.76$, so look for solutions in the interval $-140° \leq x - 140° < 220°$. The first solution is $x - 140° = -37.2°$ (1 d.p.). The tan graph repeats every $180°$, so there's another solution at $x - 140° = -37.2° + 180° = 142.8°$ (1 d.p.) (if you add on another $180°$, the answer is outside the interval). So the solutions are $x = 102.8°$ and $282.8°$ (1 d.p.).

Q9 $\sin (x + 36°) = 0.45$, so look for solutions in the interval $36° \leq x + 36° \leq 396°$. The first solution is $x + 36° = 26.7°$ (1 d.p.). This is outside the interval, so add on $360°$ to find a solution in the interval: $x + 36° = 26.7° + 360° = 386.7°$ (1 d.p.). Using a CAST diagram, the other quadrant where sin is positive is the 2nd quadrant, so there's another solution at $x + 36° = 180° - 26.7° = 153.3°$. So the solutions are $x = 117.3°$ and $350.7°$ (1 d.p.).

Q10 $\tan(x + 73°) = 1.84$, so look for solutions in the interval $73° \le x + 73° \le 433°$. The first solution is $x + 73° = 61.5°$ (1 d.p.). This is out of the interval, but use this and the pattern of the graph of $\tan x$ to find the other solutions. $\tan x$ repeats every 180°, so the next two solutions are $x + 73° = 241.5°$ and $x + 73° = 421.5°$ (1 d.p.). So the two solutions are $x = 168.5°$ and $348.5°$ (1 d.p.).

Q11 $\sin(x - \frac{\pi}{4}) = -0.25$, so look for solutions in the interval $0 - \frac{\pi}{4} \le x - \frac{\pi}{4} \le 2\pi - \frac{\pi}{4}$. The first solution is $x - \frac{\pi}{4} = -0.253$ (3 s.f.). Using a CAST diagram, the other solution is at $x - \frac{\pi}{4} = \pi + 0.253$. So adding $\frac{\pi}{4}$ to each solution gives $x = 0.533$ and $x = 4.18$ (3 s.f.).
There's no solution at $2\pi - 0.25$ because it's outside the interval.

Q12 $\cos(x + \frac{\pi}{8}) = 0.13$, so look for solutions in the interval $\frac{\pi}{8} \le x + \frac{\pi}{8} \le 2\pi + \frac{\pi}{8}$. The first solution is $x + \frac{\pi}{8} = 1.44$ (3 s.f.). Using the symmetry of the graph, there's another solution at $x + \frac{\pi}{8} = 2\pi - 1.44 = 4.84$. So the solutions are $x = 1.05$ and 4.45 (3 s.f.).

Exercise 4.4 — Using trig identities to solve equations

For all the questions in this exercise, you can either sketch a graph or use a CAST diagram.

Q1 a) This equation has already been factorised.
Either $\tan x - 5 = 0$ or $3\sin x - 1 = 0$.
$\tan x - 5 = 0 \Rightarrow \tan x = 5$
$\Rightarrow x = 78.7°$ (1 d.p.)
This is the first solution. $\tan$ repeats itself every 180°, so the other solution is $258.7°$ (1 d.p.).
$3\sin x = 1 \Rightarrow \sin x = \frac{1}{3}$
$\Rightarrow x = 19.5°$ (1 d.p.)
Using the symmetry of the graph, the other solution is $180° - 19.5° = 160.5°$ (1 d.p.).

b) $5\sin x \tan x - 4\tan x = 0$
$\tan x(5\sin x - 4) = 0$
So $\tan x = 0$ or $5\sin x - 4 = 0$.
$\tan x = 0 \Rightarrow x = 0°$ or $180°$
(from the graph of $\tan x$).
$5\sin x - 4 = 0 \Rightarrow \sin x = \frac{4}{5}$
$\Rightarrow x = 53.1°$ (1 d.p.)
Using the symmetry of the graph, the other solution is $180° - 53.1° = 126.9°$ (1 d.p.).

c) $\tan^2 x = 9 \Rightarrow \tan x = 3$ or -3.
$\tan x = 3 \Rightarrow x = 71.6°$ (1 d.p.)
Using the repetition of the tan graph, the other solution is $180° + 71.6° = 251.6°$ (1 d.p.).
$\tan x = -3 \Rightarrow x = -71.6°$ (1 d.p.)
This is outside the interval. Keep adding 180° until you've found all the solutions within the interval:
$-71.6° + 180° = 108.4°$ (1 d.p.) and
$108.4° + 180° = 288.4°$ (1 d.p.).

d) $4\cos^2 x = 3\cos x$
$4\cos^2 x - 3\cos x = 0$
$\cos x(4\cos x - 3) = 0$
So $\cos x = 0$ or $4\cos x - 3 = 0$
$\cos x = 0 \Rightarrow x = 90°$
Using the cos graph, the other solution is 270°.
$4\cos x - 3 = 0 \Rightarrow \cos x = \frac{3}{4}$
$\Rightarrow x = 41.4°$ (1 d.p.)
Using the symmetry of the graph, the other solution is $360° - 41.4° = 318.6°$ (1 d.p.).

e) $3\sin x = 5\cos x \Rightarrow \tan x = \frac{5}{3}$. The first solution is $x = 59.0°$ (1 d.p.). $\tan x$ repeats every 180°, so the other solution is $239.0°$ (1 d.p.).

f) $5\tan^2 x - 2\tan x = 0 \Rightarrow \tan x(5\tan x - 2) = 0$. So either $\tan x = 0$ or $\tan x = 0.4$. If $\tan x = 0$ then the solutions are $x = 0°$, 180° and 360°. If $\tan x = 0.4$, the first solution is $21.8°$ (1 d.p.). The graph of $\tan x$ repeats every 180°, so another solution is $x = 201.8°$ (1 d.p.).

g) $6\cos^2 x - \cos x - 2 = 0$
$\Rightarrow (3\cos x - 2)(2\cos x + 1) = 0$
So either $\cos x = \frac{2}{3}$ or $\cos x = -0.5$. If $\cos x = \frac{2}{3}$, the first solution is $48.2°$ (1 d.p.). Looking at the symmetry of the graph of $\cos x$, the other solution is $x = 360° - 48.2° = 311.8°$ (1 d.p.). If $\cos x = -0.5$, the first solution is 120°. Looking at the symmetry of the graph of $\cos x$, the other solution is $360° - 120° = 240°$.

h) $7\sin x + 3\cos x = 0 \Rightarrow 7\sin x = -3\cos x$
$\Rightarrow \tan x = -\frac{3}{7} \Rightarrow x = -23.2°$ (1 d.p.)
This is outside the required interval. Using a CAST diagram, tan is negative in the 2nd and 4th quadrants, so the solutions are $x = 180° - 23.2° = 156.8°$ (1 d.p.) and $x = 360° - 23.2° = 336.8°$ (1 d.p.).

Q2 a) $\tan x = \sin x \cos x \Rightarrow \frac{\sin x}{\cos x} - \sin x \cos x = 0$
$\Rightarrow \sin x - \sin x \cos^2 x = 0$
$\Rightarrow \sin x(1 - \cos^2 x) = 0$
$\Rightarrow \sin x(\sin^2 x) = 0$
$\Rightarrow \sin^3 x = 0$
So $\sin x = 0$. The solutions are $x = 0$, π and 2π.

b) $5\cos^2 x - 9\sin x = 3 \Rightarrow 5(1 - \sin^2 x) - 9\sin x = 3$
$\Rightarrow 5\sin^2 x + 9\sin x - 2 = 0$
$\Rightarrow (5\sin x - 1)(\sin x + 2) = 0$
So either $\sin x = 0.2$ or $\sin x = -2$. $\sin x$ can't be -2, so only $\sin x = 0.2$ will give solutions. The first solution is $x = 0.201$ (3 s.f.). The interval covers three intervals of 2π, so there will be 6 solutions. Looking at the symmetry of the sin graph and adding or subtracting 2π, the other solutions are $x = -6.08, -3.34, 2.94, 6.48$ and 9.22 (3 s.f.).
If you'd used a CAST diagram here, you'd find 0.201 and 2.94 first, then add or subtract 2π.

c) $2\sin^2 x + \sin x - 1 = 0$
$(2\sin x - 1)(\sin x + 1) = 0$
So $2\sin x - 1 = 0$ or $\sin x + 1 = 0$.

$2\sin x - 1 = 0 \Rightarrow \sin x = \frac{1}{2}$
$\Rightarrow x = \frac{\pi}{6}$. Using the symmetry of the graph, another solution is $\pi - \frac{\pi}{6} = \frac{5\pi}{6}$.
To find the other solutions in the required interval, subtract 2π from each of these:
$\frac{\pi}{6} - 2\pi = -\frac{11\pi}{6}$ and $\frac{5\pi}{6} - 2\pi = -\frac{7\pi}{6}$.
$\sin x + 1 = 0 \Rightarrow \sin x = -1$
From the graph, the solutions to this are
$x = -\frac{\pi}{2}$ and $x = \frac{3\pi}{2}$.

Q3 a) $4\sin^2 x = 3 - 3\cos x$
$4(1 - \cos^2 x) = 3 - 3\cos x$
$4 - 4\cos^2 x = 3 - 3\cos x$
$\Rightarrow 4\cos^2 x - 3\cos x - 1 = 0$, as required.

b) Solve the equation from a).
$4\cos^2 x - 3\cos x - 1 = 0$
$(4\cos x + 1)(\cos x - 1) = 0$
So $4\cos x + 1 = 0$ or $\cos x - 1 = 0$
$4\cos x + 1 = 0 \Rightarrow \cos x = -\frac{1}{4}$
$\Rightarrow x = 1.82$ (3 s.f.)
Using the symmetry of the graph, the other solution is $2\pi - 1.82 = 4.46$ (3 s.f.)
$\cos x - 1 = 0 \Rightarrow \cos x = 1$
Using the cos graph, the solutions are
$x = 0$ and $x = 2\pi$.

Q4 $9\sin^2 2x + 3\cos 2x = 7 \Rightarrow 9(1 - \cos^2 2x) + 3\cos 2x = 7$
$\Rightarrow 9 - 9\cos^2 2x + 3\cos 2x = 7$
$\Rightarrow 9\cos^2 2x - 3\cos 2x - 2 = 0$
$\Rightarrow (3\cos 2x + 1)(3\cos 2x - 2) = 0$
So either $\cos 2x = -\frac{1}{3}$ or $\cos 2x = \frac{2}{3}$. For $\cos 2x = -\frac{1}{3}$ look for solutions in the interval $0° \leq 2x \leq 720°$. The first solution is $2x = 109.47°$ (2 d.p.). Looking at the symmetry of the graph of $\cos x$, the other solutions are $2x = 250.53°$, $469.47°$ and $610.53°$ (2 d.p.). Dividing by 2 gives the solutions: $x = 54.7°$, $125.3°$, $234.7°$ and $305.3°$ (1 d.p.).

For $\cos 2x = \frac{2}{3}$, again look for solutions in the interval $0° \leq 2x \leq 720°$. The first solution is $2x = 48.19°$ (2 d.p.). Looking at the symmetry of the graph of $\cos x$, the other solutions are $2x = 311.81°$, $408.19°$, $671.81°$ (2 d.p.). Dividing by 2 gives the solutions: $x = 24.1°$, $155.9°$, $204.1°$ and $335.9°$ (1 d.p.).

Review Exercise — Chapter 3

Q1 $\cos 30° = \frac{\sqrt{3}}{2}$, $\sin 30° = \frac{1}{2}$, $\tan 30° = \frac{1}{\sqrt{3}}$
$\cos 45° = \frac{1}{\sqrt{2}}$, $\sin 45° = \frac{1}{\sqrt{2}}$, $\tan 45° = 1$
$\cos 60° = \frac{1}{2}$, $\sin 60° = \frac{\sqrt{3}}{2}$, $\tan 60° = \sqrt{3}$

Q2 a) Angle $B = 180° - 30° - 25° = 125°$.
Using the sine rule:
$\frac{c}{\sin C} = \frac{b}{\sin B} \Rightarrow c = \frac{6 \times \sin 25°}{\sin 125°} = 3.10\,\text{m}$ (3 s.f.)
$\frac{a}{\sin A} = \frac{b}{\sin B} \Rightarrow a = \frac{6 \times \sin 30°}{\sin 125°} = 3.66\,\text{m}$ (3 s.f.)

b) Area $= \frac{1}{2}ab\sin C = \frac{1}{2} \times 3.66... \times 6 \times \sin 25°$
$= 4.64\,\text{m}^2$ (3 s.f.)

Q3 a) Using the cosine rule:
$r^2 = p^2 + q^2 - 2pq\cos R$
$r = \sqrt{13^2 + 23^2 - (2 \times 13 \times 23 \times \cos 20°)}$
$= 11.664... = 11.7\,\text{km}$ (1 d.p.)
Using the sine rule:
$\frac{\sin P}{p} = \frac{\sin R}{r} \Rightarrow P = \sin^{-1}\frac{13 \times \sin 20°}{11.664...} = 22.4°$
$Q = 180° - 20° - 22.4° = 137.6°$ (1 d.p.)
Be careful here — if you didn't realise the angle was obtuse, you'd have got the wrong value for Q. If you're not sure, sketch the triangle to check.

b) Area $= \frac{1}{2}pq\sin R = 51.1\,\text{km}^2$ (1 d.p.)

Q4 Using the cosine rule:
$a^2 = b^2 + c^2 - 2bc\cos A$
$\Rightarrow A = \cos^{-1}\left(\frac{b^2 + c^2 - a^2}{2bc}\right)$
$= \cos^{-1}\left(\frac{20^2 + 25^2 - 10^2}{2 \times 20 \times 25}\right) = 22.33...°$
$= 22.3°$ (1 d.p.)
It doesn't matter which angle you start with — here it was the smallest of the 3.
$B = \cos^{-1}\left(\frac{a^2 + c^2 - b^2}{2ac}\right)$
$= \cos^{-1}\left(\frac{10^2 + 25^2 - 20^2}{2 \times 10 \times 25}\right) = 49.45...°$
$= 49.5°$ (1 d.p.)
$C = 180° - 49.45...° - 22.33...° = 108.218° = 108.2°$ (1 d.p.)

Q5 The two possible triangles are as shown:

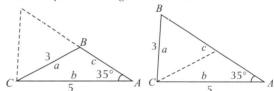

For one triangle:
$\frac{\sin A}{a} = \frac{\sin B}{b} \Rightarrow B = \sin^{-1}\left(\frac{5 \times \sin 35°}{3}\right) = 72.932...°$
$= 72.9°$ (1 d.p.)
$C = 180° - 35° - 72.932...° = 72.067...° = 72.1°$ (1 d.p.)
$\frac{c}{\sin C} = \frac{a}{\sin A} \Rightarrow c = \frac{3 \times \sin 72.06...°}{\sin 35°} = 4.976...\,\text{m}$
$= 4.98\,\text{m}$ (3 s.f.)
For the other triangle:
$B = 180° - 72.932...° = 107.067...° = 107.1°$ (1 d.p.).
This is the other solution to $\sin^{-1}\frac{5 \times \sin 35°}{3}$ for $0 \leq A \leq 180°$.
$C = 180° - 35° - 107.067...° = 37.932...° = 37.9°$ (1 d.p.)
$\frac{c}{\sin C} = \frac{a}{\sin A} \Rightarrow c = \frac{3 \times \sin 37.932...°}{\sin 35°} = 3.215...\,\text{m}$
$= 3.22\,\text{m}$ (3 s.f.)

Whenever you find an angle using the sine rule by taking $\sin^{-1}$, you need to make sure it's the right one and not '180° minus' the angle you want. You can usually tell by looking at the diagram.

Q6 $\tan x - \sin x \cos x \equiv \dfrac{\sin x}{\cos x} - \sin x \cos x$

$$\equiv \dfrac{\sin x - \sin x \cos^2 x}{\cos x}$$

$$\equiv \dfrac{\sin x(1 - \cos^2 x)}{\cos x}$$

$$\equiv \dfrac{\sin x(\sin^2 x)}{\cos x}$$

$$\equiv \sin^2 x \tan x$$

Q7 $\tan^2 x - \cos^2 x + 1 \equiv \dfrac{\sin^2 x}{\cos^2 x} - (1 - \sin^2 x) + 1$

$$\equiv \dfrac{\sin^2 x}{\cos^2 x} + \sin^2 x$$

$$\equiv \dfrac{\sin^2 x + \sin^2 x \cos^2 x}{\cos^2 x}$$

$$\equiv \dfrac{\sin^2 x(1 + \cos^2 x)}{\cos^2 x}$$

$$\equiv \tan^2 x(1 + \cos^2 x)$$

Q8 $(\sin y + \cos y)^2 + (\cos y - \sin y)^2$

$$\equiv \sin^2 y + 2\sin y \cos y + \cos^2 y$$
$$\quad + \cos^2 y - 2\sin y \cos y + \sin^2 y$$

$$\equiv 2\sin^2 y + 2\cos^2 y$$

$$\equiv 2(\sin^2 y + \cos^2 y) \equiv 2$$

Q9 $\dfrac{\sin^4 x + \sin^2 x \cos^2 x}{\cos^2 x - 1} \equiv \dfrac{\sin^4 x + \sin^2 x(1 - \sin^2 x)}{1 - \sin^2 x - 1}$

$$\equiv \dfrac{\sin^4 x + \sin^2 x - \sin^4 x}{-\sin^2 x}$$

$$\equiv \dfrac{\sin^2 x}{-\sin^2 x} \equiv -1$$

Q10 a)

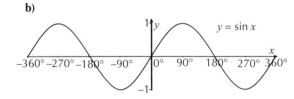

b)

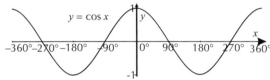

c)

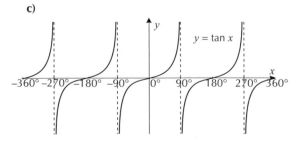

Q11 $y = -\dfrac{1}{2}\cos x$

Q12 $y = \sin 2x$

Q13 a)

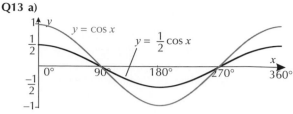

b)

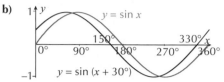

c)

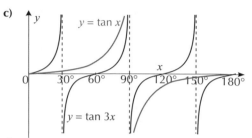

Q14 a)

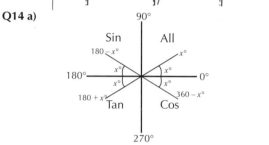

(i) The first solution is $\theta = 60°$. Looking at a CAST diagram, the other solution is $\theta = 180° - 60° = 120°$.

(ii) The first solution is $\theta = -45°$. This is out of the interval needed, but tan repeats every $180°$ so the two solutions are $135°$ and $315°$.

(iii) The first solution is $\theta = 135°$. Looking at a CAST diagram, the other solution is $180° + (180° - 135°) = 225°$.
This time you want the quadrants where cos is negative.

b) (i) The first solution is $4\theta = 131.81...°$. Using the symmetry of the graph of $\cos\theta$, the other solutions in the extended interval $-720° \le 4\theta \le 720°$ are $228.18...°$, $491.81...°$, $588.18...°$, $-131.81...°$, $-228.18...°$, $-491.81...°$ and $-588.18...°$. These are all solutions for 4θ, so the actual solutions to 1 d.p. are $\theta = 33.0°, 57.0°, 123.0°, 147.0°, -33.0°, -57.0°, -123.0°$ and $-147.0°$.

(ii) The first solution is $\theta + 35° = 17.5°$ (1 d.p.). Looking at the symmetry of the graph of $\sin\theta$, the other solution is $\theta + 35° = 180° - 17.5° = 162.5°$. These are solutions for $\theta + 35°$, so the actual solutions are $\theta = -17.5°$ and $127.5°$

(iii) The first solution is $\frac{\theta}{2} = 89.88...°$,

so $\theta = 179.77...° = 179.8°$ (1 d.p.).

As you're finding solutions for $\frac{\theta}{2}$, look in the interval $-90° \le \frac{\theta}{2} \le 90°$. Tan doesn't repeat any values within this interval, so there are no other solutions.

Q15 $6\sin^2 x = \cos x + 5 \Rightarrow 6(1 - \cos^2 x) = \cos x + 5$
$$\Rightarrow 6\cos^2 x + \cos x - 1 = 0$$
$$\Rightarrow (3\cos x - 1)(2\cos x + 1) = 0$$

So either $\cos x = \frac{1}{3}$ or $\cos x = -\frac{1}{2}$. If $\cos x = \frac{1}{3}$, the first solution is $x = 1.230....$ Looking at the symmetry of the graph of $\cos x$, the other solution is $x = 2\pi - 1.230... = 5.052...$

If $\cos x = -\frac{1}{2}$, the first solution is $x = \frac{2\pi}{3}$ (this is because $\cos\frac{\pi}{3} = \frac{1}{2}$, but you want a negative value, so as cos is negative in the second quadrant of the CAST diagram the solution is $\pi - \frac{\pi}{3} = \frac{2\pi}{3}$). Looking at the symmetry of the graph of $\cos x$, the other solution is $x = 2\pi - \frac{2\pi}{3} = \frac{4\pi}{3}$. So the solutions are $x = 1.23$ (3 s.f.), 5.05 (3 s.f.), $\frac{2\pi}{3}$ and $\frac{4\pi}{3}$.

Q16 $3\tan x + 2\cos x = 0 \Rightarrow 3\frac{\sin x}{\cos x} + 2\cos x = 0$
$$\Rightarrow 3\sin x + 2\cos^2 x = 0$$
$$\Rightarrow 3\sin x + 2(1 - \sin^2 x) = 0$$
$$\Rightarrow 3\sin x + 2 - 2\sin^2 x = 0$$
$$\Rightarrow (2\sin x + 1)(\sin x - 2) = 0$$

So either $\sin x = -\frac{1}{2}$ or $\sin x = 2$. $\sin x$ can't be 2, so look for solutions to $\sin x = -\frac{1}{2}$. The first solution is the only one within the interval: $x = -30°$.

Q17 $6\sin^2 x + \sin x - 1 = 0 \Rightarrow (3\sin x - 1)(2\sin x + 1) = 0$
So either $\sin x = \frac{1}{3}$ or $\sin x = -\frac{1}{2}$. If $\sin x = \frac{1}{3}$, the first solution is $x = 0.3398... = 0.340$ (3 s.f.). Looking at the symmetry of the graph of $\sin x$, the other solution is $x = \pi - 0.3398... = 2.801... = 2.80$ (3 s.f.).

If $\sin x = -\frac{1}{2}$, the first solution is $x = -\frac{\pi}{6}$. This is outside the interval, so look at the symmetry of the graph to find the solutions within the interval. One is $\pi + \frac{\pi}{6} = \frac{7\pi}{6}$ and another is $2\pi - \frac{\pi}{6} = \frac{11\pi}{6}$.

Q18 $\tan x - 3\sin x = 0 \Rightarrow \frac{\sin x}{\cos x} - 3\sin x = 0$
$$\Rightarrow \sin x - 3\sin x \cos x = 0$$
$$\Rightarrow \sin x(1 - 3\cos x) = 0$$

So either $\sin x = 0$ or $\cos x = \frac{1}{3}$. If $\sin x = 0$ the solutions are $x = 0°, 180°, 360°, 540°, 720°$.
If $\cos x = \frac{1}{3}$, the first solution is $x = 70.528...° = 70.5°$ (1 d.p.). Looking at the symmetry of the graph of $\cos x$, the other solutions to 1 d.p. are $x = 289.5°$, $430.5°$ and $649.5°$.

Exam-Style Questions — Chapter 3

Q1 a) Using the cosine rule:
$a^2 = b^2 + c^2 - 2bc\cos A$
If XY is a, then angle $A = 180° - 100° = 80°$.
$XY^2 = 150^2 + 250^2 - (2 \times 150 \times 250 \times \cos 80°)$
$XY^2 = 71976.3867$
$XY = \sqrt{71976.3867} = 268.28$ m (to 2 d.p.)
$= 268$ m to the nearest m.
[2 marks available — 1 mark for correct substitution into cosine rule formula, and 1 mark for correct answer.]

b) Using the sine rule:
$\frac{a}{\sin A} = \frac{b}{\sin B}$, so $\frac{250}{\sin\theta} = \frac{268.2842}{\sin 80°}$ (from (a)).
Rearranging gives:
$\frac{\sin\theta}{\sin 80°} = \frac{250}{268.2842} = 0.93$ (2 d.p.)
[2 marks available — 1 mark for correct substitution into sine rule formula and 1 mark for correct final answer.]

Q2 $2 - \sin x = 2\cos^2 x$, and $\cos^2 x \equiv 1 - \sin^2 x$
$\Rightarrow 2 - \sin x = 2(1 - \sin^2 x)$
$\Rightarrow 2 - \sin x = 2 - 2\sin^2 x$
$\Rightarrow 2\sin^2 x - \sin x = 0$
$\sin x(2\sin x - 1) = 0 \Rightarrow \sin x = 0$ or $\sin x = \frac{1}{2}$.
For $\sin x = 0$, $x = 0, \pi$ and 2π.
For $\sin x = \frac{1}{2}$, $x = \frac{\pi}{6}$ and $\pi - \frac{\pi}{6} = \frac{5\pi}{6}$.

[6 marks available — 1 mark for correct substitution using trig identity, 1 mark for factorising quadratic in sin x, 1 mark for finding correct values of sin x, 1 mark for all three solutions when sin x = 0, 1 mark for each of the other 2 correct solutions.]

Q3 a) $(1 + 2\cos x)(3\tan^2 x - 1) = 0$
$\Rightarrow 1 + 2\cos x = 0 \Rightarrow \cos x = -\frac{1}{2}$.
OR:
$3\tan^2 x - 1 = 0 \Rightarrow \tan^2 x = \frac{1}{3}$
$\Rightarrow \tan x = \frac{1}{\sqrt{3}}$ or $-\frac{1}{\sqrt{3}}$.
For $\cos x = -\frac{1}{2}$, $x = \frac{2\pi}{3}$ and $-\frac{2\pi}{3}$.

Drawing the cos x graph helps you find the second one here, and don't forget the limits are $-\pi \le x \le \pi$.

For $\tan x = \frac{1}{\sqrt{3}}$,
$x = \frac{\pi}{6}$ and $-\pi + \frac{\pi}{6} = -\frac{5\pi}{6}$.
For $\tan x = -\frac{1}{\sqrt{3}}$,
$x = -\frac{\pi}{6}$ and $-\frac{\pi}{6} + \pi = \frac{5\pi}{6}$.

Again, look at the graph of tan x if you're unsure.

[6 marks available — 1 mark for each correct solution.]

b) $3\cos^2 x = \sin^2 x \Rightarrow 3 = \frac{\sin^2 x}{\cos^2 x} \Rightarrow 3 = \tan^2 x$
$\Rightarrow \tan x = \pm\sqrt{3}$.
For $\tan x = \sqrt{3}$, $x = \frac{\pi}{3}$ and $\frac{\pi}{3} - \pi = -\frac{2\pi}{3}$.
For $\tan x = -\sqrt{3}$, $x = -\frac{\pi}{3}$ and $-\frac{\pi}{3} + \pi = \frac{2\pi}{3}$.
[4 marks available — 1 mark for each correct solution.]

Q4 a) $3\cos x = 2\sin x$, and $\tan x \equiv \dfrac{\sin x}{\cos x}$,

You need to substitute tan in somewhere, so look at how you can rearrange to get sin/cos in the equation...

Divide through by $\cos x$ to give:
$3\dfrac{\cos x}{\cos x} = 2\dfrac{\sin x}{\cos x} \Rightarrow 3 = 2\tan x$
$\Rightarrow \tan x = \dfrac{3}{2}$ (or = 1.5).

[2 marks available — 1 mark for correct substitution of tan x, 1 mark for correct final answer.]

b) Using $\tan x = 1.5$, the first solution is $x = 56.3°$ (1 d.p.) **[1 mark]** and a 2nd solution can be found from $x = 180° + 56.3° = 236.3°$ (1 d.p.) **[1 mark]**. *Remember tan x repeats every 180°.*

Q5 a) (i)
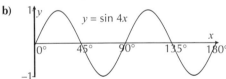

[2 marks available — 1 mark for correct shape of cos x graph, 1 mark for shift of 60° to the left.]

(ii) $\cos x$ crosses the x-axis at 90° and 270°. Shifting these left by 60° gives the solutions $x = 30°$ and $210°$.
[2 marks available — 1 mark for each correct solution.]

b)

[graph of $y = \sin 4x$ with axis marks at 0°, 45°, 90°, 135°, 180°]

[2 marks available — 1 mark for correct shape, 1 mark for stretch of scale factor ¼.]

c) The first solution is $4x = 30°$ so $x = 7.5°$ **[1 mark]**. The graph in (b) shows there are 4 solutions between 0° and 180°, and looking at the symmetry of the graph, they lie 7.5° from where the graph cuts the x-axis as follows:

$x = 45° - 7.5° = 37.5°$ **[1 mark]**
$x = 90° + 7.5° = 97.5°$ **[1 mark]**
$x = 135° - 7.5° = 127.5°$ **[1 mark]**

If you hadn't sketched the graph, you could solve it by extending the range.

Q6 a) $\tan\left(x + \dfrac{\pi}{6}\right) = \sqrt{3} \Rightarrow x + \dfrac{\pi}{6} = \dfrac{\pi}{3}$ **[1 mark]**

tan is positive in the 3rd quadrant of a CAST diagram, so you can work out the other solution as follows...

2nd solution can be found from:
$x + \dfrac{\pi}{6} = \pi + \dfrac{\pi}{3} = \dfrac{4}{3}\pi$ **[1 mark]**. So subtracting $\dfrac{\pi}{6}$ from each solution gives: $x = \dfrac{\pi}{6}$ **[1 mark]**,
$x = \dfrac{7\pi}{6}$ **[1 mark]**.

b) For this one, change the interval and look for solutions in the interval $-\dfrac{\pi}{4} \le x - \dfrac{\pi}{4} \le 2\pi - \dfrac{\pi}{4}$.
$2\cos\left(x - \dfrac{\pi}{4}\right) = \sqrt{3}$
so $\cos\left(x - \dfrac{\pi}{4}\right) = \dfrac{\sqrt{3}}{2}$.

Solving this gives $x - \dfrac{\pi}{4} = \dfrac{\pi}{6}$, which is in the interval — so it's a solution **[1 mark]**. From the symmetry of the cos graph there's another solution at $2\pi - \dfrac{\pi}{6} = \dfrac{11\pi}{6}$. But this is outside the interval for $x - \dfrac{\pi}{4}$, so you can ignore it **[1 mark]**. Using symmetry again, there's also a solution at $-\dfrac{\pi}{6}$ **[1 mark]** — and this one is in your interval. So solutions for $x - \dfrac{\pi}{4}$ are $-\dfrac{\pi}{6}$ and $\dfrac{\pi}{6}$
$\Rightarrow x = \dfrac{\pi}{12}$ and $\dfrac{5\pi}{12}$. **[1 mark]**

You might find it useful to sketch the graph for this one — or you could use the CAST diagram if you prefer.

c) $\sin 2x = -\dfrac{1}{2}$, so look for solutions in the interval $0 \le 2x \le 4\pi$.
It's easier to see what's going on by drawing a graph.

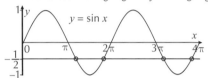

The graph shows there are 4 solutions between 0 and 4π. Putting $\sin 2x = -\dfrac{1}{2}$ into your calculator gives you the solution $2x = -\dfrac{\pi}{6}$, but this is outside the interval. From the graph, you can see that the solutions within the interval occur at

$\pi + \dfrac{\pi}{6}, 2\pi - \dfrac{\pi}{6}, 3\pi + \dfrac{\pi}{6}$ and $4\pi - \dfrac{\pi}{6}$ **[1 mark]**, so
$2x = \dfrac{7\pi}{6}, \dfrac{11\pi}{6}, \dfrac{19\pi}{6}$ and $\dfrac{23\pi}{6}$ **[1 mark]**. Dividing by 2 gives: $x = \dfrac{7\pi}{12}, \dfrac{11\pi}{12}, \dfrac{19\pi}{12}$ and $\dfrac{23\pi}{12}$

[1 mark for each correct solution]

Q7 a) $2(1 - \cos x) = 3\sin^2 x$, and $\sin^2 x \equiv 1 - \cos^2 x$.
$\Rightarrow 2(1 - \cos x) = 3(1 - \cos^2 x)$ **[1 mark]**.
You need to get the whole equation into either sin or cos to get something useful at the end, so get used to spotting places to use the trig identities.
$\Rightarrow 2 - 2\cos x = 3 - 3\cos^2 x$
$\Rightarrow 3\cos^2 x - 2\cos x - 1 = 0$ **[1 mark]**.

b) From (a), the equation can be written as:
$3\cos^2 x - 2\cos x - 1 = 0$
Now this looks like a quadratic equation, which can be factorised:
$(3\cos x + 1)(\cos x - 1) = 0$
$\Rightarrow \cos x = -\dfrac{1}{3}$ **[1 mark]** or $\cos x = 1$ **[1 mark]**
For $\cos x = -\dfrac{1}{3}$, $x = 109.5°$ (to 1 d.p.) **[1 mark]**, and a 2nd solution can be found from $x = (360° - 109.5°) = 250.5°$ **[1 mark]**.
For $\cos x = 1$, $x = 0°$ **[1 mark]** and $360°$ **[1 mark]**.

Chapter 4:
Logarithms and Exponentials

1. Logs
Exercise 1.1 — Logs

Q1 a) $\log_2 8 = 3$ **b)** $\log_5 625 = 4$
c) $\log_{49} 7 = \frac{1}{2}$ **d)** $\log_8 4 = \frac{2}{3}$
e) $\log_{10} \frac{1}{100} = -2$ **f)** $\log_2 0.125 = -3$

Q2 a) $\log_4 9 = x$ **b)** $\log_x 40 = 3$
c) $\log_8 x = 11$

Q3 a) $5^3 = 125$ **b)** $10^4 = 10\,000$
c) $\left(\frac{1}{2}\right)^{-2} = 4$ **d)** $7^6 = a$
e) $5^{0.2} = t$ **f)** $4^1 = m$
g) $\left(\frac{1}{4}\right)^{\frac{1}{2}} = p$ **h)** $10^5 = k$
i) $x^m = a$

Q4 a) 3 **b)** −2 **c)** 0
d) 0.301 **e)** 0.477 **f)** 0.778
Use the 'log' button on your calculator to work these out.

Q5 a) 2 **b)** 3 **c)** −1

Q6 a) $x^2 = 49 \Rightarrow x = 7$
b) $x^3 = 8 \Rightarrow x = 2$
c) $x^5 = 100\,000 \Rightarrow x = 10$
d) $x^{\frac{1}{2}} = 3 \Rightarrow x = 3^2 \Rightarrow x = 9$
e) $x^{\frac{1}{3}} = 7 \Rightarrow x = 7^3 \Rightarrow x - 343$
f) $x^{\frac{1}{5}} = 2 \Rightarrow x = 2^5 \Rightarrow x = 32$
g) $3^4 = x \Rightarrow x = 81$
h) $2^6 = x \Rightarrow x = 64$
i) $7^1 = x \Rightarrow x = 7$
j) $9^{\frac{1}{2}} = x \Rightarrow x = 3$
k) $64^{\frac{1}{3}} = x \Rightarrow x = 4$
l) $27^{\frac{2}{3}} = x \Rightarrow x = 9$

Q7 a) $a^2 = x$ and $a^4 = y$, so $y = x^2$.
b) $a^3 = x$ and $(2a)^3 = y \Rightarrow 8a^3 = y$, so $y = 8x$.
c) $a^5 = x$ and $a^{20} = y$, $(a^5)^4 = y$, so $y = x^4$.

Exercise 1.2 — Laws of logs

Q1 a) $\log_a 2 + \log_a 5 = \log_a (2 \times 5) = \log_a 10$
b) $\log_m 8 + \log_m 7 = \log_m (8 \times 7) = \log_m 56$
c) $\log_b 8 - \log_b 4 = \log_b (8 \div 4) = \log_b 2$
d) $\log_m 15 - \log_m 5 = \log_m (15 \div 5) = \log_m 3$
e) $3 \log_n 4 = \log_n (4^3) = \log_n 64$
f) $2 \log_a 7 = \log_a (7^2) = \log_a 49$
g) $\frac{1}{2} \log_b 16 = \log_b (16^{\frac{1}{2}}) = \log_b 4$
h) $\frac{2}{3} \log_a 125 = \log_a (125^{\frac{2}{3}}) = \log_a 25$

i) $\log_3 a - \log_2 a$ can't be simplified any further as the bases are different.
The words 'where possible' in the question are a clue that there might be some parts which can't be simplified.

Q2 a) $2 \log_a 5 + \log_a 4 = \log_a (5^2) + \log_a 4$
$= \log_a (25 \times 4) = \log_a 100$
b) $3 \log_m 2 - \log_m 4 = \log_m (2^3) - \log_m 4$
$= \log_m (8 \div 4) = \log_m 2$
c) $3 \log_n 4 - 2 \log_n 8 = \log_n (4^3) - \log_n (8^2)$
$= \log_n (64 \div 64) = \log_n 1 = 0$
d) $\frac{2}{3} \log_b 216 - 2 \log_b 3 = \log_b (216^{\frac{2}{3}}) - \log_b (3^2)$
$= \log_b (36 \div 9) = \log_b 4$
e) $1 + \log_a 6 = \log_a a + \log_a 6 = \log_a 6a$
f) $2 - \log_b 5 = 2 \log_b b - \log_b 5 = \log_b b^2 - \log_b 5$
$= \log_b \left(\frac{b^2}{5}\right)$

Q3 a) $\log_a 6 = \log_a (2 \times 3) = \log_a 2 + \log_a 3 = x + y$
b) $\log_a 16 = \log_a 2^4 = 4 \log_a 2 = 4x$
c) $\log_a 60 = \log_a (2 \times 2 \times 3 \times 5)$
$= \log_a 2^2 + \log_a 3 + \log_a 5 = 2x + y + z$

Q4 a) $\log_b b^3 = 3 \log_b b = 3$
b) $\log_a \sqrt{a} = \log_a a^{\frac{1}{2}} = \frac{1}{2} \log_a a = \frac{1}{2}$
c) $\log_m 4m - 2 \log_m 2 = \log_m 4 + \log_m m - \log_m 2^2$
$= \log_m 4 + 1 - \log_m 4 = 1$
d) $\log_{2b} \left(\frac{4 \times b}{2}\right) = \log_{2b} 2b - 1$

Q5 a) $\log_2 4^x = x \log_2 4 = x \log_2 2^2 = 2x \log_2 2 = 2x$
b) $\frac{\log_a 54 - \log_a 6}{\log_a 3} = \frac{\log_a (54 \div 6)}{\log_a 3} = \frac{\log_a 9}{\log_a 3}$
$= \frac{\log_a 3^2}{\log_a 3} = \frac{2 \log_a 3}{\log_a 3} = 2$

Q6 a) $\log_6 3 = \frac{\log_{10} 3}{\log_{10} 6} = 0.613$ (3 s.f.)
b) $\log_9 2 = \frac{\log_{10} 2}{\log_{10} 9} = 0.315$ (3 s.f.)
c) $\log_3 13 = \frac{\log_{10} 13}{\log_{10} 3} = 2.33$ (3 s.f.)
d) $\log_5 4 = \frac{\log_{10} 4}{\log_{10} 5} = 0.861$ (3 s.f.)

2. Exponentials
Exercise 2.1 — Exponentials and logs

Q1 a) Take logs of both sides:
$\log 2^x = \log 3 \Rightarrow x \log 2 = \log 3$
$\Rightarrow x = \frac{\log 3}{\log 2} = 1.584... = 1.58$ (3 s.f.)

b) $4^x = 16 \Rightarrow \log 4^x = \log 16 \Rightarrow x \log 4 = \log 16$
$\Rightarrow x = \frac{\log 16}{\log 4} = \frac{4 \log 2}{2 \log 2} = 2$
This one's actually pretty easy if you spot that $16 = 4^2$ so $x = 2$.

c) $7^x = 2 \Rightarrow \log 7^x = \log 2 \Rightarrow x \log 7 = \log 2$

$\Rightarrow x = \dfrac{\log 2}{\log 7} = 0.3562... = 0.356 \,(3 \text{ s.f.})$

d) $1.8^x = 0.4 \Rightarrow \log 1.8^x = \log 0.4$
$\Rightarrow x \log 1.8 = \log 0.4$

$\Rightarrow x = \dfrac{\log 0.4}{\log 1.8} = -1.558... = -1.56 \,(3 \text{ s.f.})$

Notice that this solution is negative, because log 0.4 is negative.

e) $0.7^x = 3 \Rightarrow \log 0.7^x = \log 3 \Rightarrow x \log 0.7 = \log 3$

$\Rightarrow x = \dfrac{\log 3}{\log 0.7} = -3.080... = -3.08 \,(3 \text{ s.f.})$

f) $0.5^x = 0.2 \Rightarrow \log 0.5^x = \log 0.2$
$\Rightarrow x \log 0.5 = \log 0.2$

$\Rightarrow x = \dfrac{\log 0.2}{\log 0.5} = 2.321... = 2.32 \,(3 \text{ s.f.})$

g) Take logs of both sides:
$\log 2^{3x-1} = \log 5 \Rightarrow (3x - 1)\log 2 = \log 5$
$\Rightarrow 3x \log 2 = \log 5 + \log 2$
$\qquad\qquad = \log(5 \times 2)$

$\Rightarrow x = \dfrac{\log 10}{3 \log 2} = 1.11 \,(3 \text{ s.f.})$

h) $10^{3-x} = 8 \Rightarrow \log 10^{3-x} = \log 8$
$\Rightarrow (3 - x)\log 10 = \log 8$
$\Rightarrow 3 - x = \log 8$
$\Rightarrow x = 3 - \log 8 = 2.10 \;(3 \text{ s.f.})$

Remember that $\log_{10} 10 = 1$.

i) $0.4^{5x-4} = 2 \Rightarrow \log 0.4^{5x-4} = \log 2$
$\Rightarrow (5x - 4)\log 0.4 = \log 2$
$\Rightarrow 5x \log 0.4 = \log 2 + 4 \log 0.4$

$\Rightarrow x = \dfrac{\log 2 + 4 \log 0.4}{5 \log 0.4} = 0.649 \;(3 \text{ s.f.})$

Q2 a) Take exponentials of both sides using base 10 (since the logarithm is base 10):
$10^{\log 5x} = 10^3 \Rightarrow 5x = 1000 \Rightarrow x = 200$

b) Take exponentials of both sides (using base 2):
$\Rightarrow 2^{\log_2 (x+3)} = 2^4 \Rightarrow x + 3 = 16 \Rightarrow x = 13$

c) Take exponentials of both sides (using base 3):
$\Rightarrow 3^{\log_3 (5-2x)} = 3^{2.5} \Rightarrow 5 - 2x = 3^{2.5}$

$\Rightarrow x = \dfrac{5 - 3^{2.5}}{2} = -5.294... = -5.29 \,(3 \text{ s.f.})$

Q3 a) $4^{x+1} = 3^{2x} \Rightarrow \log 4^{x+1} = \log 3^{2x}$
$\Rightarrow (x + 1) \log 4 = 2x \log 3$

Multiply out the brackets:
$\Rightarrow x \log 4 + \log 4 = 2x \log 3$

Collect x-terms on one side:
$\Rightarrow \log 4 = 2x \log 3 - x \log 4 = x (2 \log 3 - \log 4)$

$\Rightarrow x = \dfrac{\log 4}{2 \log 3 - \log 4} = 1.709... = 1.71 (3 \text{ s.f.})$

b) $2^{5-x} = 4^{x+3} \Rightarrow \log 2^{5-x} = \log 4^{x+3}$
$\Rightarrow (5 - x) \log 2 = (x + 3) \log 4$
$\Rightarrow 5 \log 2 - x \log 2 = x \log 4 + 3 \log 4$
$\Rightarrow 5 \log 2 - 3 \log 4 = x \log 4 + x \log 2$
$\Rightarrow 5 \log 2 - 3 \log 4 = x \,(\log 4 + \log 2)$

$\Rightarrow x = \dfrac{5 \log 2 - 3 \log 4}{\log 4 + \log 2} = -\dfrac{1}{3}$

c) $3^{2x-1} = 6^{3-x} \Rightarrow \log 3^{2x-1} = \log 6^{3-x}$
$\Rightarrow (2x - 1) \log 3 = (3 - x) \log 6$
$\Rightarrow 2x \log 3 - \log 3 = 3 \log 6 - x \log 6$
$\Rightarrow 2x \log 3 + x \log 6 = 3 \log 6 + \log 3$
$\Rightarrow x \,(2 \log 3 + \log 6) = 3 \log 6 + \log 3$

$\Rightarrow x = \dfrac{3 \log 6 + \log 3}{2 \log 3 + \log 6} = 1.622... = 1.62 \,(3 \text{ s.f.})$

Q4 a) $\log 2x = \log(x + 1) - 1 \Rightarrow \log 2x - \log(x + 1) = -1$

$\Rightarrow \log\left(\dfrac{2x}{x+1}\right) = -1 \Rightarrow \dfrac{2x}{x+1} = 10^{-1} = \dfrac{1}{10}$

$\Rightarrow x + 1 = 20x \Rightarrow x = \dfrac{1}{19} = 0.0526 \text{ (to 3 s.f.)}$

b) $\log_2 2x = 3 - \log_2 (9 - 2x)$
$\Rightarrow \log_2 2x + \log_2 (9 - 2x) = 3$
$\Rightarrow \log_2 2x (9 - 2x) = 3$

Take exponentials of base 2 of both sides to get:
$\Rightarrow 2x(9 - 2x) = 2^3 \Rightarrow 18x - 4x^2 = 8$
$\Rightarrow 4x^2 - 18x + 8 = 0 \Rightarrow 2x^2 - 9x + 4 = 0$
$\Rightarrow (2x - 1)(x - 4) = 0$

So, $x = \dfrac{1}{2}$ or $x = 4$

c) $\log_6 x = 1 - \log_6 (x + 1)$
$\Rightarrow \log_6 x + \log_6 (x + 1) = 1 \Rightarrow \log_6 x(x + 1) = 1$

Take exponentials of base 6 of both sides to get:
$\Rightarrow x(x + 1) = 6^1 \Rightarrow x^2 + x - 6 = 0$
$\Rightarrow (x + 3)(x - 2) = 0 \Rightarrow x = 2$

x = −3 is not a solution because logarithms of negative numbers don't exist.

d) $\log_2 (2x + 1) = 3 + 2 \log_2 x$
$\Rightarrow \log_2 (2x + 1) = 3 + \log_2 x^2$

$\Rightarrow \log_2 (2x + 1) - \log_2 x^2 = 3 \Rightarrow \log_2 \dfrac{2x + 1}{x^2} = 3$

Take exponentials of base 2 of both sides to get:
$\Rightarrow \dfrac{2x + 1}{x^2} = 2^3 \Rightarrow 2x + 1 = 8x^2$

$\Rightarrow 8x^2 - 2x - 1 = 0 \Rightarrow (4x + 1)(2x - 1) = 0$

So $x = \dfrac{1}{2}$.

Q5 $2^{x+y} = 8 \Rightarrow 2^{x+y} = 2^3 \Rightarrow x + y = 3$.

$\log_2 x - \log_2 y = 1 \Rightarrow \log_2 \dfrac{x}{y} = 1$

Take exponentials of base 2 of both sides to get:
$\Rightarrow \dfrac{x}{y} = 2^1 = 2 \Rightarrow x = 2y$

Solve $x + y = 3$ and $x = 2y$ simultaneously:

$x = 2y$ so put this into $x + y = 3 \Rightarrow 2y + y = 3$
$\Rightarrow 3y = 3$
$\Rightarrow y = 1$ and $x = 2$.

Q6 $9^{x-2} = 3^y \Rightarrow (3^2)^{x-2} = 3^y \Rightarrow 3^{(2(x-2))} = 3^y$ so $2(x-2) = y$

$\log_3 2x = 1 + \log_3 y \Rightarrow \log_3 2x - \log_3 y = 1$

$\Rightarrow \log_3 \frac{2x}{y} = 1 \Rightarrow \frac{2x}{y} = 3^1 \Rightarrow 2x = 3y$

Solve $2(x-2) = y$ and $2x = 3y$ simultaneously:

$2x = 3y$ so put this into $2(x-2) = y \Rightarrow 3y - 4 = y \Rightarrow$
$2y = 4 \Rightarrow y = 2$ and $x = 3$.

Q7 a) Let $y = 2^x$, then $2^{2x} - 5(2^x) + 4 = 0$ is equivalent to
the quadratic equation $y^2 - 5y + 4 = 0$.
$\Rightarrow (y-1)(y-4) = 0$, so $y = 1$ or $y = 4$.
So $2^x = 4$ or $2^x = 1 \Rightarrow x = 2$ or $x = 0$.
The y^2 in the quadratic equation comes from $2^{2x} = (2^x)^2$.

b) Let $y = 4^x$, then $4^{2x} - 17(4^x) + 16 = 0$ is equivalent
to the quadratic equation $y^2 - 17y + 16 = 0$.
$\Rightarrow (y-1)(y-16) = 0$, so $y = 1$ or $y = 16$.
So $4^x = 1$ or $4^x = 16 \Rightarrow x = 0$ or $x = 2$.

c) Let $y = 3^x$, then $3^{2x+2} = 3^{2x} \times 3^2 = y^2 \times 9 = 9y^2$.
So $3^{2x+2} - 82(3^x) + 9 = 0$ is equivalent to the
quadratic equation $9y^2 - 82y + 9 = 0$.
$\Rightarrow (9y-1)(y-9) = 0$, so $y = \frac{1}{9}$ or $y = 9$.
So $3^x = \frac{1}{9}$ or $3^x = 9 \Rightarrow x = -2$ or $x = 2$.

d) Let $y = 2^x$, then $2^{2x+3} = 2^{2x} \times 2^3 = y^2 \times 8 = 8y^2$.
So $2^{2x+3} - 9(2^x) + 1 = 0$ is equivalent to the
quadratic equation $8y^2 - 9y + 1 = 0$.
$\Rightarrow (8y-1)(y-1) = 0$, so $y = \frac{1}{8}$ or $y = 1$.
So $2^x = \frac{1}{8}$ or $2^x = 1 \Rightarrow x = -3$ or $x = 0$.

Q8 After n years, the investment is worth 500×1.08^n.
So, $500 \times 1.08^n = 1500 \Rightarrow 1.08^n = 3$.
Take logs of both sides:
$\log 1.08^n = \log 3 \Rightarrow n \log 1.08 = \log 3$
$\Rightarrow n = \dfrac{\log 3}{\log 1.08} = 14.3$

So it will take 15 years for Howard's investment to
exceed £1500.
*The question asks for full years, so round your answer up —
after 14 years the value won't be quite high enough.*

Review Exercise — Chapter 4

Q1 a) $\log_4 16 = 2$

b) $\log_{216} 6 = \frac{1}{3}$

c) $\log_3 \frac{1}{81} = -4$

Q2 a) $3^3 = 27$ so $\log_3 27 = 3$

b) To get fractions you need negative powers
$3^{-3} = \frac{1}{27}$
$\log_3 \left(\frac{1}{27} \right) = -3$

c) Logs are subtracted so divide
$\log_3 18 - \log_3 2 = \log_3 (18 \div 2)$
$= \log_3 9$
$= 2$ (as $3^2 = 9$)

Q3 a) Logs are added so multiply
(remember $2 \log 5 = \log 5^2$).
$\log 3 + 2 \log 5 = \log (3 \times 5^2)$
$= \log 75$

b) Logs are subtracted so divide
$\frac{1}{2} \log 36 - \log 3 = \log (36^{\frac{1}{2}} \div 3)$
$= \log (6 \div 3)$
$= \log 2$

c) Logs are subtracted so divide
$\log 2 - \frac{1}{4} \log 16 = \log (2 \div 16^{\frac{1}{4}})$
$= \log (2 \div 2)$
$= \log 1 = 0$

Q4 $\log_b(x^2 - 1) - \log_b(x - 1) = \log_b \dfrac{x^2 - 1}{x - 1}$
$= \log_b \dfrac{(x+1)(x-1)}{x-1} = \log_b(x + 1)$
This uses the difference of two squares.

Q5 a) $\log_7 12 = \dfrac{\log_{10} 12}{\log_{10} 7} = 1.276... = 1.28$ (3 s.f.)

b) $\log_5 8 = \dfrac{\log_{10} 8}{\log_{10} 5} = 1.292... = 1.29$ (3 s.f.)

c) $\log_{16} 125 = \dfrac{\log_{10} 125}{\log_{10} 16} = 1.741... = 1.74$ (3 s.f.)

Q6 E.g. $\dfrac{2 + \log_a 4}{\log_a 2a} - \dfrac{2 + 2\log_a 2}{\log_a 2 + \log_a a} = \dfrac{2(1 + \log_a 2)}{\log_a 2 + 1} = 2$

Q7 a) Filling in the answers is just a case of using a
calculator:

x	-3	-2	-1	0	1	2	3
y	0.0156	0.0625	0.25	1	4	16	64

b)

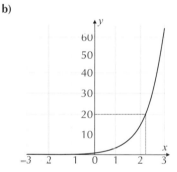

c) From the graph 2.2 is a good estimate.

d) First take logs of base 10 of both sides:
$\log_{10} 4^x = \log_{10} 20 \Rightarrow x \log_{10} 4 = \log_{10} 20$
$\Rightarrow x = \dfrac{\log_{10} 20}{\log_{10} 4} = 2.160... = 2.16$ (3 s.f.),
so the estimate in part c) was correct to 1 d.p.

Q8 a) Take logs of base 10 of both sides:
$\log_{10} 10^x = \log_{10} 240 \Rightarrow x \log_{10} 10 = \log_{10} 240$
$\Rightarrow x = \log_{10} 240 = 2.380... = 2.38$ (to 3 s.f.)
This uses the fact that $\log_{10} 10 = 1$.

b) Take exponentials of both sides using base 10:
$10^{\log_{10} x} = 10^{5.3}$
$\Rightarrow x = 10^{5.3} = 199\,526.2... = 200\,000$ (to 3 s.f.)

c) Take logs of base 10 of both sides:
$\log_{10} 10^{2x+1} = \log_{10} 1500$
$\Rightarrow 2x + 1 = \log_{10} 1500 = 3.176...$, so $2x = 2.176...$,
so $x = 1.088... = 1.09$ (to 3 s.f.)

d) Take logs of base 10 of both sides:
$\log 4^{x-1} = \log 200 \Rightarrow (x-1)\log 4 = \log 200$
$\Rightarrow x = \dfrac{\log 200}{\log 4} + 1 = 4.821... = 4.82$ (3 s.f.)

Q9 Using $y = 10^x$, the equation becomes $2y^2 - 7y + 5 = 0$, which factorises to give $(2y - 5)(y - 1) = 0$. So $y = 2.5$ or $y = 1$, that is, $10^x = 2.5$ or $10^x = 1$.
Solving these for x gives $x = \log 2.5$ or $x = 0$.
The question asks for exact solutions so leave it in log form.

Q10 First solve for $1.5^P = 1\,000\,000$
$P\log_{10}1.5 = \log_{10}1\,000\,000 \Rightarrow P\log_{10}1.5 = 6$
so $P = \dfrac{6}{\log_{10}1.5}$, $P = 34.07...$
We need the next biggest integer, so $P = 35$.

Q11 In n years' time, the population will be 2000×0.75^n, so we need to find the value of n such that $2000 \times 0.75^n = 200$. Solving for n, $0.75^n = 0.1$
$\Rightarrow n\log 0.75 = \log 0.1 \Rightarrow n = \dfrac{\log 0.1}{\log 0.75} \Rightarrow n = 8.003...,$
so the spiders are in danger of extinction after 8 years.
This time you need to round down, as the spiders will face extinction in the 8th year.

Exam-Style Questions — Chapter 4

Q1 a) $2^x = 9$, so taking logs of both sides gives
$\log 2^x = \log 9$ *[1 mark]*
$\Rightarrow x\log 2 = \log 9$ *[1 mark]*
$\Rightarrow x = \dfrac{\log 9}{\log 2} = 3.17$ to 2 d.p. *[1 mark]*

b) $2^{2x} = (2^x)^2$ (from the power laws),
so let $y = 2^x$ and $y^2 = 2^{2x}$. This gives a quadratic in y: $y^2 - 13y + 36 = 0$ *[1 mark]*
$(y - 9)(y - 4) = 0$, so $y = 9$ or $y = 4$
$\Rightarrow 2^x = 9$ *[1 mark]* or $2^x = 4$ *[1 mark]*
From a), $2^x = 9 \Rightarrow x = 3.17$ to 2 d.p. *[1 mark]*
and for $2^x = 4$, $x = 2$ (since $2^2 = 4$) *[1 mark]*.

Q2 a) $\log_3 x = -\dfrac{1}{2}$, so take exponentials of base 3 of each side to remove the log:
$x = 3^{-\frac{1}{2}}$ *[1 mark]*
$\Rightarrow x = \dfrac{1}{3^{\frac{1}{2}}}$ *[1 mark]* $\Rightarrow x = \dfrac{1}{\sqrt{3}}$ *[1 mark]*.

b) $2\log_3 x = -4$
$\Rightarrow \log_3 x = -2$, so take exponentials of base 3 of each side to remove the log:
$x = 3^{-2}$ *[1 mark]*
$\Rightarrow x = \dfrac{1}{9}$ *[1 mark]*

Q3 a) $6^{(3x+2)} = 9$, so taking logs of both sides gives:
$(3x + 2)\log 6 = \log 9$ *[1 mark]*
$\Rightarrow 3x\log 6 = \log 9 - 2\log 6$
$\Rightarrow x = \dfrac{\log 9 - 2\log 6}{3\log 6}$ *[1 mark]*
$= -0.2579... = -0.258$ (3 s.f.) *[1 mark]*

b) $3^{y^2-4} = 7^{(y+2)}$, so taking logs of both sides gives:
$(y^2 - 4)\log 3 = (y + 2)\log 7$ *[1 mark]*
$\Rightarrow (y + 2)(y - 2)\log 3 - (y + 2)\log 7 = 0$ *[1 mark]*
$\Rightarrow (y + 2)[(y - 2)\log 3 - \log 7] = 0$ *[1 mark]*

$\Rightarrow y + 2 = 0$ or $(y - 2)\log 3 - \log 7 = 0$
$\Rightarrow y = -2$ or $y = \dfrac{\log 7 + 2\log 3}{\log 3}$
$\Rightarrow y = -2$ *[1 mark]* or $y = 3.77$ (3 s.f.) *[1 mark]*

Q4 a) (i) $\log_a 20 - 2\log_a 2$
$= \log_a 20 - \log_a 2^2$ *[1 mark]*
$= \log_a (20 \div 2^2)$ *[1 mark]*
$= \log_a 5$ *[1 mark]*

(ii) $\dfrac{1}{2}\log_a 16 + \dfrac{1}{3}\log_a 27$
$= \log_a (16^{\frac{1}{2}}) + \log_a (27^{\frac{1}{3}})$ *[1 mark]*
$= \log_a (16^{\frac{1}{2}} \times 27^{\frac{1}{3}})$ *[1 mark]*
$= \log_a (4 \times 3) = \log_a 12$ *[1 mark]*

b) (i) $\log_2 64 = 6$ *[1 mark]* (since $2^6 = 64$)

(ii) $2\log_3 9 = \log_3 9^2 = \log_3 81$ *[1 mark]*
$\log_3 81 = 4$ *[1 mark]* (since $3^4 = 81$)

c) (i) $\log_6 25 = \dfrac{\log 25}{\log 6} = 1.7965$ to 4 d.p. *[1 mark]*

(ii) $\log_3 10 + \log_3 2 = \log_3 (10 \times 2) = \log_3 20$
[1 mark]
$\log_3 20 = \dfrac{\log 20}{\log 3} = 2.7268$ to 4 d.p. *[1 mark]*

Q5 $\log_7 (y + 3) + \log_7 (2y + 1) = 1$
$\Rightarrow \log_7 ((y + 3)(2y + 1)) = 1$
To remove the $\log_7$, take exponentials of base 7:
$(y + 3)(2y + 1) = 7^1 = 7$
Multiply out, rearrange, and factorise:
$2y^2 + 7y + 3 = 7$
$\Rightarrow 2y^2 + 7y - 4 = 0$
$\Rightarrow (2y - 1)(y + 4) = 0$
$\Rightarrow y = \dfrac{1}{2}$ or $y = -4$,
but since $y + 3 > 0$ and $2y + 1 > 0$, $y = \dfrac{1}{2}$ is the only solution.

[5 marks available — 1 mark for combining the two logs, 1 mark for 7 to the power of each side, 1 mark for the correct factorisation of the quadratic, 1 mark for correct solutions and 1 mark for stating that only $y = \frac{1}{2}$ is a valid solution.]

Q6 a) $\log_4 p - \log_4 q = \dfrac{1}{2}$, so using the log laws:
$\log_4 \left(\dfrac{p}{q}\right) = \dfrac{1}{2}$
Taking exponentials of base 4 gives:
$\dfrac{p}{q} = 4^{\frac{1}{2}} = \sqrt{4} = 2$
$\Rightarrow p = 2q$
[3 marks available — 1 mark for combining the two logs, 1 mark for taking exponentials of base 4 of each side, 1 mark for the correct final working.]

b) Since $p = 2q$ (from a)), the equation can be written: $\log_2 (2q) + \log_2 q = 7$ *[1 mark]*
This simplifies to:
$\log_2 (2q^2) = 7$ *[1 mark]*
Taking exponentials of base 2 gives:
$2q^2 = 2^7 = 128$ *[1 mark]*
$\Rightarrow q^2 = 64, \Rightarrow q = 8$ (since p and q are positive)
[1 mark] $p = 2q \Rightarrow p = 16$ *[1 mark]*

Chapter 5:
Sequences and Series

1. Geometric Sequences and Series
Exercise 1.1 — Geometric sequences

Q1 Common ratio $r = \dfrac{\text{second term}}{\text{first term}} = \dfrac{3}{2} = 1.5$.
Then:

    4th term $= 3^{rd}$ term $\times\ 1.5 = 4.5 \times 1.5 = 6.75$
    5th term $= 4^{th}$ term $\times\ 1.5 = 6.75 \times 1.5 = 10.125$,
    6th term $= 5^{th}$ term $\times\ 1.5 = 10.125 \times 1.5$
                              $= 15.1875$,
    7th term $= 6^{th}$ term $\times\ 1.5 = 15.1875 \times 1.5$
                              $= 22.78125$.

This method isn't as slow as it looks because you can use a scientific calculator to get the terms of the series quickly: press '2 =' then 'x 1.5 =' to get the second term. Pressing '=' repeatedly will give you the following terms. Even so, the method below is quicker, so unless you're asked to find the term after one you've already got, you're better off doing this:

Or: First term $a = 2$, common ratio $r = 1.5$
    n^{th} term $= ar^{n-1} = 2 \times (1.5)^{n-1}$
    7th term $= 2 \times (1.5)^6 = 2 \times 11.390625 = 22.78125$

Q2 Common ratio $r = \dfrac{7^{th}\ \text{term}}{6^{th}\ \text{term}} = \dfrac{6561}{2187} = 3$.
The 6th term is $2187 = ar^5 = a \times 3^5 \Rightarrow a = \dfrac{2187}{3^5} = 9$.
So the first term is 9.

Q3 Common ratio: $r = \dfrac{\text{second term}}{\text{first term}} = \dfrac{12}{24} = 0.5$.
First term: $a = 24$.
n^{th} term: $u_n = ar^{n-1} = 24 \times (0.5)^{n-1}$
9th term: $u_9 = ar^8 = 24 \times (0.5)^8 = 0.09375$

Q4 First term: $a = 1.125$
n^{th} term: $u_n = ar^{n-1} = 1.125r^{n-1}$
14th term: $9216 = u_{14} = 1.125r^{13}$

$9216 = 1.125r^{13} \Rightarrow r^{13} = \dfrac{9216}{1.125} = 8192$
                    $\Rightarrow r = \sqrt[13]{8192} = 2$

Q5 Common ratio: $r = \dfrac{1.1}{1} = 1.1$, first term: $a = 1$.
n^{th} term: $u_n = ar^{n-1} = 1 \times 1.1^{n-1} = 1.1^{n-1}$
So to find the number of terms in the sequence that are less than 4, solve: $\quad u_n = 1.1^{n-1} < 4$
                                    $1.1^{n-1} < 4$
             $\Rightarrow \quad \log 1.1^{n-1} < \log 4$
Use the log law, $\log x^n = n(\log x)$:
             $\Rightarrow (n-1)\log 1.1 < \log 4$
log 1.1 > 0 so dividing through by log 1.1 doesn't change the direction of the inequality:
             $\Rightarrow \quad\quad n-1 < \dfrac{\log 4}{\log 1.1}$
             $\Rightarrow \quad\quad n-1 < 14.54...$
             $\Rightarrow \quad\quad\quad n < 15.54...$
so u_n is less than 4 when n is less than 15.54..., therefore u_{15} is the last term that's less than 4, so there are 15 terms that are less than 4.
You could solve this as an equation instead, finding the value of n such that u_n = 4 and rounding down.

Q6 $a = 5$, $r = 0.6$, n^{th} term: $u_n = ar^{n-1} = 5 \times (0.6)^{n-1}$
10th term: $u_{10} = 5 \times (0.6)^9 = 0.050388$ (6 d.p.)
15th term: $u_{15} = 5 \times (0.6)^{14} = 0.003918$ (6 d.p.)
Difference: $0.003918 - 0.050388 = -0.04647$ (5 d.p.)
You could also have:
Difference: 0.050388 − 0.003918 = 0.04647 (5 d.p.)

Q7 $a = 25\,000$, $r = 0.8$
n^{th} term: $u_n = ar^{n-1} = 25\,000 \times (0.8)^{n-1}$
to find the first term in the sequence less than 1000,
solve: $\quad u_n = 25\,000 \times (0.8)^{n-1} < 1000$

          $25\,000 \times (0.8)^{n-1} < 1000$
          $\Rightarrow (0.8)^{n-1} < \dfrac{1000}{25\,000} = 0.04$
          $\Rightarrow \quad \log(0.8)^{n-1} < \log 0.04$
          $\Rightarrow (n-1)\log 0.8 < \log 0.04$
          $\Rightarrow \quad\quad\quad n-1 > \dfrac{\log 0.04}{\log 0.8}$
          $\Rightarrow \quad\quad\quad n-1 > 14.425...$
          $\Rightarrow \quad\quad\quad\quad n > 15.425...$
so u_n is less than 1000 when n is greater than 15.425..., therefore u_{16} is the first term that's less than 1000.
0.8 < 1 so log 0.8 < 0 and dividing through by log 0.8 changes the direction of the inequality because log 0.8 is negative. Again, you could solve this as an equation by finding n such that u_n = 1000 and rounding up.

Q8 Divide consecutive terms to find the common ratio r:
e.g. $r = \dfrac{\text{second term}}{\text{first term}} = \dfrac{-5}{5} = -1$

Q9 a) Common ratio $r = \dfrac{\text{second term}}{\text{first term}} = \dfrac{\frac{3}{16}}{\frac{1}{4}} = \dfrac{3}{4}$

    b) First term: $a = \dfrac{1}{4}$
        n^{th} term: $u_n = ar^{n-1} = \dfrac{1}{4} \times \left(\dfrac{3}{4}\right)^{n-1}$
        8th term: $u_8 = \dfrac{1}{4} \times \left(\dfrac{3}{4}\right)^7 = \dfrac{1}{4} \times \dfrac{2187}{16384} = \dfrac{2187}{65536}$
                                  $(= 0.03337$ to 5 d.p.$)$

Q10 $r = 0.8$, n^{th} term: $u_n = ar^{n-1} = a(0.8)^{n-1}$
7th term: $196.608 = u_7 = a(0.8)^6$
$196.608 = a(0.8)^6 \Rightarrow a = \dfrac{196.608}{0.8^6} = 750$

Q11 a) Common ratio $r = \dfrac{\text{second term}}{\text{first term}} = \dfrac{-2.4}{3} = -0.8$
    b) Continuing the sequence gives -1.536, 1.2288, -0.98304. So there are 5 terms in the series before a term has modulus less than 1.
    You could also answer this part of the question by writing a new series where each term is the modulus of the old series, then using logs to find the first term less than 1. But in this case it's much easier to just find the next few terms.

Exercise 1.2 — Geometric series

Q1 The sum of the first n terms is $S_n = \dfrac{a(1-r^n)}{(1-r)}$, $a = 8$ and $r = 1.2$, so the sum of the first 15 terms is:
$$S_{15} = \dfrac{a(1-r^{15})}{(1-r)} = \dfrac{8(1-(1.2)^{15})}{(1-1.2)} = 576.28 \text{ to 2 d.p.}$$

Q2 For a geometric series with first term a and common ratio r: $\sum_{k=0}^{n-1} ar^k = \frac{a(1-r^n)}{1-r}$,

so: $\sum_{k=0}^{9} ar^k = \sum_{k=0}^{9} 25\,(0.7)^k = \frac{25\,(1-(0.7)^{10})}{1-0.7} = 80.98$ (to 2 d.p.)

Q3 $a = 3$ and $r = 2$, the sum of the first n terms is:

$\frac{3\,(1-2^n)}{(1-2)} = -3\,(1-2^n)$

$196\,605 = S_n = -3\,(1-2^n) \Rightarrow -65\,535 = 1 - 2^n$

$\Rightarrow 65\,536 = 2^n \Rightarrow \log 65\,536 = \log 2^n$

$\Rightarrow \log 65\,536 = n \log 2 \Rightarrow n = \frac{\log 65\,536}{\log 2} = 16$

Q4 The first term is $a = 4$,

the common ratio is $r = \frac{\text{second term}}{\text{first term}} = \frac{5}{4} = 1.25$

The sum of the first x terms is

$S_x = \frac{a(1-r^x)}{(1-r)} = \frac{4\,(1-(1.25)^x)}{(1-1.25)} = -16\,(1-(1.25)^x)$

So: $103.2 = -16\,(1-(1.25)^x) \Rightarrow -6.45 = 1-(1.25)^x$

$\Rightarrow 7.45 = 1.25^x \Rightarrow \log 7.45 = x \log 1.25$

$\Rightarrow x = \frac{\log 7.45}{\log 1.25} = 9.00$ to 2 d.p.

So $x = 9$ (as it must be an integer)

Q5 a) 3rd term $= ar^2 = 6$, 8th term $= ar^7 = 192$.
Dividing the two equations gives:

$\frac{ar^7}{ar^2} = r^5 = \frac{192}{6} = 32 \Rightarrow r = \sqrt[5]{32} = 2$

b) 3rd term $= ar^2 = 6$ and $r = 2$, so $a = \frac{6}{r^2} = \frac{6}{2^2} = 1.5$

c) The sum of the first 15 terms is:

$S_{15} = \frac{a(1-r^{15})}{(1-r)} = \frac{1.5\,(1-2^{15})}{(1-2)} = 49\,150.5$

Q6 a) Common ratio:

$\frac{\text{2nd term}}{\text{1st term}} = \frac{\text{3rd term}}{\text{2nd term}}$, so:

$\frac{k}{k+10} = \frac{2k-21}{k}$

$\Rightarrow k^2 = (k+10)(2k-21)$

$\Rightarrow k^2 = 2k^2 - 21k + 20k - 210$

$\Rightarrow 0 = k^2 - k - 210$

b) Factorising $k^2 - k - 210 = 0$ gives:
$(k-15)(k+14) = 0$, so $k = 15$ or $k = -14$,
since $k > 0$, $k = 15$.

c) $k = 15$ gives the first three terms 25, 15, 9.
Common ratio $= \frac{\text{second term}}{\text{first term}} = \frac{15}{25} = 0.6$

d) $a = 25$ and $r = 0.6$, so sum of first 10 terms is:

$S_{10} = \frac{a(1-r^{10})}{(1-r)} = \frac{25\,(1-0.6^{10})}{(1-0.6)} = 62.12$ to 2 d.p.

Q7 a) $1 + x + x^2 = 3 \Rightarrow x^2 + x - 2 = 0$

$\Rightarrow (x-1)(x+2) = 0$

$\Rightarrow x = 1$ or $x = -2$.

Since the terms are all different $x \neq 1$ (as 1 is the first term and $x = 1 \Rightarrow x^2 = 1^2 = 1$), hence $x = -2$.

b) $a = 1$ and $r = \frac{\text{second term}}{\text{first term}} = \frac{-2}{1} = -2$,
so the sum of the first 7 terms is:

$S_7 = \frac{a(1-r^7)}{(1-r)} = \frac{1\,(1-(-2)^7)}{(1-(-2))} = 43$

Q8 $a = 7.2$ and $r = 0.38$, so:

$\sum_{k=0}^{9} ar^k = \frac{a(1-r^{10})}{(1-r)} = \frac{7.2\,(1-0.38^{10})}{(1-0.38)} = 11.61$ to 2 d.p.

Q9 $1.2 = S_8 = \frac{a(1-r^8)}{(1-r)} = \frac{a\left(1-\left(-\frac{1}{3}\right)^8\right)}{\left(1-\left(-\frac{1}{3}\right)\right)} = a(0.749...)$

$\Rightarrow a = 1.60$ to 2 d.p.

Q10 The geometric sequence a, $-2a$, $4a$, ... has first term a and common ratio $r = -2$, so $\sum_{k=0}^{12} a(-2)^k$ is the sum of the first 13 terms of the sequence. Therefore:

$\sum_{k=0}^{12} a(-2)^k = -5735.1$

$\sum_{k=0}^{12} a(-2)^k = \frac{a(1-r^{13})}{(1-r)} = \frac{a(1-(-2)^{13})}{(1-(-2))} = 2731a$

$\Rightarrow -5735.1 = 2731a \Rightarrow a = -2.1$

Exercise 1.3 —
Convergent geometric series

Q1 a) $r = \frac{1.1}{1} = 1.1$, $|r| = |1.1| = 1.1 > 1$,
so the sequence does not converge.

b) $r = \frac{0.8^2}{0.8} = 0.8$, $|r| = |0.8| = 0.8 < 1$,
so the sequence converges.

c) $r = \frac{\frac{1}{4}}{1} = \frac{1}{4}$, $|r| = \left|\frac{1}{4}\right| = \frac{1}{4} < 1$,
so the sequence converges.

d) $r = \frac{\frac{9}{2}}{3} = \frac{3}{2}$, $|r| = \left|\frac{3}{2}\right| = \frac{3}{2} > 1$,
so the sequence does not converge.

e) $r = \frac{-\frac{1}{2}}{1} = -\frac{1}{2}$, $|r| = \left|-\frac{1}{2}\right| = \frac{1}{2} < 1$,
so the sequence converges.

f) $r = \frac{5}{5} = 1$, $|r| = |1| = 1$ (and 1 is not less than 1),
so the sequence does not converge.

Q2 Find the common ratio: $r = \frac{2^{\text{nd}} \text{ term}}{1^{\text{st}} \text{ term}} = \frac{8.1}{9} = 0.9$.
The first term is $a = 9$.
The sum to infinity is:

$S_\infty = \frac{a}{1-r} = \frac{9}{1-0.9} = \frac{9}{0.1} = 90$

Q3 $S_\infty = \frac{a}{1-r} = 2a \Rightarrow \frac{1}{1-r} = 2 \Rightarrow 1 = 2 - 2r$

$\Rightarrow 2r = 1 \Rightarrow r = 0.5$

Q4 a) Sum to infinity is $13.5 = S_\infty = \frac{a}{1-r}$
Sum of the first three terms is $13 = S_3 = \frac{a(1-r^3)}{(1-r)}$

Divide S_3 by S_∞: $\frac{13}{13.5} = \frac{a(1-r^3)}{1-r} \div \frac{a}{1-r}$

$= \frac{a(1-r^3)}{1-r} \times \frac{1-r}{a} = 1 - r^3$

You can cancel the $1 - r$ because r can't be 1 (as the series converges), so $1 - r \neq 0$.
So $1 - r^3 = \frac{13}{13.5} = \frac{26}{27} \Rightarrow r^3 = \frac{1}{27} \Rightarrow r = \frac{1}{3}$.

b) $13.5 = S_\infty = \frac{a}{1-r} \Rightarrow a = 13.5(1-r)$

$= 13.5 \times \frac{2}{3} = 9$.

Q5 $ar = 3, 12 = S_\infty = \dfrac{a}{1-r} \Rightarrow 12 - 12r = a$

The first equation gives $a = \dfrac{3}{r}$,

plugging this into the second equation gives:

$12 - 12r = \dfrac{3}{r} \Rightarrow 12r - 12r^2 = 3$

$\Rightarrow 12r^2 - 12r + 3 = 0$

$\Rightarrow 4r^2 - 4r + 1 = 0$

This factorises to $(2r - 1)(2r - 1) = 0$

Hence $2r - 1 = 0 \Rightarrow r = 0.5$

Then $a = \dfrac{3}{r} = \dfrac{3}{0.5} = 6$.

Q6 **a)** $a = 6, 10 = S_\infty = \dfrac{a}{1-r} = \dfrac{6}{1-r}$

$\Rightarrow 1 - r = \dfrac{6}{10} = 0.6 \Rightarrow r = 0.4$

b) 5th term: $u_5 = ar^4 = 6 \times 0.4^4 = 0.1536$.

Q7 **a)** Second term $= ar = -48$, 5th term $= ar^4 = 0.75$.

Dividing gives: $r^3 = \dfrac{ar^4}{ar} = \dfrac{0.75}{-48} = -0.015625$

$\Rightarrow r = -0.25$

b) $ar = -48 \Rightarrow a = \dfrac{-48}{r} = \dfrac{-48}{-0.25} = 192$

c) $|r| < 1$ so you can find the sum to infinity:

$S_\infty = \dfrac{a}{1-r} = \dfrac{192}{1-(-0.25)} = \dfrac{192}{1.25} = 153.6$.

Q8 The sum of terms after the 10th is $S_\infty - S_{10}$. So the question tells you that $S_\infty - S_{10} < \dfrac{1}{100}S_\infty$

$\Rightarrow \dfrac{99}{100}S_\infty - S_{10} < 0 \Rightarrow \dfrac{99}{100}S_\infty < S_{10}$

Then:

$0.99(S_\infty) = \dfrac{0.99a}{1-r} < \dfrac{a(1-r^{10})}{(1-r)} = S_{10}$

You can cancel and keep the inequality sign because the series is convergent so $|r| < 1 \Rightarrow 1 - r > 0$, and you know $a > 0$ from the question:

$\Rightarrow 0.99 < 1 - r^{10} \Rightarrow r^{10} < 0.01$

$\Rightarrow |r| < \sqrt[10]{0.01} \Rightarrow |r| < 0.631$ (to 3 s.f.)

Exercise 1.4 — Real-life problems

Q1 The heights form a geometric sequence (starting from the smallest doll), with $a = 3$, $r = 1.25$.
The nth term is $ar^{n-1} = 3 \times 1.25^{n-1}$,
The first term is the height of the first doll, so the height of the 8th doll is the 8th term.
The 8th term is: $3 \times 1.25^7 = 14.3$ cm (to 1 d.p.)

Q2 The value decreases by 15% each year so multiply by $1 - 0.15 = 0.85$ to get from one term to the next, so the common ratio $r = 0.85$.
So the nth term $u_n = ar^{n-1} = a(0.85)^{n-1}$
The price when new is the first term in the series and the value after 10 years is the 11th term.
11th term: $2362 = u_{11} = a(0.85)^{10}$
$2362 = a(0.85)^{10} \Rightarrow a = \dfrac{2362}{0.85^{10}} = 11\,997.496...$
$= 11\,997.50$ to 2 d.p.
When new, the car cost £11 997.50

Q3 **a)** The cost increases by 3% each year so multiply by $1 + 0.03 = 1.03$ to get from one term to the next, so 2006 cost $= 1.03 \times$ (2005 cost)
$= 1.03 \times$ £120 $=$ £123.60

b) The costs each year form a geometric sequence with common ratio $r = 1.03$ and first term $a = 120$, so the total cost between 2005 and 2010 (including 2005 and 2010) is the sum of the first 6 terms:
$S_6 = \dfrac{a(1-r^6)}{(1-r)} = \dfrac{120(1-1.03^6)}{(1-1.03)} = 776.21$ to 2 d.p.
Nigel paid £776.21 (the cost to the nearest penny).

Q4 The annual amounts he earns form a geometric sequence:
$2000, 2000 \times 1.04, 2000 \times 1.04^2, ..., 2000 \times 1.04^7$.
The common ratio is 1.04 because each year the amount increases by 4%. The first term is 2000.
The total amount he earns in the first 8 years is the sum of the first 8 terms of the sequence:
Using $S_n = \dfrac{a(1-r^n)}{(1-r)}$, where $a = 2000$, $r = 1.04$, $n = 8$,
$S_8 = \dfrac{2000(1-1.04^8)}{(1-1.04)} = 18428.45...$
To the nearest pound he received £18 428.

Q5 4 weeks is 28 days.

The claim is that the leeks' height increases by 15% every 2 days.

$28 \div 2 = 14$, so there are 14 lots of 2 days in 28 days

In 28 days, the height should increase 14 times by 15%.

So if the claim is correct, the height of the leeks after 28 days will be the 15th term of a geometric progression with first term 5 and common ratio 1.15.

The first term is the initial height, the second term is the height after one 15% increase and so on, so the 15th term is the height after fourteen 15% increases.

The common ratio is 1.15 because multiplying something by 1.15 is the same as increasing it by 15%.

So $a = 5$ and $r = 1.15$.

The nth term is: $u_n = ar^{n-1} = 5(1.15)^{n-1}$

So the 15th term is: $5 \times 1.15^{14} = 35.3785...$ cm.

So if the claim were true, the leeks would be 35.4 cm tall (to 1 d.p.). Since the leeks only reach a height of 25 cm, the claim on the compost is not justified.

Q6 **a)** The gnome value goes up by 2% each year, so the price after 1 year is 102% of £80 000:
$80\,000 \times 1.02 = 81\,600$
The value after 1 year is £ 81 600.

b) To get from one term to the next you multiply by 1.02 (to increase by 2% each time), so the common ratio $r = 1.02$.

c) Price at start $= a = 80\,000$, $r = 1.02$
n^{th} term $= u_n = ar^{n-1} = 80\,000 \times (1.02)^{n-1}$
The value at the start is the 1st term,
the value after 1 year is the 2nd term and so on,
so the value after 10 years is the 11th term:
11th term $= u_{11} = 80\,000 \times (1.02)^{10} = 97\,519.55$
(to 2 d.p.)
The value after 10 years is £97 519.55
(to the nearest penny)

d) The value after k years is the $(k + 1)^{th}$ term:
$(k + 1)^{th}$ term $= u_{k+1} = ar^k = 80\,000 \times (1.02)^k$
After k years the value is more than 120 000, so:
$u_{k+1} = 80\,000 \times (1.02)^k > 120\,000$

$\Rightarrow \qquad (1.02)^k > \dfrac{120\,000}{80\,000} = 1.5$

$\Rightarrow \qquad \log(1.02)^k > \log 1.5$

$\Rightarrow \qquad k\log(1.02) > \log 1.5$

e) $k\log(1.02) > \log 1.5$ (from part d))

$\Rightarrow \quad k > \dfrac{\log 1.5}{\log 1.02} = 20.475...$

So u_{k+1} (the value after k years) is more than 120 000 when $k > 20.475...$, therefore the value exceeds 120 000 after 21 years.

Q7 The thickness of the paper doubles every time you fold it in half.
So the paper thickness forms a geometric progression:
After 1 fold, thickness $= 0.01 \times 2$ cm.
After 2 folds, thickness $= 0.01 \times 2^2$ cm
After n folds, thickness $= 0.01 \times 2^n$ cm
Distance to the moon:
384 000 km $= 3.84 \times 10^5$ km
$= (3.84 \times 10^5 \times 1000)$ m
$= (3.84 \times 10^5 \times 1000 \times 100)$ cm
$= 3.84 \times 10^{10}$ cm
Therefore the paper reaches the moon when:
$0.01 \times 2^n = 3.84 \times 10^{10}$
$2^n = 3.84 \times 10^{12}$.
Taking logs: $n \log 2 = \log(3.84 \times 10^{12})$
$\Rightarrow n = \log(3.84 \times 10^{12}) \div \log 2 = 41.80...$
So after approximately 42 folds the paper would reach the moon.

Q8 a) The distances she runs form a geometric sequence with $a = 12$, $r = 1.03$.
The nth term is $ar^{n-1} = 12 \times 1.03^{n-1}$,
so the 10th term is: $12 \times 1.03^9 = 15.7$ miles (to 1 d.p.).

b) The total distance she runs in 20 days is the sum of the first 20 terms of the sequence.
Using: $S_n = \dfrac{a(1 - r^n)}{(1 - r)}$, where $a = 12$, $r = 1.03$, $n = 20$,

$S_{20} = \dfrac{12(1 - 1.03^{20})}{(1 - 1.03)} = 322.44...$

In 20 days she runs a total of 322 miles (to the nearest mile).

Q9 After 0 years, $u_1 = a$ and after 1 year, $u_2 = ar$.
So after 10 years, $u_{11} = ar^{10}$. She wants her investment to double so $u_{11} = ar^{10} = 2a \Rightarrow r^{10} = 2$
$\Rightarrow |r| = \sqrt[10]{2} = 1.071773$.
So the interest rate needed is 7.17 % (3 s.f.).

2. Binomial Expansions

Exercise 2.1 —
Binomial expansions — $(1 + x)^n$

Q1 Pascal's triangle is
$$1$$
$$1 \quad 1$$
$$1 \quad 2 \quad 1$$
$$1 \quad 3 \quad 3 \quad 1$$
$$1 \quad 4 \quad 6 \quad 4 \quad 1$$

The expansion of $(1 + x)^4$ takes its coefficients of each term, in ascending powers of x, from the 5th row:
$(1 + x)^4 = 1 + 4x + 6x^2 + 4x^3 + x^4$

Q2 a) $^6C_2 = 15$

b) $\dbinom{12}{5} = \,^{12}C_5 = 792$

c) $\dfrac{30!}{4!26!} = \,^{30}C_4 = 27\,405$

d) $^8C_8 = 1$

Q3 a) $\dfrac{9!}{4!5!} = \dfrac{9 \times 8 \times 7 \times 6 \times 5 \times 4 \times 3 \times 2 \times 1}{(4 \times 3 \times 2 \times 1)(5 \times 4 \times 3 \times 2 \times 1)}$
$= \dfrac{9 \times 8 \times 7 \times 6}{4 \times 3 \times 2 \times 1} = 3 \times 7 \times 6 = 126$

b) $^{10}C_3 = \dfrac{10!}{3!(10 - 3)!} = \dfrac{10 \times 9 \times 8}{3 \times 2 \times 1}$
$= 10 \times 3 \times 4 = 120$

c) $\dfrac{15!}{11!4!} = \dfrac{15 \times 14 \times 13 \times 12}{4 \times 3 \times 2 \times 1} = 15 \times 7 \times 13 = 1365$

d) $\dbinom{8}{6} = \dfrac{8!}{6!(8 - 6)!} = \dfrac{8 \times 7}{2 \times 1} = 4 \times 7 = 28$

Q4 $(1 + x)^{10} = 1 + \,^{10}C_1x + \,^{10}C_2x^2 + \,^{10}C_3x^3 + ...$

You can work out the coefficients ^nC_r using a calculator, or using the method below. Using the notation $\dbinom{n}{r}$ instead of ^nC_r for the coefficients is fine too. The formula for ^nC_r only works when n and r are positive integers — which is fine in C2, but in C4 you'll need to use the formula on p.85 instead.

$^nC_r = \dfrac{n!}{r!(n - r)!}$

$^{10}C_1 = \dfrac{10!}{1!(10 - 1)!} = \dfrac{10 \times 9 \times 8 \times ...1}{1 \times 9 \times 8 \times ...1} = 10$

$^{10}C_2 = \dfrac{10!}{2!(10 - 2)!} = \dfrac{10 \times 9 \times 8 \times 7 \times ...1}{2 \times 1 \times 8 \times 7 \times ...1}$
$= \dfrac{10 \times 9}{2} = 45$

$^{10}C_3 = \dfrac{10!}{3!(10 - 3)!} = \dfrac{10 \times 9 \times 8 \times 7 \times ...1}{3 \times 2 \times 1 \times 7 \times ...1}$
$= \dfrac{10 \times 9 \times 8}{3 \times 2}$
$= 10 \times 3 \times 4 = 120$

$(1 + x)^{10} = 1 + 10x + 45x^2 + 120x^3 + ...$

For the rest of this exercise you can use one of the methods shown in question 1 or 4 to find the coefficients ^nC_r, or you can use a calculator.

Q5 $(1 + x)^6 = 1 + \,^6C_1x + \,^6C_2x^2 + \,^6C_3x^3 + \,^6C_4x^4 + \,^6C_5x^5 + \,^6C_6x^6$
$= 1 + 6x + 15x^2 + 20x^3 + 15x^4 + 6x^5 + x^6$

Q6 $(1 + x)^7 = 1 + \,^7C_1x + \,^7C_2x^2 + \,^7C_3x^3 + ...$
$= 1 + 7x + 21x^2 + 35x^3 + ...$

Exercise 2.2 —
Binomial expansions — $(1 + ax)^n$

Q1 a) $(1 - x)^6 = 1 + {}^6C_1(-x) + {}^6C_2(-x)^2 + {}^6C_3(-x)^3$
$\qquad\qquad\qquad + {}^6C_4(-x)^4 + {}^6C_5(-x)^5 + {}^6C_6(-x)^6$
$\qquad = 1 - 6x + 15x^2 - 20x^3 + 15x^4 - 6x^5 + x^6$

For part a) you could just use the formula for the expansion
of $(1 - x)^n$: $(1 - x)^n = 1 - {}^nC_1x + {}^nC_2x^2 - {}^nC_3x^3 + ...$

b) $(1 + x)^9$
$\qquad = 1 + {}^9C_1x + {}^9C_2x^2 + {}^9C_3x^3 + {}^9C_4x^4 + {}^9C_5x^5 + ...$
$(1 - x)^9$
$\qquad = 1 - {}^9C_1x + {}^9C_2x^2 - {}^9C_3x^3 + {}^9C_4x^4 - {}^9C_5x^5 + ...$
So:
$(1 + x)^9 - (1 - x)^9 = 2({}^9C_1x + {}^9C_3x^3 + {}^9C_5x^5 + ...)$

The even powers cancel out, so only the terms with odd
powers appear (and they're doubled because one term
comes from each expansion).

$\qquad = 2(9x + 84x^3 + 126x^5 + 36x^7 + x^9)$
$\qquad = 18x + 168x^3 + 252x^5 + 72x^7 + 2x^9$

Q2 The first 3 terms will include 1 and the terms in x and
x^2 so expand each bracket up to and including the
term in x^2:
$(1 + x)^3(1 - x)^4$
$\qquad = (1 + 3x + 3x^2 + ...)(1 - 4x + 6x^2 - ...)$
$\qquad = 1 - 4x + 6x^2 + ... + 3x - 12x^2$
$\qquad\qquad\qquad + ... + 3x^2 + (\text{higher power terms})$
$\qquad = 1 - x - 3x^2 + ...$

Q3 The expansion of $(1 + x)^5(1 + y)^7$ is the expansions of
$(1 + x)^5$ and $(1 + y)^7$ multiplied together. We need the
x^3 term from $(1 + x)^5$ and the y^2 term from $(1 + y)^7$.
Multiplying the coefficients gives the x^3y^2 coefficient:

x^3 coefficient: $\dfrac{5!}{3!(5 - 3)!} = \dfrac{5 \times 4}{2 \times 1} = 10$

y^2 coefficient: $\dfrac{7!}{2!(7 - 2)!} = \dfrac{7 \times 6}{2 \times 1} = 21$

x^3y^2 coefficient: $10 \times 21 = 210$

Q4 $(1 - 2x)^5 = 1 + {}^5C_1(-2x) + {}^5C_2(-2x)^2 + {}^5C_3(-2x)^3$
$\qquad\qquad\qquad + {}^5C_4(-2x)^4 + {}^5C_5(-2x)^5$
$\qquad = 1 + 5(-2x) + 10(4x^2) + 10(-8x^3) + 5(16x^4) + 1(-32x^5)$
$\qquad = 1 - 10x + 40x^2 - 80x^3 + 80x^4 - 32x^5$

Q5 a) $(1 - 3x)^6 = 1 + {}^6C_1(-3x) + {}^6C_2(-3x)^2 + {}^6C_3(-3x)^3 + ...$
$\qquad\qquad = 1 - 18x + 135x^2 - 540x^3 + ...$

b) $(1 + x)(1 - 3x)^6 = (1 + x)(1 - 18x + ...)$
$\qquad\qquad = 1 - 18x + ... + x - 18x^2 + ...$
$\qquad\qquad \approx 1 - 17x$

You're told you can ignore x^2 and higher terms.

Q6 $(1 + kx)^8$
$\qquad = 1 + {}^8C_1kx + {}^8C_2(kx)^2 + {}^8C_3(kx)^3 + ...$
$\qquad = 1 + 8kx + 28k^2x^2 + 56k^3x^3 + ...$

Q7 a) $\left(1 + \dfrac{x}{2}\right)^{12} = 1 + {}^{12}C_1\left(\dfrac{x}{2}\right) + {}^{12}C_2\left(\dfrac{x}{2}\right)^2$
$\qquad\qquad\qquad + {}^{12}C_3\left(\dfrac{x}{2}\right)^3 + {}^{12}C_4\left(\dfrac{x}{2}\right)^4 + ...$

$\qquad = 1 + 6x + \dfrac{33}{2}x^2 + \dfrac{55}{2}x^3 + \dfrac{495}{16}x^4 + ...$

b) $1 + \left(\dfrac{x}{2}\right) = 1.005$ when $x = 0.01$.
Substitute this value into the expansion:
$1.005^{12} \approx 1 + 6(0.01) + \dfrac{33}{2}(0.01)^2$
$\qquad\qquad + \dfrac{55}{2}(0.01)^3 + \dfrac{495}{16}(0.01)^4$
$1.005^{12} \approx 1.061677809 = 1.0616778$ to 7 d.p.

Exercise 2.3 —
Binomial expansions — $(a + b)^n$

Q1 Using the formula for the expansion of $(a + b)^n$:
$(a + b)^n = a^n + \dbinom{n}{1}a^{n-1}b + \dbinom{n}{2}a^{n-2}b^2 + ... + b^n$

In this case $a = 3$ and $b = x$:
$(3 + x)^6 = 3^6 + {}^6C_13^5x + {}^6C_23^4x^2 + {}^6C_33^3x^3 + ...$
$\qquad = 729 + 6(243x) + 15(81x^2) + 20(27x^3) + ...$
$\qquad = 729 + 1458x + 1215x^2 + 540x^3 + ...$

Q2 In this case $a = 2$ and $b = x$:
$(2 + x)^4 = 2^4 + {}^4C_12^3x + {}^4C_22^2x^2 + {}^4C_32x^3 + {}^4C_4x^4$
$\qquad = 16 + 4(8x) + 6(4x^2) + 4(2x^3) + x^4$
$\qquad = 16 + 32x + 24x^2 + 8x^3 + x^4$

Q3 a) The term in x^5 is ${}^8C_5(\lambda x)^5 = 56\lambda^5x^5$
Therefore $56\lambda^5 = 57\ 344$
$\qquad\qquad \Rightarrow \lambda^5 = 1024 \Rightarrow \lambda = \sqrt[5]{1024} = 4$

b) $(1 + 4x)^8 = 1 + {}^8C_1(4x) + {}^8C_2(4x)^2 + ...$
$\qquad\qquad = 1 + 32x + 448x^2 + ...$

Q4 a) $(2 + x)^8$
$\qquad = 2^8 + {}^8C_12^7x + {}^8C_22^6x^2 + {}^8C_32^5x^3 + {}^8C_42^4x^4 + ...$
$\qquad = 256 + 1024x + 1792x^2 + 1792x^3 + 1120x^4 + ...$

b) $2 + x = 2.01$ when $x = 0.01$
Hence: $2.01^8 = 256 + 1024(0.01) + 1792(0.01)^2$
$\qquad\qquad + 1792(0.01)^3 + 1120(0.01)^4 + ...$
$\qquad\qquad \approx 266.4210032$
An approximation to 2.01^8 is: 266.42100
(to 5 d.p.)

Q5 $(3 + 5x)^7$
$\qquad = 3^7 + {}^7C_13^6(5x) + {}^7C_23^5(5x)^2 + {}^7C_33^4(5x)^3 + ...$
$\qquad = 2187 + 25515x + 127575x^2 + 354375x^3 + ...$

Q6 a) $(3 + 2x)^6 = 3^6 + {}^6C_13^5(2x) + {}^6C_23^4(2x)^2$
$\qquad\qquad\qquad + {}^6C_33^3(2x)^3 + {}^6C_43^2(2x)^4 + ...$
$\qquad\qquad = 729 + 2916x + 4860x^2$
$\qquad\qquad\qquad + 4320x^3 + 2160x^4 + ...$

b) $(1 + x)(3 + 2x)^6 = (3 + 2x)^6 + x(3 + 2x)^6$
$\qquad = (729 + 2916x + 4860x^2$
$\qquad\qquad\qquad + 4320x^3 + 2160x^4 + ...)$
$\qquad + (729x + 2916x^2 + 4860x^3$
$\qquad\qquad\qquad + 4320x^4 + 2160x^5 + ...)$
$\qquad = 729 + 3645x + 7776x^2 + 9180x^3$
$\qquad\qquad\qquad + 6480x^4 + ...$

(The term in x^5 is the 6th term.)

Q7 a) Expansion of $(1 + x)^n = 1 + nx + \dfrac{n(n - 1)}{2}x^2 + ...$
So:
$\qquad\qquad \dfrac{n(n - 1)}{2} = 231$
$\Rightarrow \qquad\qquad n(n - 1) = 462$
$\Rightarrow \qquad n^2 - n - 462 = 0$
$\Rightarrow (n + 21)(n - 22) = 0$
$\Rightarrow n = -21$ or $n = 22$

Hence $n = 22$ since $n > 0$.
This factorisation was a bit tricky, but you know that
462 is the product of two consecutive numbers.
So to give you an idea of the factors, try square rooting
462. $\sqrt{462} = 21.49...$ so the roots are 21 and 22.

b) Coefficient of term in x^3 is:
$$\frac{22!}{3!(22-3)!} = \frac{22 \times 21 \times 20}{3 \times 2 \times 1} = 1540$$
So the term in x^3 is: $1540x^3$

Q8 The coefficient of x^2 is ${}^8C_2 a^6 3^2$
The coefficient of x^5 is ${}^8C_5 a^3 3^5$
Therefore:
$$28 \times a^6 \times 3^2 = \frac{32}{27} \times 56 \times a^3 \times 3^5$$
$$\Rightarrow 28a^3 = \frac{32}{27} \times 56 \times 3^3$$
$$\Rightarrow a^3 = \frac{32 \times 56 \times 27}{27 \times 28} = 64$$
$$\Rightarrow a = \sqrt[3]{64} = 4.$$

Q9 Expand each bracket up to the term in x^3:
$(1 + 2x)^5 = 1 + 5(2x) + 10(2x)^2 + 10(2x)^3 + ...$
$= 1 + 10x + 40x^2 + 80x^3 + ...$
$(3 - x)^4 = 3^4 + 4(3)^3(-x) + 6(3)^2(-x)^2 + 4(3)(-x)^3 + ...$
$= 81 - 108x + 54x^2 - 12x^3 + ...$
Multiply the terms that will give a result in x^3:
$(1 \times -12x^3) + (10x \times 54x^2) + (40x^2 \times -108x)$
$\qquad\qquad + (80x^3 \times 81) = 2688x^3$
So the coefficient of x^3 is 2688.

Q10 a) The coefficient of x^3 is:
$$\frac{n!}{3!(n-3)!} = \frac{n(n-1)(n-2)}{3 \times 2 \times 1}$$
The coefficient of x^2 is:
$$\frac{n!}{2!(n-2)!} = \frac{n(n-1)}{2}$$
The coefficient of x^3 is three times the coefficient of x^2, so:
$$\frac{n(n-1)(n-2)}{3 \times 2 \times 1} = 3 \times \frac{n(n-1)}{2}.$$
$$\Rightarrow \qquad \frac{n-2}{3} = 3$$
$$\Rightarrow \qquad n = 11$$

b) $(1 + x)^{11} = 1 + 11x + 55x^2 + ...$
The coefficient of x^2 is $a \times$ (coefficient of x),
so $55 = 11a \Rightarrow a = 5$.

Q11 a) $(2 + \mu x)^8 = 2^8 + {}^8C_1 2^7(\mu x) + {}^8C_2 2^6(\mu x)^2 + ...$
$= 256 + (8 \times 128)(\mu x) + (28 \times 64)(\mu x)^2 + ...$
$= 256 + 1024\mu x + 1792\mu^2 x^2 + ...$

b) The coefficient of x^2 is:
$87\,808 = 1792\mu^2 \Rightarrow \mu^2 = 49 \Rightarrow \mu = 7$ or -7.

Review Exercise — Chapter 5

Q1 First term $= a$, second term $= ar$
$$r = \frac{\text{second term}}{\text{first term}} = \frac{1875}{3125} = 0.6$$

Q2 First term is $a = 3$,
common ratio is $r = \dfrac{\text{second term}}{\text{first term}} = \dfrac{-9}{3} = -3$
n^{th} term of a geometric sequence is: $u_n = ar^{n-1}$
So the nth term of this sequence is:
$u_n = 3(-3)^{n-1} = -(-3)^n$

Q3 a) $a = 2$, $r = \dfrac{\text{second term}}{\text{first term}} = \dfrac{-6}{2} = -3$
10th term, $u_{10} = ar^9 = 2 \times (-3)^9 = -39366$

b) The sum of the first 10 terms is:
$$S_{10} = \frac{a(1 - r^{10})}{(1-r)} = \frac{2(1 - (-3)^{10})}{1 - (-3)} = -29524$$

For the rest of this exercise the common ratio r is found by dividing
two consecutive terms, unless another method is given.

Q4 a) $a = 2$, $r = 4$, so $S_{12} = \dfrac{2(1 - 4^{12})}{1 - 4} = 11\,184\,810$

b) $a = 30$, $r = 0.5$, so $S_{12} = \dfrac{30(1 - (0.5)^{12})}{1 - 0.5} = 59.985$
$\qquad\qquad\qquad\qquad\qquad\qquad\qquad$ to 3 d.p.

Q5 For $a = 7$, $r = 0.6$
$$\sum_{k=0}^{5} 7(0.6)^k = S_6 = \frac{a(1 - r^6)}{(1-r)}$$
$$= \frac{7(1 - (0.6)^6)}{(1 - 0.6)} = 16.68 \text{ to 2 d.p.}$$

Q6 a) $r = \dfrac{2}{1} = 2$, $|2| > 1$, so series is divergent.

b) $r = \dfrac{27}{81} = \dfrac{1}{3}$, $\left|\dfrac{1}{3}\right| < 1$, so series is convergent.

c) $r = \dfrac{\frac{1}{3}}{1} = \dfrac{1}{3}$, $\left|\dfrac{1}{3}\right| < 1$, so series is convergent.

d) $r = \dfrac{1}{4}$, $\left|\dfrac{1}{4}\right| < 1$, so series is convergent.

Q7 a) $r = \dfrac{\text{second term}}{\text{first term}} = \dfrac{12}{24} = \dfrac{1}{2}$

b) 7th term $= ar^6 = 24 \times \left(\dfrac{1}{2}\right)^6 = 0.375$ or $\dfrac{3}{8}$

c) $S_{10} = \dfrac{a(1 - r^{10})}{1 - r} = \dfrac{24\left(1 - \left(\frac{1}{2}\right)^{10}\right)}{1 - \frac{1}{2}} = 47.953$
$\qquad\qquad\qquad\qquad\qquad\qquad\qquad\qquad$ to 3 d.p.

d) $S_\infty = \dfrac{a}{1-r} = \dfrac{24}{1 - \frac{1}{2}} = 48$

Q8 $a = 2$, $r = \dfrac{6}{2} = 3$
Need to find n so that $ar^{n-1} = 1458$,
$2 \times 3^{n-1} = 1458 \Rightarrow 3^{n-1} = 729$.
Then, use logs to find that: $\log 3^{n-1} = \log 729$
$$(n-1)\log 3 = \log 729$$
$$n - 1 = \frac{\log 729}{\log 3}$$
$\Rightarrow n - 1 = 6 \Rightarrow n = 7$, so the 7th term equals 1458.

Q9 a) The Year 1 donations are £20 000.
The donations increase by 8% each year, so the
second year the donations will be:
$$1.08 \times 20\,000 = £21600$$

b) The donations increase by 8% each year, so you
multiply by 1.08 to get from one year's total to
the next, so the common ratio is 1.08.

c) The first term of the sequence is $a = 20\,000$.
The common ratio is $r = 1.08$.
The donations in Year n are the nth term of the
sequence: $u_n = ar^{n-1} = 20\,000(1.08)^{n-1}$

d) The total of the donations from Year 1 to Year 10
is the sum of the first 10 terms of the sequence:
$$S_{10} = \frac{a(1 - r^{10})}{1 - r} = \frac{20\,000(1 - (1.08)^{10})}{1 - 1.08}$$
$$= £289\,731 \qquad \text{(to the nearest £)}$$

Q10 The sum of the first n terms of a geometric series is S_n:

$$S_n = a + ar + ar^2 + ar^3 + ... + ar^{n-1}$$

Then: $rS_n = ar + ar^2 + ar^3 + ar^4... + ar^n$

Subtracting rS_n from S_n gives:

$$(1 - r)S_n = a - ar^n$$
$$= a(1 - r^n) \implies S_n = \frac{a(1 - r^n)}{1 - r}$$

Q11 $\sum\limits_{k=0}^{\infty} ar^k = S_\infty = \dfrac{a}{1 - r}$

In this case $a = 33$ and $r = 0.25$, so:

$$\sum\limits_{k=0}^{\infty} ar^k = \sum\limits_{k=0}^{\infty} 33(0.25)^k$$
$$= \frac{33}{1 - 0.25} = \frac{33}{0.75} = 44$$

Q12 $(1 + x)^{12} = 1 + {}^{12}C_1 x + {}^{12}C_2 x^2 + {}^{12}C_3 x^3 + ...$
$$= 1 + 12x + 66x^2 + 220x^3 + ...$$

Q13 $(1 - x)^{20} = 1 - {}^{20}C_1 x + {}^{20}C_2 x^2 - ...$
$$(\text{or} = 1 + {}^{20}C_1(-x) + {}^{20}C_2(-x)^2 + ...)$$
$$= 1 - 20x + 190x^2 - ...$$

Q14 ${}^{16}C_4(-2x)^4 = 1820 \times 16x^4 = 29120x^4$

Q15 a) $\left(1 + \frac{x}{3}\right)^9 = 1 + {}^9C_1\left(\frac{x}{3}\right) + {}^9C_2\left(\frac{x}{3}\right)^2 + {}^9C_3\left(\frac{x}{3}\right)^3 + ...$
$$= 1 + 9\left(\frac{x}{3}\right) + 36\left(\frac{x}{3}\right)^2 + 84\left(\frac{x}{3}\right)^3 + ...$$
$$= 1 + 3x + 4x^2 + \frac{28}{9}x^3 + ...$$

b) $\left(1 + \frac{x}{3}\right)^9 = 1.003^9$ when $x = 0.009$,

so find an approximation by putting $x = 0.009$ into the expansion of $\left(1 + \frac{x}{3}\right)^9$.

Terms with higher powers of 0.009 can be ignored because 0.009 is small:

$$1.003^9 \approx 1 + 3(0.009) + 4(0.009)^2 + \frac{28}{9}(0.009)^3$$
$$\approx 1.027326 \text{ to 6 d.p.}$$

Q16 $(1 + 3x)^5 = 1 + {}^5C_1(3x) + {}^5C_2(3x)^2 + {}^5C_3(3x)^3$
$$+ {}^5C_4(3x)^4 + {}^5C_5(3x)^5$$
$$= 1 + 5(3x) + 10(3x)^2 + 10(3x)^3 + 5(3x)^4 + (3x)^5$$
$$= 1 + 15x + 90x^2 + 270x^3 + 405x^4 + 243x^5$$

Q17 a) $(1 + ax)^8 = 1 + {}^8C_1(ax) + {}^8C_2(ax)^2 + {}^8C_3(ax)^3$
$$+ {}^8C_4(ax)^4 + ...$$
$$= 1 + 8ax + 28a^2x^2 + 56a^3x^3 + 70a^4x^4 + ...$$

b) The coefficient of x^2 is double the coefficient of x^3, so $28a^2 = 2 \times 56a^3 = 112a^3$
$$\implies \frac{28}{112} = a \implies a = \frac{1}{4}$$

c) The coefficient of x is $8a = 8 \times \frac{1}{4} = 2$

Q18 $(4 - 5x)^7 = 4^7 + {}^7C_1 4^6(-5x) + {}^7C_2 4^5(-5x)^2 + ...$
$$= 16\,384 + (7 \times 4096)(-5x) + (21 \times 1024)(-5x)^2 + ...$$
$$= 16\,384 - 143\,360x + 537\,600x^2 + ...$$

This method uses the $(a + b)^n$ formula. You could also rearrange the bracket so you can expand it using the $(1 + x)^n$ formula — if you do it this way make sure you multiply through by 4^7 at the end. For the rest of this exercise the $(a + b)^n$ formula is used, but you can use the other method for all of these questions.

Q19 $(2 + 3x)^5 = 2^5 + {}^5C_1 2^4(3x) + {}^5C_2 2^3(3x)^2 + ...$

So the term in x^2 is ${}^5C_2 2^3(3x)^2 = 720x^2$, and the coefficient of x^2 is 720.

You can go straight to the term in x^2 without writing the expansion out, but you can write the first few terms out like this to make sure you get it right.

Q20 a) $\left(3 + \frac{x}{4}\right)^{11} = 3^{11} + {}^{11}C_1 3^{10}\left(\frac{x}{4}\right) + {}^{11}C_2 3^9\left(\frac{x}{4}\right)^2$
$$+ {}^{11}C_3 3^8\left(\frac{x}{4}\right)^3 + {}^{11}C_4 3^7\left(\frac{x}{4}\right)^4 + ...$$
$$= 177\,147 + (11 \times 59\,049)\left(\frac{x}{4}\right)$$
$$+ (55 \times 19\,683)\left(\frac{x}{4}\right)^2 + (165 \times 6561)\left(\frac{x}{4}\right)^3$$
$$+ (330 \times 2187)\left(\frac{x}{4}\right)^4 + ...$$
$$= 177\,147 + \frac{649\,539}{4}x + \frac{1\,082\,565}{16}x^2$$
$$+ \frac{1\,082\,565}{64}x^3 + \frac{721\,710}{256}x^4 + ...$$

b) $\left(3 + \frac{x}{4}\right)^{11} = 3.002^{11}$ when $x = 0.008$

Put $x = 0.008$ into the expansion:

$$3.002^{11} = 177\,147 + \frac{649\,539}{4}(0.008)$$
$$+ \frac{1\,082\,565}{16}(0.008)^2 + \frac{1\,082\,565}{64}(0.008)^3$$
$$+ \frac{721\,710}{256}(0.008)^4 + ...$$
$$\approx 178\,450.417 \text{ to 3 d.p.}$$

(higher powers can be ignored as they're small)

Q21 a) $(2 + kx)^{13} = 2^{13} + {}^{13}C_1 2^{12}(kx) + {}^{13}C_2 2^{11}(kx)^2 + ...$
$$= 8192 + 53\,248kx + 159\,744k^2x^2 + ...$$

b) The coefficient of x is $\frac{1}{6}$ of the coefficient of x^2:

$$6(53\,248k) = 159\,744k^2$$
$$\implies k = 6 \times \frac{53\,248}{159\,744} = 2$$

Exam Questions — Chapter 5

Q1 $(4 + 3x)^{10} = 4^{10} + {}^{10}C_1 4^9(3x) + {}^{10}C_2 4^8(3x)^2$
$$+ {}^{10}C_3 4^7(3x)^3 + {}^{10}C_4 4^6(3x)^4 + ...$$

So: x coefficient $= 10 \times 4^9 \times 3 = 7\,864\,320$

[1 mark]

x^2 coefficient $= 45 \times 4^8 \times 9 = 26\,542\,080$

[1 mark]

x^3 coefficient $= 120 \times 4^7 \times 27 = 53\,084\,160$

[1 mark]

x^4 coefficient $= 210 \times 4^6 \times 81 = 69\,672\,960$

[1 mark]

Q2 a) The series is defined by $u_{n+1} = 12 \times 1.3^n$ so the common ratio = 1.3, which is greater than 1, so the sequence is divergent. *[1 mark]*

b) $u_3 = 12 \times 1.3^2 = 20.28$ *[1 mark]*

$u_{10} = 12 \times 1.3^9 = 127.25$ to 2 d.p. *[1 mark]*

Q3 a) $S_\infty = \dfrac{a}{1 - r} = \dfrac{20}{1 - \frac{3}{4}} = \dfrac{20}{\frac{1}{4}} = 80$

[2 marks available — 1 mark for formula, 1 mark for correct answer]

b) $u_{15} = ar^{14} = 20 \times \left(\frac{3}{4}\right)^{14} = 0.356$ (to 3 sig. fig.)

[2 marks available — 1 mark for formula, 1 mark for correct answer]

c) Use the formula for the sum of a geometric series to write an expression for S_n:

$$S_n = \frac{a(1 - r^n)}{1 - r} = \frac{20\left(1 - \left(\frac{3}{4}\right)^n\right)}{1 - \frac{3}{4}} \quad \textbf{\textit{[1 mark]}}$$

so $\dfrac{20\left(1 - \left(\frac{3}{4}\right)^n\right)}{1 - \frac{3}{4}} > 79.76$

Now rearrange and use logs to get n on its own:

$$\frac{20\left(1 - \left(\frac{3}{4}\right)^n\right)}{1 - \frac{3}{4}} > 79.76 \Rightarrow 20\left(1 - \left(\frac{3}{4}\right)^n\right) > 19.94$$

$\Rightarrow 1 - \left(\frac{3}{4}\right)^n > 0.997 \Rightarrow 0.003 > 0.75^n$ *[1 mark]*

$\Rightarrow \log 0.003 > n \log 0.75$ *[1 mark]*

$\Rightarrow \dfrac{\log 0.003}{\log 0.75} < n$ *[1 mark]*

Remember — if x < 1, then log x has a negative value.

$\dfrac{\log 0.003}{\log 0.75} = 20.1929....$

so $n > 20.1929....$

But n must be an integer, so $n = 21$ *[1 mark]*

Q4 a) $u_n = ar^{n-1}$ where $a = 1$ and $r = 1.5$,

so $u_5 = 1 \times (1.5)^4$ *[1 mark]*

$= 5.06$ km to the nearest 10 m *[1 mark]*

b) $a = 2$ and $r = 1.2$ *[1 mark]*

$u_9 = 2 \times (1.2)^8 = 8.60$ km (2 d.p.)

$u_{10} = 2 \times (1.2)^9 = 10.32$ km (2 d.p.) *[1 mark]*

$u_9 < 10$ km and $u_{10} > 10$ km

Day 10 is the first day Chris runs > 10 km. *[1 mark]*

You could also use logs to answer Q4 b) — the method is the same as in Q3 c).

c) In 10 days, Alex ran a total of 30 km *[1 mark]*

Use the formula for the sum of first n terms:

$$S_n = \frac{a(1 - r^n)}{1 - r}.$$

Chris ran a total of:

$\dfrac{2(1 - 1.2^{10})}{1 - 1.2} = 51.917$ km (nearest m) *[1 mark]*

Heather ran a total of:

$\dfrac{1(1 - 1.5^{10})}{1 - 1.5} = 113.330$ km (nearest m) *[1 mark]*

So they raised:

$30 + 51.917 + 113.330$
$= £195.25$ (nearest penny) *[1 mark]*

Q5 a) $S_\infty = \dfrac{a}{1 - r}$ and $u_2 = ar$ *[1 mark]*

So $36 = \dfrac{a}{1 - r}$ i.e. $36 - 36r = a$ *[1 mark]*

and $5 = ar$. *[1 mark]*

Substituting for a gives:

$5 = (36 - 36r)r = 36r - 36r^2$

i.e. $36r^2 - 36r + 5 = 0$ *[1 mark]*

b) Factorising gives: $(6r - 1)(6r - 5) = 0$

So: $r = \dfrac{1}{6}$ or $r = \dfrac{5}{6}$.

[1 mark for each correct value]

If $r = \dfrac{1}{6}$ and $ar = 5$ then $\dfrac{a}{6} = 5$ i.e. $a = 30$

If $r = \dfrac{5}{6}$ and $ar = 5$ then $\dfrac{5a}{6} = 5$ i.e. $a = 6$

[1 mark for each correct value]

Q6 a) $\left(1 + \frac{x}{3}\right)^8 = 1 + {}^8C_1\left(\frac{x}{3}\right) + {}^8C_2\left(\frac{x}{3}\right)^2 + ...$

$= 1 + 8\left(\frac{x}{3}\right) + 28\left(\frac{x}{3}\right)^2 + ...$

$= 1 + \dfrac{8}{3}x + \dfrac{28}{9}x^2 + ...$

[1 mark for each correct term]

b) $\left(1 + \frac{x}{3}\right)^8 = 1.002^8$ when $x = 0.006$, so find an approximation by putting $x = 0.006$ *[1 mark]* into the expansion of $\left(1 + \frac{x}{3}\right)^8$.

Terms with higher powers of 0.006 can be ignored because 0.006 is small:

$1.002^8 \approx 1 + \dfrac{8}{3}(0.006) + \dfrac{28}{9}(0.006)^2$ *[1 mark]*

$= 1 + 0.016 + 0.000112$

$= 1.0161$ to 4 d.p. *[1 mark]*

Q7 a) Common ratio:

$$r = \frac{\text{second term}}{\text{first term}} = \frac{5}{2} = 2.5$$

First term $a = 2$

$$S_8 = \frac{a(1 - r^8)}{1 - r} = \frac{2(1 - 2.5^8)}{1 - 2.5} = 2033.17 \text{ to 2 d.p.}$$

[3 marks available — 1 mark for correct values of a and r, 1 mark for putting values into formula, 1 mark for correct answer]

b) An infinite geometric series with common ratio r is convergent when $|r| < 1$. *[1 mark]*

You could also have 'when −1 < r < 1'.

c) $S_\infty = \dfrac{a}{1 - r} = \dfrac{8}{1 - \left(-\frac{3}{4}\right)} = \dfrac{8}{1.75} = 4.57$ to 2 d.p.

[2 marks available — 1 mark for using formula, 1 mark for correct answer]

Q8 a) $(3 + kx)^9 = 3^9 + {}^9C_1 3^8(kx) + {}^9C_2 3^7(kx)^2 + ...$

Giving the binomial coefficients in another form is also fine — using factorials, writing out the numbers, e.g.
$\dfrac{9 \times 8 \times ... \times 1}{(2 \times 1)(7 \times 6 \times ... \times 1)}$ *or using* $\binom{n}{r}$ *notation.*

$= 19\,683 + (9 \times 6561 \times kx) + (36 \times 2187 \times k^2x^2)$

$= 19\,683 + 59\,049kx + 78\,732k^2x^2$

[1 mark for each correct term]

b) $59\,049k = \dfrac{3}{4}(78\,732k^2)$

$\Rightarrow 59\,049 = \dfrac{3}{4}(78\,732k)$

It's OK to divide by k here, because you know from the question that k ≠ 0.

$\Rightarrow 59\,049 = 59\,049k \Rightarrow k = 1$ *[1 mark]*

Q9 a) $S_\infty = \dfrac{a}{1 - r} = -9 \Rightarrow a = -9(1 - r)$ *[1 mark]*

and $ar = -2 \Rightarrow a = \dfrac{-2}{r}$

$\Rightarrow \dfrac{-2}{r} = -9(1 - r)$ *[1 mark]*

$\Rightarrow -2 = -9r + 9r^2 \Rightarrow 9r^2 - 9r + 2 = 0$ *[1 mark]*

b) $9r^2 - 9r + 2 = 0$,

factorising: $(3r - 1)(3r - 2) = 0$ *[1 mark]*

$\Rightarrow r = \frac{1}{3}$ or $r = \frac{2}{3}$ *[1 mark]*

c) $ar = -2 \Rightarrow a = \frac{-2}{r}$

$r = \frac{1}{3} \Rightarrow a = \frac{-2}{\frac{1}{3}} = -6$ *[1 mark]*

$r = \frac{2}{3} \Rightarrow a = \frac{-2}{\frac{2}{3}} = -3$ *[1 mark]*

d) r takes its smallest possible value,

so $r = \frac{1}{3} \Rightarrow a = -6$.

The 7th term is: $u_7 = ar^6$ *[1 mark]*

$= -6 \times \left(\frac{1}{3}\right)^6 = -0.0082$ to 4 d.p. *[1 mark]*

e) $S_5 = \frac{a(1 - r^5)}{1 - r} = \frac{-6\left(1 - \left(\frac{1}{3}\right)^5\right)}{1 - \frac{1}{3}}$ *[1 mark]*

$= -8.96$ to 2 d.p. *[1 mark]*

Q10 a) $(1 + 3x)^6 = 1 + {}^6C_1(3x) + {}^6C_2(3x)^2 + \dots$

$= 1 + 6(3x) + 15(3x)^2 + \dots$

$= 1 + 18x + 135x^2 + \dots$

[1 mark for each correct term]

You could answer this question by using the formula for expansions of $(1 + x)^n$ and replacing x with 3x, or by getting coefficients from Pascal's triangle — if you do this make sure you don't forget to multiply each binomial coefficient by the right power of 3.

b) $(1 - 2x)(1 + 3x)^6$

$= (1 - 2x)(1 + 18x + 135x^2 + \dots)$

$\approx (1 - 2x)(1 + 18x + 135x^2)$

$= 1 + 18x + 135x^2 - 2x - 36x^2 - 270x^3$ *[1 mark]*

$\approx 1 + 18x + 135x^2 - 2x - 36x^2$

$= 1 + 16x + 99x^2$ *[1 mark]*

Chapter 6 — Differentiation

1. Differentiation

Exercise 1.1 — Differentiation

Q1 a) $f'(x) = 6x$ **b)** $f'(x) = 20x^3 + 2x$

c) $f(x) = (x - 3)^2 = x^2 - 6x + 9$

$f'(x) = 2x - 6$

d) $f(x) = (3x + 1)(x - 2) = 3x^2 - 5x - 2$

$f'(x) = 6x - 5$

Q2 a) $f'(x) = 6x^5 + 6x + 2 \Rightarrow f'(0) = 0 + 0 + 2 = 2$

b) $f(x) = (x + 3)^2 = x^2 + 6x + 9$

$f'(x) = 2x + 6 \Rightarrow f'(0) = 0 + 6 = 6$

c) $f(x) = (2x + 1)(2x - 1) = 4x^2 - 1$

$f'(x) = 8x \Rightarrow f'(0) = 0$

d) $f'(x) = 20x^3 + 3 \Rightarrow f'(0) = 0 + 3 = 3$

Q3 a) $y = \frac{2}{x} = 2x^{-1} \Rightarrow \frac{dy}{dx} = -2x^{-2} = -\frac{2}{x^2}$

b) $y = x^2 + \frac{1}{x} = x^2 + x^{-1}$

$\frac{dy}{dx} = 2x - x^{-2} = 2x - \frac{1}{x^2}$

c) $y = \frac{2x + 3}{x^3} = 2x^{-2} + 3x^{-3}$

$\frac{dy}{dx} = -4x^{-3} - 9x^{-4} = -\frac{4x + 9}{x^4}$

d) $y = x^{-3} + 2x^{-4}$

$\frac{dy}{dx} = -3x^{-4} - 8x^{-5}$

Q4 a) $y = 4\sqrt{x} = 4x^{\frac{1}{2}}$

$\frac{dy}{dx} = 2x^{-\frac{1}{2}} = \frac{2}{\sqrt{x}}$

b) $y = x^2 + 7\sqrt{x} = x^2 + 7x^{\frac{1}{2}}$

$\frac{dy}{dx} = 2x + \frac{7}{2}x^{-\frac{1}{2}} = 2x + \frac{7}{2\sqrt{x}}$

c) $y = (x - \sqrt{x})^2 = x^2 - 2x^{\frac{3}{2}} + x$

$\frac{dy}{dx} = 2x - 3x^{\frac{1}{2}} + 1 = 2x - 3\sqrt{x} + 1$

d) $y = \sqrt{x}(1 + \sqrt{x}) = \sqrt{x} + x = x^{\frac{1}{2}} + x$

$\frac{dy}{dx} = \frac{1}{2}x^{-\frac{1}{2}} + 1 = \frac{1}{2\sqrt{x}} + 1$

Q5 a) $y = \sqrt{x}(x + \sqrt{x})^2 = \sqrt{x}(x^2 + 2x^{\frac{3}{2}} + x)$

$= x^{\frac{5}{2}} + 2x^2 + x^{\frac{3}{2}}$

$\frac{dy}{dx} = \frac{5}{2}x^{\frac{3}{2}} + 4x + \frac{3}{2}x^{\frac{1}{2}} = \frac{1}{2}(5(\sqrt{x})^3 + 8x + 3\sqrt{x})$

b) $y = (2 + \frac{1}{x})(x + 1) = 2x + 2 + 1 + x^{-1}$

$\frac{dy}{dx} = 2 - \frac{1}{x^2}$

Q6 a) $f(x) = \sqrt{x}(x^2 + 3)(x^2 - 3) = \sqrt{x}(x^4 - 9) = x^{\frac{9}{2}} - 9x^{\frac{1}{2}}$

$f'(x) = \frac{9}{2}x^{\frac{7}{2}} - \frac{9}{2}x^{-\frac{1}{2}} = \frac{9}{2x^{\frac{1}{2}}}(x^4 - 1) = \frac{9}{2\sqrt{x}}(x^4 - 1)$

b) If $f'(x) = 0$ then either $\frac{9}{2\sqrt{x}} = 0$ or $x^4 - 1 = 0$. The first part can't be equal to 0, so $x^4 = 1 \Rightarrow x = \pm1$. x can't be -1 as there's a $\sqrt{x}$, so the only solution is $x = 1$.

Q7 $f'(x) = 3x^2 + 2kx + k$.

When $x = 0$, $3(0)^2 + 2k(0) + k = 3$, so $k = 3$.

2. Using Differentiation

Exercise 2.1 — Stationary points

Q1 a) The graph has 2 stationary points — a minimum and a point of inflexion.

b) The graph has 3 stationary points — a maximum, a minimum and a point of inflexion.

Q2 a) $\frac{dy}{dx} = 2x + 3$. When $\frac{dy}{dx} = 0$, $2x + 3 = 0 \Rightarrow x = -\frac{3}{2}$

b) $y = (3 - x)(4 + 2x) = 12 + 2x - 2x^2$

$\frac{dy}{dx} = 2 - 4x$. When $\frac{dy}{dx} = 0$, $2 - 4x = 0 \Rightarrow x = \frac{1}{2}$

Q3 a) $\frac{dy}{dx} = 4x - 5$. When $\frac{dy}{dx} = 0$, $4x - 5 = 0 \Rightarrow x = \frac{5}{4}$

When $x = \frac{5}{4}$, $y = 2(\frac{5}{4})^2 - 5(\frac{5}{4}) + 2 = -\frac{9}{8}$

So the coordinates are $(\frac{5}{4}, -\frac{9}{8})$.

b) $\frac{dy}{dx} = -2x + 3$. When $\frac{dy}{dx} = 0$, $-2x + 3 = 0$
$\Rightarrow x = \frac{3}{2}$. When $x = \frac{3}{2}$, $y = -(\frac{3}{2})^2 + 3(\frac{3}{2}) - 4 = -\frac{7}{4}$.
So the coordinates are $(\frac{3}{2}, -\frac{7}{4})$.

c) $\frac{dy}{dx} = -6 - 6x$. When $\frac{dy}{dx} = 0$, $-6 - 6x = 0 \Rightarrow x = -1$
When $x = -1$, $y = 7 - 6(-1) - 3(-1)^2 = 10$.
So the coordinates are $(-1, 10)$.

d) $y = (x - 1)(2x + 3) = 2x^2 + x - 3$
$\frac{dy}{dx} = 4x + 1$. When $\frac{dy}{dx} = 0$, $4x + 1 = 0 \Rightarrow x = -\frac{1}{4}$
When $x = -\frac{1}{4}$, $y = (-\frac{1}{4} - 1)(2(-\frac{1}{4}) + 3) = -\frac{25}{8}$.
So the coordinates are $(-\frac{1}{4}, -\frac{25}{8})$.

Q4 a) $\frac{dy}{dx} = 3x^2 - 3$. When $\frac{dy}{dx} = 0$, $3x^2 - 3 = 0$
$\Rightarrow x = \pm 1$. When $x = 1$, $y = 1^3 - 3(1) + 2 = 0$.
When $x = -1$, $y = (-1)^3 - 3(-1) + 2 = 4$.
So the coordinates are $(1, 0)$ and $(-1, 4)$.

b) $\frac{dy}{dx} = 12x^2$. When $\frac{dy}{dx} = 0$, $12x^2 = 0 \Rightarrow x = 0$
When $x = 0$, $y = 4(0)^3 + 5 = 5$.
So the coordinates are $(0, 5)$.

Q5 $f'(x) = 5x^4 + 3$. When $f'(x) = 0$, $5x^4 + 3 = 0 \Rightarrow x^4 = -\frac{3}{5}$.
Finding a solution would involve finding the fourth root of a negative number. But $x^4 = (x^2)^2$, so x^4 is always positive and so there are no stationary points.

Q6 a) $\frac{dy}{dx} = 3x^2 - 14x - 5$

b) When $\frac{dy}{dx} = 0$, $3x^2 - 14x - 5 = 0$
$\Rightarrow (3x + 1)(x - 5) = 0$, so either $x = -\frac{1}{3}$ or $x = 5$.
When $x = -\frac{1}{3}$, $y = (-\frac{1}{3})^3 - 7(-\frac{1}{3})^2 - 5(-\frac{1}{3}) + 2$
$= \frac{77}{27}$.
When $x = 5$, $y = 5^3 - 7(5)^2 - 5(5) + 2 = -73$.
So the coordinates are $(-\frac{1}{3}, \frac{77}{27})$ and $(5, -73)$.

Q7 For stationary points to occur, $f'(x)$ must equal zero, so $f'(x) = 3x^2 + k = 0 \Rightarrow -\frac{k}{3} = x^2$. For this equation to have a solution, k can't be positive (or it would be taking the square root of a negative number), so $k \leq 0$. Therefore, if the graph has no stationary points, $k > 0$.

Exercise 2.2 —
Maximum and minimum points

Q1 a) negative   **b)** positive   **c)** negative
    d) negative   **e)** positive

Q2 a) $\frac{dy}{dx} = 3x^2 + 4x + 5$ $\frac{d^2y}{dx^2} = 6x + 4$

b) $\frac{dy}{dx} = 12x^3$ $\frac{d^2y}{dx^2} = 36x^2$

c) $\frac{dy}{dx} = 28x^6 - 3$ $\frac{d^2y}{dx^2} = 168x^5$

d) $\frac{dy}{dx} = 5x^4 - 30x^2$ $\frac{d^2y}{dx^2} = 20x^3 - 60x$

e) $\frac{dy}{dx} = -7 + 2x$ $\frac{d^2y}{dx^2} = 2$

f) $\frac{dy}{dx} = -2$ $\frac{d^2y}{dx^2} = 0$

g) $y = (x + 2)(3x - 4) = 3x^2 + 2x - 8$
$\frac{dy}{dx} = 6x + 2$ $\frac{d^2y}{dx^2} = 6$

h) $y = \frac{1}{x^2} + 2x^3 + x = x^{-2} + 2x^3 + x$
$\frac{dy}{dx} = -2x^{-3} + 6x^2 + 1$ $\frac{d^2y}{dx^2} = 6x^{-4} + 12x$

Q3 a) $(1, 3)$
All the clues are in the question — the derivative when $x = 1$ is zero so you know it's a stationary point, and the y-value when $x = 1$ is 3.

b) The second derivative at $x = 1$ is positive, so it's a minimum.

Q4 a) $\frac{dy}{dx} = -2x$. When $\frac{dy}{dx} = 0$, $x = 0$. When $x = 0$, $y = 5 - 0 = 5$. So the coordinates are $(0, 5)$.
$\frac{d^2y}{dx^2} = -2$, so it's a maximum turning point.

b) $\frac{dy}{dx} = 6x^2 - 6$. When $\frac{dy}{dx} = 0$, $6x^2 = 6 \Rightarrow x = \pm 1$
When $x = 1$, $y = 2 - 6 + 2 = -2$. When $x = -1$, $y = -2 + 6 + 2 = 6$. So the coordinates are $(1, -2)$ and $(-1, 6)$. $\frac{d^2y}{dx^2} = 12x$.
At $(1, -2)$, $\frac{d^2y}{dx^2} = 12$, so it's a minimum.
At $(-1, 6)$, $\frac{d^2y}{dx^2} = -12$ so it's a maximum.

c) $\frac{dy}{dx} = 3x^2 - 6x - 24$. When $\frac{dy}{dx} = 0$, $x^2 - 2x - 8 = 0$
$\Rightarrow (x - 4)(x + 2) = 0 \Rightarrow x = 4$ or -2.
When $x = 4$, $y = 64 - 48 - 96 + 15 = -65$.
When $x = -2$, $y = -8 - 12 + 48 + 15 = 43$.
So the coordinates are $(4, -65)$ and $(-2, 43)$.
$\frac{d^2y}{dx^2} = 6x - 6$. At $(4, -65)$, $\frac{d^2y}{dx^2} = 24 - 6 = 18$, so it's a minimum.
At $(-2, 43)$, $\frac{d^2y}{dx^2} = -12 - 6 = -18$, so it's a maximum.

d) $\frac{dy}{dx} = 4x^3 + 12x^2 + 8x$. When $\frac{dy}{dx} = 0$,
$x^3 + 3x^2 + 2x = 0 \Rightarrow x(x + 2)(x + 1) = 0$, so $x = 0$, -1 or -2. When $x = 0$, $y = 0 + 0 + 0 - 10 = -10$.
When $x = -1$, $y = 1 - 4 + 4 - 10 = -9$. When $x = -2$, $y = 16 - 32 + 16 - 10 = -10$. So the stationary points are $(0, -10)$, $(-1, -9)$ and $(-2, -10)$.
$\frac{d^2y}{dx^2} = 12x^2 + 24x + 8$. At $(0, -10)$,
$\frac{d^2y}{dx^2} = 0 + 0 + 8 = 8$, so it's a minimum.
At $(-1, -9)$, $\frac{d^2y}{dx^2} = 12 - 24 + 8 = -4$,
so it's a maximum.
At $(-2, -10)$, $\frac{d^2y}{dx^2} = 48 - 48 + 8 = 8$, so it's a minimum.

Q5 a) $f'(x) = 24x^2 + 32x + 8$. When $f'(x) = 0$,
$3x^2 + 4x + 1 = 0 \Rightarrow (3x + 1)(x + 1) = 0$, so $x = -1$
or $-\frac{1}{3}$. When $x = -1$, $f(x) = -8 + 16 - 8 + 1 = 1$.
When $x = -\frac{1}{3}$, $f(x) = -\frac{8}{27} + \frac{16}{9} - \frac{8}{3} + 1 = -\frac{5}{27}$.
So the coordinates are $(-1, 1)$ and $(-\frac{1}{3}, -\frac{5}{27})$.
$f''(x) = 48x + 32$. At $(-1, 1)$ $f''(x) = -48 + 32 = -16$,
so it's a maximum.
At $(-\frac{1}{3}, -\frac{5}{27})$, $f''(x) = -\frac{48}{3} + 32 = 16$,
so it's a minimum.

b) $f(x) = \frac{27}{x^3} + x = 27x^{-3} + x \Rightarrow f'(x) = -81x^{-4} + 1$.
When $f'(x) = 0$, $x^4 = 81 \Rightarrow x = \pm 3$. When $x = 3$,
$f(x) = \frac{27}{27} + 3 = 4$. When $x = -3$, $f(x) = -\frac{27}{27} - 3$
$= -4$. So the coordinates are $(3, 4)$ and $(-3, -4)$.
$f''(x) = 324x^{-5}$. At $(3, 4)$ $f''(x) = \frac{4}{3}$,
so it's a minimum.
At $(-3, -4)$ $f''(x) = -\frac{4}{3}$, so it's a maximum.

Q6 a) $f'(x) = 3x^2 - 6x$. $f''(x) = 6x - 6$.

b) When $f'(x) = 0$, $3x^2 - 6x = 0 \Rightarrow x(x - 2) = 0$,
so $x = 0$ or $x = 2$. When $x = 0$, $f(x) = 0 - 0 + 4 = 4$.
When $x = 2$, $f(x) = 8 - 12 + 4 = 0$.
So the coordinates are $(0, 4)$ and $(2, 0)$.
At $(0, 4)$ $f''(x) = 0 - 6 = -6$, so it's a maximum.
At $(2, 0)$ $f''(x) = 12 - 6 = 6$, so it's a minimum.

Q7 a) $V = r^2 + \frac{2000}{r} = r^2 + 2000r^{-1} \Rightarrow \frac{dV}{dr} = 2r - \frac{2000}{r^2}$
When $\frac{dV}{dr} = 0$, $2r = \frac{2000}{r^2} \Rightarrow r^3 = 1000 \Rightarrow r = 10$

b) $\frac{d^2V}{dr^2} = 2 + \frac{4000}{r^3}$. When $r = 10$,
$\frac{d^2V}{dr^2} = 2 + 4 = 6$, so it's a minimum.

Q8 $f(x) = x^3 + ax^2 + bx + c \Rightarrow f'(x) = 3x^2 + 2ax + b$.
$\Rightarrow f''(x) = 6x + 2a$. At the point $(3, 10)$:
$10 = 3^3 + a(3^2) + b(3) + c \Rightarrow 10 = 27 + 9a + 3b + c$
As $(3, 10)$ is a stationary point, $0 = 3(3^2) + 2a(3) + b$
$\Rightarrow 0 = 27 + 6a + b$. We know that $f''(3) = 0$,
so $0 = 6(3) + 2a \Rightarrow 0 = 18 + 2a \Rightarrow a = -9$.
Then $0 = 27 + 6a + b = 27 + 6(-9) + b \Rightarrow b = 27$
And $10 = 27 + 9a + 3b + c = 27 + 9(-9) + 3(27) + c$
$\Rightarrow c = -17$. So $f(x) = x^3 - 9x^2 + 27x - 17$.

Q9 a) $\frac{dy}{dx} = 4x^3 + 3kx^2 + 2x$. Stationary points occur
when $\frac{dy}{dx} = 0$, so $4x^3 + 3kx^2 + 2x = 0 \Rightarrow$
$x(4x^2 + 3kx + 2) = 0$ so $x = 0$ or $4x^2 + 3kx + 2 = 0$.
As you know the only stationary point occurs at
$x = 0$, the part in brackets can't have any
solutions. This gives you information about the
discriminant of the quadratic equation:
$b^2 - 4ac < 0 \Rightarrow 9k^2 < 32 \Rightarrow k^2 < \frac{32}{9}$.

b) When $x = 0$, $y = 0 + 0 + 0 + 17 = 17$, so the
coordinates are $(0, 17)$.
$\frac{d^2y}{dx^2} = 12x^2 + 6kx + 2$. When $x = 0$, $\frac{d^2y}{dx^2} = 2$, so
it's a minimum.

Exercise 2.3 —
Increasing and decreasing functions

Q1 a) $\frac{dy}{dx} = 2x + 7$. If the function is increasing, $\frac{dy}{dx} > 0$
$\Rightarrow 2x > -7 \Rightarrow x > -\frac{7}{2}$

b) $\frac{dy}{dx} = 10x + 3$. If the function is increasing,
$\frac{dy}{dx} > 0 \Rightarrow 10x > -3 \Rightarrow x > -\frac{3}{10}$.

c) $\frac{dy}{dx} = -18x$. If the function is increasing, $\frac{dy}{dx} > 0$
$\Rightarrow -18x > 0 \Rightarrow x < 0$.
*Be careful with the direction of the inequality sign if
you're dividing by a negative number.*

Q2 a) $f'(x) = -3 - 4x$. If the function is decreasing,
$f'(x) < 0 \Rightarrow -4x < 3 \Rightarrow x > -\frac{3}{4}$.

b) $f(x) = (6 - 3x)(6 + 3x) = 36 - 9x^2$
$f'(x) = -18x$. If the function is decreasing, $f'(x) < 0$
$\Rightarrow -18x < 0 \Rightarrow x > 0$.

c) $f(x) = (1 - 2x)(7 - 3x) = 7 - 17x + 6x^2$
$f'(x) = -17 + 12x$. If the function is decreasing,
$f'(x) < 0 \Rightarrow 12x < 17 \Rightarrow x < \frac{17}{12}$.

Q3 a) $\frac{dy}{dx} = 3x^2 - 12x - 15$. If the function is increasing,
$\frac{dy}{dx} > 0 \Rightarrow 3x^2 - 12x - 15 > 0$
$\Rightarrow x^2 - 4x - 5 > 0 \Rightarrow (x - 5)(x + 1) > 0$
For this expression to be > 0, both brackets must
be positive or both brackets must be negative.
So either $x > 5$ and $x > -1$ or $x < 5$ and $x < -1$.
So the function is increasing when $x < -1$ and
when $x > 5$.

b) $\frac{dy}{dx} = 3x^2 + 12x + 12$. If the function is
increasing, $\frac{dy}{dx} > 0 \Rightarrow 3x^2 + 12x + 12 > 0$
$\Rightarrow x^2 + 4x + 4 > 0 \Rightarrow (x + 2)(x + 2) > 0$
$\Rightarrow (x + 2)^2 > 0 \Rightarrow x \neq -2$.
*Remember that you can use a different method, e.g.
sketching the quadratic, to solve the inequality if you
prefer.*

Q4 a) $f'(x) = 3x^2 - 6x - 9$. If the function is decreasing,
$f'(x) < 0 \Rightarrow 3x^2 - 6x - 9 < 0 \Rightarrow x^2 - 2x - 3 < 0$
$\Rightarrow (x - 3)(x + 1) < 0$. For the expression to be < 0,
one bracket must be positive and one negative.
So either $x < 3$ and $x > -1$ or $x > 3$ and $x < -1$.
The second situation is impossible, so $-1 < x < 3$.

b) $f'(x) = 3x^2 - 8x + 4$. If the function is decreasing,
$f'(x) < 0 \Rightarrow 3x^2 - 8x + 4 < 0 \Rightarrow (3x - 2)(x - 2) < 0$.
For the expression to be < 0, either $x < \frac{2}{3}$ and
$x > 2$ or $x > \frac{2}{3}$ and $x < 2$. The first situation is
impossible, so $\frac{2}{3} < x < 2$.

Q5 $f'(x) = 3x^2 + 1$. x^2 can't be negative ($x^2 \geq 0$), so $f'(x)$
must always be positive and so $f(x)$ is an increasing
function for all real values of x.

Q6 $f'(x) = -3 - 3x^2$. x^2 can't be negative ($x^2 \geq 0$), so $f'(x)$
must always be ≤ -3 (hence negative), so $f(x)$ is a
decreasing function.

Q7 a) $\frac{dy}{dx} = 8x^3 + 1$. If the function is decreasing,

$\frac{dy}{dx} < 0 \Rightarrow 8x^3 + 1 < 0 \Rightarrow x^3 < -\frac{1}{8} \Rightarrow x < -\frac{1}{2}$

b) $\frac{dy}{dx} = 4x^3 - 6x^2 - 10x$. If the function is decreasing, $\frac{dy}{dx} < 0 \Rightarrow 4x^3 - 6x^2 - 10x < 0$

$\Rightarrow x(2x - 5)(x + 1) < 0$. For this to be true, there are 4 possibilities — all are less than zero, or one is less than zero and the other two are not:

Either $x < 0$ and $x < \frac{5}{2}$ and $x < -1$, so $x < -1$

Or $x < 0$ and $x > \frac{5}{2}$ and $x > -1$ (impossible)

Or $x > 0$ and $x < \frac{5}{2}$ and $x > -1$, so $0 < x < \frac{5}{2}$

Or $x > 0$ and $x > \frac{5}{2}$ and $x < -1$ (impossible)

This gives the ranges $x < -1$ and $0 < x < \frac{5}{2}$.

Remember, if x must be smaller than 0, 5/2 and –1, you can just dismiss the two higher numbers and simplify it to x being smaller than –1.

Q8 a) $y = x^2 + \sqrt{x} = x^2 + x^{\frac{1}{2}} \Rightarrow \frac{dy}{dx} = 2x + \frac{1}{2\sqrt{x}}$

$\frac{dy}{dx} > 0$ for all $x > 0$, so the function is increasing for all $x > 0$.

b) $y = 4x^2 + \frac{1}{x} = 4x^2 + x^{-1} \Rightarrow \frac{dy}{dx} = 8x - \frac{1}{x^2}$

The function is increasing when $\frac{dy}{dx} > 0$

$\Rightarrow 8x - \frac{1}{x^2} > 0 \Rightarrow x^3 > \frac{1}{8} \Rightarrow x > \frac{1}{2}$

Q9 If the function is decreasing, $\frac{dy}{dx} < 0$. $\frac{dy}{dx} = -3 - 5ax^4$

$\Rightarrow -3 - 5ax^4 < 0 \Rightarrow ax^4 > -\frac{3}{5}$. The right-hand side is negative, so as $x^4 \geq 0$, a must also be positive to make the LHS > RHS for all x. So $a > 0$.

Q10 If the function is increasing, $\frac{dy}{dx}$ will always be greater than 0. $\frac{dy}{dx} = kx^{k-1} + 1 \Rightarrow kx^{k-1} + 1 > 0$.

When $k = 1$, $x^0 + 1 > 0$ — true for all x

When $k = 2$, $2x^1 + 1 > 0$ — not true for all x

When $k = 3$, $3x^2 + 1 > 0$ — true for all x

When $k = 4$, $4x^3 + 1 > 0$ — not true for all x, etc.

So k must be an odd number greater than zero.

Exercise 2.4 — Curve sketching

Q1 a) When $x = 0$, $y = 0^3 - 2(0)^2 = 0$, so the curve crosses the axes at $(0, 0)$. When $y = 0$, $x^3 - 2x^2 = 0 \Rightarrow x^2(x - 2) = 0 \Rightarrow x = 0$ or $x = 2$. So the curve also crosses the axes at $(2, 0)$.

You already knew it crossed the x-axis at x = 0, so you can ignore that one.

b) $\frac{dy}{dx} = 3x^2 - 4x$. When $\frac{dy}{dx} = 0$, $3x^2 - 4x = 0$

$\Rightarrow x(3x - 4) = 0 \Rightarrow x = 0$ or $x = \frac{4}{3}$. When $x = \frac{4}{3}$, $y = (\frac{4}{3})^3 - 2(\frac{4}{3})^2 = -\frac{32}{27}$. So the coordinates are $(0, 0)$ and $(\frac{4}{3}, -\frac{32}{27})$.

c) $\frac{d^2y}{dx^2} = 6x - 4$. At $x = 0$, $\frac{d^2y}{dx^2} = -4$, so it's a maximum. At $x = \frac{4}{3}$, $\frac{d^2y}{dx^2} = 4$, so it's a minimum.

d) It's a positive cubic, so it'll go from bottom left to top right:

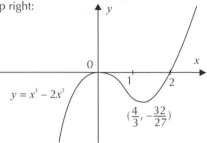

Q2 a) $x^3 + x^2 = 0 \Rightarrow x^2(x + 1) = 0 \Rightarrow x = 0$ or $x = -1$.

b) $f'(x) = 3x^2 + 2x$. When $f'(x) = 0$, $3x^2 + 2x = 0$ $\Rightarrow x(3x + 2) = 0 \Rightarrow x = 0$ or $x = -\frac{2}{3}$. When $x = 0$, $y = 0$. When $x = -\frac{2}{3}$, $y = \frac{4}{27}$, so the stationary points are at $(0, 0)$ and $(-\frac{2}{3}, \frac{4}{27})$.

$f''(x) = 6x + 2$. At $(0, 0)$, $f''(x) = 2$, so it's a minimum. At $(-\frac{2}{3}, \frac{4}{27})$, $f''(x) = -2$, so it's a maximum.

c)

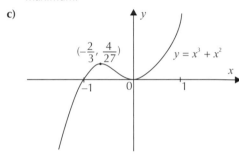

This is another positive cubic, so it goes from bottom left to top right.

Q3 a) $f'(x) = 4x^3 - 3x^2$, $f''(x) = 12x^2 - 6x$.

b) $f(x)$ is increasing for $f'(x) > 0 \Rightarrow 4x^3 - 3x^2 > 0$ $\Rightarrow x^2(4x - 3) > 0$, so either $x^2 > 0$ and $x > \frac{3}{4}$ $\Rightarrow x > \frac{3}{4}$ or $x^2 < 0$ and $x < \frac{3}{4}$. x^2 can't be less than 0, so this situation is impossible.

$f(x)$ is decreasing for $f'(x) < 0 \Rightarrow 4x^3 - 3x^2 < 0$ $\Rightarrow x^2(4x - 3) < 0$, so either $x^2 < 0$ and $x > \frac{3}{4}$ or $x^2 > 0$ and $x < \frac{3}{4}$. The first situation is impossible, so it's decreasing when $x < \frac{3}{4}$, $x \neq 0$.

c) When $x = 0$, $f(x) = 0$. When $f(x) = 0$, $x^4 - x^3 = 0$ $\Rightarrow x^3(x - 1) = 0$, so $x = 0$ or $x = 1$. So the curve crosses the axes at $(0, 0)$ and $(1, 0)$.

Stationary points occur when $f'(x) = 0$. $f'(x) = 4x^3 - 3x^2 = 0 \Rightarrow x^2(4x - 3) = 0$, so $x = 0$ or $x = \frac{3}{4}$. When $x = 0$, $y = 0$ and when $x = \frac{3}{4}$, $y = -\frac{27}{256} = -0.11$ (2 d.p.).

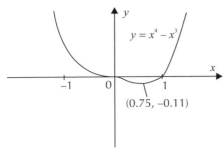

$y = x^4 - x^3$

(0.75, −0.11)

f''(0) = 0 so we cannot say whether the point (0, 0) is a maximum or a minimum — but the function is decreasing for x < 0.75 so (0, 0) must be a point of inflexion.

Q4 **a)** When $x = 0$, $y = 0$. When $y = 0$, $3x^3 + 3x^2 = 0$
$\Rightarrow 3x^2(x + 1) = 0$, so $x = 0$ or -1.

When $\dfrac{dy}{dx} = 0$, $9x^2 + 6x = 0 \Rightarrow 3x(3x + 2) = 0$

so $x = 0$ or $x = -\dfrac{2}{3}$. When $x = -\dfrac{2}{3}$, $y = \dfrac{4}{9}$.

$\dfrac{d^2y}{dx^2} = 18x + 6$. When $x = 0$, $\dfrac{d^2y}{dx^2} = 6$,

so it's a minimum. When $x = -\dfrac{2}{3}$, $\dfrac{d^2y}{dx^2} = -6$,
so it's a maximum.

It's a positive cubic, so it'll go from bottom left to top right.

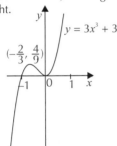

$y = 3x^3 + 3x^2$

$\left(-\dfrac{2}{3}, \dfrac{4}{9}\right)$

b) When $x = 0$, $y = 0$. When $y = 0$, $-x^3 + 9x = 0$
$\Rightarrow x(9 - x^2) = 0$, so $x = 0$ or $x = \pm3$.

When $\dfrac{dy}{dx} = 0$, $-3x^2 + 9 = 0 \Rightarrow x = \pm\sqrt{3}$.

When $x = \sqrt{3}$, $y = 6\sqrt{3}$ and when $x = -\sqrt{3}$,
$y = -6\sqrt{3}$

$\dfrac{d^2y}{dx^2} = -6x$. when $x = \sqrt{3}$, $\dfrac{d^2y}{dx^2} = -6\sqrt{3}$,

so it's a maximum, When $x = -\sqrt{3}$, $\dfrac{d^2y}{dx^2} = 6\sqrt{3}$,
so it's a minimum.

It's a negative cubic, so it'll go from top left to bottom right.

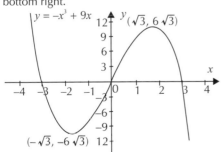

$y = -x^3 + 9x$

$(\sqrt{3}, 6\sqrt{3})$

$(-\sqrt{3}, -6\sqrt{3})$

c) When $x = 0$, $y = 0$. When $y = 0$, $x^4 - x^2 = 0$
$\Rightarrow x^2(x^2 - 1) = 0$, so $x = 0$ or ±1.

When $\dfrac{dy}{dx} = 0$, $4x^3 - 2x = 0 \Rightarrow x(2x^2 - 1) = 0$.

So $x = 0$ or $\pm\dfrac{1}{\sqrt{2}}$. When $x = \dfrac{1}{\sqrt{2}}$, $y = -\dfrac{1}{4}$

and when $x = -\dfrac{1}{\sqrt{2}}$, $y = -\dfrac{1}{4}$.

$\dfrac{d^2y}{dx^2} = 12x^2 - 2$. When $x = 0$, $\dfrac{d^2y}{dx^2} = -2$,

so it's a maximum. When $x = \dfrac{1}{\sqrt{2}}$, $\dfrac{d^2y}{dx^2} = 4$,

so it's a minimum. When $x = -\dfrac{1}{\sqrt{2}}$, $\dfrac{d^2y}{dx^2} = 4$,
so it's a minimum.

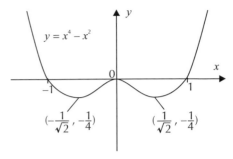

$y = x^4 - x^2$

$\left(-\dfrac{1}{\sqrt{2}}, -\dfrac{1}{4}\right)$ $\left(\dfrac{1}{\sqrt{2}}, -\dfrac{1}{4}\right)$

d) When $x = 0$, $y = 0$. When $y = 0$, $x^4 + x^2 = 0$
$\Rightarrow x^2(x^2 + 1) = 0$, so $x = 0$ ($x^2 = -1$ has no solutions).

When $\dfrac{dy}{dx} = 0$, $4x^3 + 2x = 0 \Rightarrow x(2x^2 + 1) = 0$,

so $x = 0$.

$\dfrac{d^2y}{dx^2} = 12x^2 + 2$. When $x = 0$, $\dfrac{d^2y}{dx^2} = 2$,
so it's a minimum.

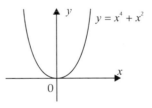

$y = x^4 + x^2$

Q5 When $x = 0$, $y = 1$. When $y = 0$, $(x + 1)(x - 1)^2 = 0$,
so $x = 1$ or -1.

When $\dfrac{dy}{dx} = 0$, $3x^2 - 2x - 1 = 0 \Rightarrow (3x + 1)(x - 1) = 0$

so $x = 1$ or $-\dfrac{1}{3}$. When $x = 1$, $y = 0$ and when $x = -\dfrac{1}{3}$,

$y = \dfrac{32}{27}$.

$\dfrac{d^2y}{dx^2} = 6x - 2$. When $x = 1$, $\dfrac{d^2y}{dx^2} = 4$,

so it's a minimum. When $x = -\dfrac{1}{3}$, $\dfrac{d^2y}{dx^2} = -4$,
so it's a maximum.

It's a positive cubic, so it'll go from bottom left to top right.

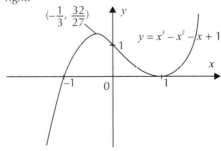

$\left(-\dfrac{1}{3}, \dfrac{32}{27}\right)$

$y = x^3 - x^2 - x + 1$

Q6 a) When $x = 2$, $x^3 - 4x = 8 - 8 = 0$.
When $x = -2$, $x^3 - 4x = -8 - (-8) = 0$
When $x = 0$, $x^3 - 4x = 0 - 0 = 0$.

b) $\frac{dy}{dx} = 3x^2 - 4$, $\frac{d^2y}{dx^2} = 6x$.
When $\frac{dy}{dx} = 0$, $3x^2 - 4 = 0 \Rightarrow x = \pm\frac{2\sqrt{3}}{3} \approx \pm 1.2$.
When $x = \frac{2\sqrt{3}}{3}$, $y \approx -3.1$ and when $x = -\frac{2\sqrt{3}}{3}$,
$y \approx 3.1$. So the coordinates of the stationary
points are roughly $(\frac{2\sqrt{3}}{3}, -3.1)$ and $(-\frac{2\sqrt{3}}{3}, 3.1)$.
At $(\frac{2\sqrt{3}}{3}, -3.1)$, $\frac{d^2y}{dx^2} = 6 \times \frac{2\sqrt{3}}{3} = 4\sqrt{3}$, so it's a
minimum. At $(-\frac{2\sqrt{3}}{3}, 3.1)$, $\frac{d^2y}{dx^2} = 6 \times -\frac{2\sqrt{3}}{3} = -4\sqrt{3}$, so it's a maximum.

c)

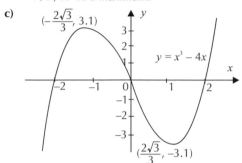

Q7 a) $f'(x) = 1 - \frac{1}{x^2}$. When $f'(x) = 0$, $1 - \frac{1}{x^2} = 0$
$\Rightarrow x^2 = 1 \Rightarrow x = \pm 1$. So the graph of $f(x) = x + \frac{1}{x}$
has stationary points at $x = 1$ and $x = -1$.

b) When $x = 1$, $y = 1 + 1 = 2$, and when $x = -1$,
$y = -1 - 1 = -2$. So the coordinates are $(1, 2)$
and $(-1, -2)$.
$f''(x) = \frac{2}{x^3}$. At $(1, 2)$, $f''(x) = 2$, so it's a minimum.
At $(-1, -2)$, $f''(x) = -2$, so it's a maximum.

c) $f(x) = x + \frac{1}{x}$. As x tends to 0 from below (x is
negative), $f(x)$ tends to $-\infty$. As $x \rightarrow 0$ from above
(x is positive), $f(x) \rightarrow \infty$.

d) As x tends to ∞, $f(x)$ tends to x
i.e. the graph tends towards the line $y = x$.
As x tends to $-\infty$, $f(x)$ tends to x
i.e. the graph tends towards the line $y = x$.

e)

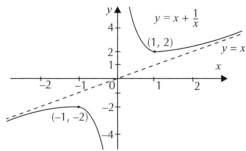

Q8 a) $\frac{dy}{dx} = 4x^3 - \frac{4}{\sqrt{x^3}}$. When $x = 1$, $\frac{dy}{dx} = 4 - 4 = 0$.

b) When $x = 1$, $y = 9$
As $x \rightarrow 0$, $y \rightarrow \infty$ and as $x \rightarrow \infty$, $y \rightarrow x^4$.
When $y = 0$, $x^4 + \frac{8}{\sqrt{x}} = 0 \Rightarrow \sqrt{x^9} = -8$. This has
no solutions, so the curve doesn't cross the x-axis.
$\frac{d^2y}{dx^2} = 12x^2 + \frac{6}{\sqrt{x^5}}$. When $x = 1$, $\frac{d^2y}{dx^2} = 18$,
so it's a minimum.

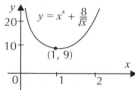

3. Real-Life Problems
Exercise 3.1 — Differentiation in real-life problems

Q1 Let the length of fence parallel to the wall be x and
the lengths perpendicular to the wall be y.
We want to maximise area, i.e $A = xy$.
Total length of fence = $66 = x + 2y \Rightarrow y = \frac{66 - x}{2}$
So we want to maximise $A = x(\frac{66 - x}{2}) = 33x - \frac{x^2}{2}$.
$\frac{dA}{dx} = 33 - x$, so when $\frac{dA}{dx} = 0$, $x = 33$.
Check that this gives a maximum value by
differentiating again: $\frac{d^2A}{dx^2} = -1$, so it's a maximum.
When $x = 33$, $y = \frac{66 - 33}{2} = 16.5$.
Area = $33 \times 16.5 = 544.5$ m².
*If you'd labelled the sides the other way round, your working
would be different but you'd still get the same answer.*

Q2 Total length of fence for a rectangular area of length
x m and width y m = $2x + 2y$.
Total area = $100 = xy \Rightarrow y = \frac{100}{x}$
Sub this into equation for length: $f(x) = 2x + \frac{200}{x}$
You want to minimise the length, so find $f'(x)$:
$f'(x) = 2 - \frac{200}{x^2}$. When $f'(x) = 0$, $2 - \frac{200}{x^2} = 0$
$\Rightarrow x^2 = 100 \Rightarrow x = 10$ (length can't be negative).
Check to see if this gives a minimum value by
differentiating again:
$f''(x) = \frac{400}{x^3} \Rightarrow f''(10) = 0.4$, so it's a minimum.
Now find the value of y when $x = 10$:
$100 = xy \Rightarrow y = 10$.
Finally, length = $2x + 2y = 20 + 20 = 40$ m.

Q3 $\frac{dh}{dt} = 30 - 9.8t$. When $\frac{dh}{dt} = 0$, $30 - 9.8t = 0$
$\Rightarrow t = 3.061... = 3.06$ (3 s.f.).
Check this gives a maximum value:
$\frac{d^2h}{dt^2} = -9.8$, so it's a maximum. So the maximum
value of $h = 30(3.061...) - 4.9(3.061...)^2$
$= 45.9$ m (3 s.f.).

Q4 a) Surface area = area of top and bottom plus area of curved face $= \pi r^2 + \pi r^2 + (2\pi r \times h)$
$= 2\pi r^2 + 2\pi r h$.
To match the question, find an expression for h by thinking about the volume of the tin:
Volume = area of base × height $= \pi r^2 h = 500$
$\Rightarrow h = \dfrac{500}{\pi r^2}$. So surface area $= 2\pi r^2 + 2\pi r\dfrac{500}{\pi r^2}$
$= 2\pi r^2 + \dfrac{1000}{r}$.

b) $\dfrac{dA}{dr} = 4\pi r - \dfrac{1000}{r^2}$. When $\dfrac{dA}{dr} = 0$,
$4\pi r - \dfrac{1000}{r^2} = 0 \Rightarrow r = \sqrt[3]{\dfrac{250}{\pi}} = 4.30$ cm (3 s.f.).

Check that this gives a minimum value:
$\dfrac{d^2A}{dr^2} = 4\pi + \dfrac{2000}{r^3}$. When $r = 4.30$,
$\dfrac{d^2A}{dr^2} = 37.7$, so it's a minimum.
You don't really need to work out $\dfrac{d^2A}{dr^2}$ when r = 4.30, you can tell it is positive straight away.

c) Surface area $= 2\pi r^2 + \dfrac{1000}{r} = 349$ cm^2 (3 s.f.).

Q5 a) Volume of the box = length × width × height
$= (40 - 2x) \times (40 - 2x) \times x = 4x^3 - 160x^2 + 1600x$

b) $\dfrac{dV}{dx} = 12x^2 - 320x + 1600$. When $\dfrac{dV}{dx} = 0$,
$12x^2 - 320x + 1600 = 0 \Rightarrow 3x^2 - 80x + 400 = 0$
$\Rightarrow (3x - 20)(x - 20) = 0 \Rightarrow x = 20$ or $x = \dfrac{20}{3}$.
Differentiate again to find which of these is a maximum: $\dfrac{d^2V}{dx^2} = 24x - 320$. When $x = 20$,
$\dfrac{d^2V}{dx^2} = 160$, and when $x = \dfrac{20}{3}$, $\dfrac{d^2V}{dx^2} = -160$,
so V is a maximum when $x = \dfrac{20}{3}$.
Note that if x = 20, then the volume of the box V = 0 so this can't be the maximum of V — it's always worth checking that the values of x are sensible in the context of the question.

So the maximum volume is:
$4(\tfrac{20}{3})^3 - 160(\tfrac{20}{3})^2 + 1600(\tfrac{20}{3}) = 4740$ cm^3 (3 s.f.).

Q6 a) The prism is made up of 5 shapes: 2 triangles with base x and height x (area $= \tfrac{1}{2}x^2$), 2 rectangles with width x and length l (area $= xl$) and 1 rectangle with width h and length l (area $= hl$). So the total surface area is given by:
$A = x^2 + 2xl + hl$. To get rid of the l, find an expression for l by looking at the volume:
$300 = \tfrac{1}{2}x^2 l \Rightarrow l = \dfrac{600}{x^2}$
To get rid of the h, form an expression for it in terms of x. It's the hypotenuse of a right-angled triangle, so $h^2 = x^2 + x^2 \Rightarrow h = \sqrt{2x^2} = \sqrt{2}\,x$.
Now put these into the original formula for A:
$A = x^2 + 2x(\dfrac{600}{x^2}) + \sqrt{2}\,x(\dfrac{600}{x^2}) = x^2 + \dfrac{600(2 + \sqrt{2})}{x}$

b) $\dfrac{dA}{dx} = 2x - \dfrac{600(2 + \sqrt{2})}{x^2}$. When $\dfrac{dA}{dx} = 0$,
$2x - \dfrac{600(2 + \sqrt{2})}{x^2} = 0 \Rightarrow x^3 = 300(2 + \sqrt{2})$
$\Rightarrow x = \sqrt[3]{600 + 300\sqrt{2}}$

Check that this gives a minimum value:
$\dfrac{d^2A}{dx^2} = 2 + \dfrac{1200(2 + \sqrt{2})}{x^3}$.
When $x = \sqrt[3]{600 + 300\sqrt{2}}$, $\dfrac{d^2A}{dx^2} = 6$, so it's a minimum.

Review Exercise — Chapter 6

Q1 $\dfrac{dy}{dx} = 3x^2 - 12x - 63$
When $\dfrac{dy}{dx} = 0$, $3x^2 - 12x - 63 = 0$
$\Rightarrow x^2 - 4x - 21 = 0 \Rightarrow (x - 7)(x + 3) = 0$
So $x = 7$ or $x = -3$. When $x = 7$,
$y = 7^3 - 6(7)^2 - 63(7) + 21 = -371$. When $x = -3$,
$y = (-3)^3 - 6(-3)^2 - 63(-3) + 21 = 129$. So the stationary points are $(7, -371)$ and $(-3, 129)$.

Q2 a) $y = x^3 + \dfrac{3}{x} = x^3 + 3x^{-1}$
$\dfrac{dy}{dx} = 3x^2 - 3x^{-2} = 3x^2 - \dfrac{3}{x^2}$.
When $\dfrac{dy}{dx} = 0$, $3x^2 - \dfrac{3}{x^2} = 0 \Rightarrow x^4 = 1 \Rightarrow x = \pm 1$.
When $x = 1$, $y = 1^3 + \dfrac{3}{1} = 4$. When $x = -1$,
$y = (-1)^3 + \dfrac{3}{-1} = -4$. So the stationary points are $(1, 4)$ and $(-1, -4)$.

b) $\dfrac{d^2y}{dx^2} = 6x + 6x^{-3}$. At $(1, 4)$, $\dfrac{d^2y}{dx^2} = 6 + 6 = 12$, so it's a minimum. At $(-1, -4)$, $\dfrac{d^2y}{dx^2} = -6 + (-6)$
$= -12$, so it's a maximum.

Q3 $\dfrac{dy}{dx} = 8x^3 - 2x$. When $\dfrac{dy}{dx} = 0$, $8x^3 - 2x = 0$
$\Rightarrow x(4x^2 - 1) = 0$ so either $x = 0$ or $4x^2 - 1 = 0$
$\Rightarrow x^2 = \dfrac{1}{4}$
$\Rightarrow x = \pm\dfrac{1}{2}$. When $x = 0$, $y = 0 - 0 + 4 = 4$.
When $x = \dfrac{1}{2}$, $y = 2(\tfrac{1}{2})^4 - (\tfrac{1}{2})^2 + 4 = \dfrac{31}{8}$.
When $x = -\dfrac{1}{2}$, $y = 2(-\tfrac{1}{2})^4 - (-\tfrac{1}{2})^2 + 4 = \dfrac{31}{8}$.
So the coordinates are $(0, 4)$, $(\tfrac{1}{2}, \tfrac{31}{8})$ and $(-\tfrac{1}{2}, \tfrac{31}{8})$.
$\dfrac{d^2y}{dx^2} = 24x^2 - 2$. At $(0, 4)$, $\dfrac{d^2y}{dx^2} = -2$, so it's a maximum. At $(\tfrac{1}{2}, \tfrac{31}{8})$ and $(-\tfrac{1}{2}, \tfrac{31}{8})$, $\dfrac{d^2y}{dx^2} = 4$, so they're minimums.

Q4 a) $y = 6(x + 2)(x - 3) = 6x^2 - 6x - 36$
$\dfrac{dy}{dx} = 12x - 6$. It's increasing when $\dfrac{dy}{dx} > 0$, i.e. when $12x - 6 > 0 \Rightarrow x > \dfrac{1}{2}$ and it's decreasing when $12x - 6 < 0 \Rightarrow x < \dfrac{1}{2}$.

b) $y = \dfrac{1}{x^2} = x^{-2}$, $\dfrac{dy}{dx} = -2x^{-3} = -\dfrac{2}{x^3}$
It's increasing when $-\dfrac{2}{x^3} > 0 \Rightarrow x < 0$ and it's decreasing when $-\dfrac{2}{x^3} < 0 \Rightarrow x > 0$.

Q5 First find where it crosses the axes:
When $x = 0$, $y = 0 - 0 = 0$. When $y = 0$,
$3x^3 - 16x = 0 \Rightarrow x(3x^2 - 16) = 0$, so $x = 0$ and
$x = \pm\dfrac{4}{\sqrt{3}}$ ($\approx \pm 2.3$).

Then differentiate to find the stationary points:

$\frac{dy}{dx} = 9x^2 - 16 = 0 \Rightarrow x^2 = \frac{16}{9} \Rightarrow x = \pm\frac{4}{3}$ ($\approx \pm1.3$)

When $x = \frac{4}{3}$, $y = -\frac{128}{9}$ (≈ -14.2) and when

$x = -\frac{4}{3}$, $y = \frac{128}{9}$ (≈ 14.2).

Differentiate again to find if they're maximum or minimum points:

$\frac{d^2y}{dx^2} = 18x$, so when $x = \frac{4}{3}$ it's a minimum and when $x = -\frac{4}{3}$ it's a maximum.

It's a positive cubic, so will go from bottom left to top right.

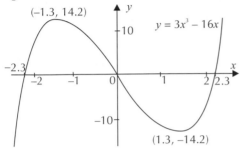

$y = 3x^3 - 16x$

Q6 First find where it crosses the axes:

When $x = 0$, $y = 0 + 0 = 0$.

When $y = 0$, $-3x^3 + 6x^2 = 0 \Rightarrow 3x^2(2 - x) = 0$, so $x = 0$ and $x = 2$.

Then differentiate to find the stationary points:

$\frac{dy}{dx} = -9x^2 + 12x = 0 \Rightarrow 3x(4 - 3x) = 0 \Rightarrow x = 0$ and $x = \frac{4}{3}$. When $x = 0$, $y = 0$ and when $x = \frac{4}{3}$, $y = \frac{32}{9}$.

Differentiate again to find if they're maximum or minimum points:

$\frac{d^2y}{dx^2} = -18x + 12$. At $(0, 0)$ $\frac{d^2y}{dx^2} = 12$, so it's a minimum. At $(\frac{4}{3}, \frac{32}{9})$ $\frac{d^2y}{dx^2} = -12$, so it's a maximum. It's a negative cubic, so will go from top left to bottom right.

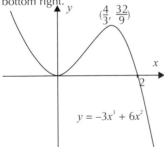

$y = -3x^3 + 6x^2$

Q7 $xy = 20 \Rightarrow y = \frac{20}{x} \Rightarrow f(x) = x^2 + \frac{400}{x^2}$

We want to minimise $f(x)$ so differentiate and find when $f'(x) = 0$. $f'(x) = 2x - \frac{800}{x^3} = 0 \Rightarrow x^4 = 400$ $\Rightarrow x^2 = 20 \Rightarrow x = \sqrt{20}$.

To find if it's a minimum, differentiate again: $f''(x) = 2 + \frac{2400}{x^4}$. When $x = \sqrt{20}$, $f''(x) = 8$, so it's a minimum. When $x = \sqrt{20}$, $y = \frac{20}{\sqrt{20}} = \sqrt{20}$.

So the least possible value of $x^2 + y^2$ is $(\sqrt{20})^2 + (\sqrt{20})^2 = 40$.

Q8 We need to find when h is at its maximum value, i.e. when $\frac{dh}{dm} = 0$, so differentiate:

$\frac{dh}{dm} = \frac{m}{5} - \frac{3m^2}{800} = 0 \Rightarrow \frac{1}{5}m(1 - \frac{3m}{160}) = 0$

So $m = 0$ and $m = \frac{160}{3}$. The answer can't be $m = 0$ or there'd be no fuel, so the max height is reached when $m = \frac{160}{3}$, $h = \frac{1}{10}(\frac{160}{3})^2 - \frac{1}{800}(\frac{160}{3})^3 = 94.8$ m (3 s.f.).

You can check that this is in fact the maximum by finding the second derivative when $m = \frac{160}{3}$.

Q9 a) There are 6 rectangular sides of the box which make up 3 identical pairs of sides. The total surface area is $A = 2(xy + 2x^2 + 2xy) = 4x^2 + 6xy$. To get rid of y, rearrange the formula for the volume of the box: $2x^2y = 200 \Rightarrow y = \frac{100}{x^2}$ so $A = 4x^2 + 6x(\frac{100}{x^2}) = 4x^2 + \frac{600}{x}$.

b) $\frac{dA}{dx} = 8x - \frac{600}{x^2} = 0 \Rightarrow x^3 = 75$

$\Rightarrow x = 4.217... = 4.22$ cm (3 s.f.)

Check this is a minimum by differentiating again:

$\frac{d^2A}{dx^2} = 8 + \frac{1200}{x^3}$. When $x = 4.217...$, $\frac{d^2A}{dx^2} = 24$, so it's a minimum.

c) Surface area $= 4(4.217...)^2 + \frac{600}{4.21...}$ $= 213$ cm^2 (3 s.f.).

Exam-Style Questions — Chapter 6

Q1 a) $y = 6 + \frac{4x^3 - 15x^2 + 12x}{6} = 6 + \frac{2}{3}x^3 - \frac{5}{2}x^2 + 2x$

$\frac{dy}{dx} = 2x^2 - 5x + 2$

[1 mark for each correct term]

b) Stationary points occur when $2x^2 - 5x + 2 = 0$ *[1 mark]*. Factorising the equation gives: $(2x - 1)(x - 2) = 0$, so stationary points occur when $x = 2$ *[1 mark]* and $x = \frac{1}{2}$ *[1 mark]*.

When $x = 2$:

$y = 6 + \frac{4(2^3) - 15(2^2) + (12 \times 2)}{6} = \frac{16}{3}$ *[1 mark]*

When $x = \frac{1}{2}$:

$y = 6 + \frac{4(\frac{1}{2})^3 - 15(\frac{1}{2})^2 + 12(\frac{1}{2})}{6} = \frac{155}{24}$ *[1 mark]*

So coordinates of the stationary points on the curve are $(2, \frac{16}{3})$ and $(\frac{1}{2}, \frac{155}{24})$.

c) Differentiate again to find $\frac{d^2y}{dy^2} = 4x - 5$ *[1 mark]*.

When $x = 2$ this gives $4(2) - 5 = 3$, which is positive, therefore the curve has a minimum at $(2, \frac{16}{3})$. *[1 mark]*

When $x = \frac{1}{2}$ this gives $4(\frac{1}{2}) - 5 = -3$, which is negative, so the maximum is at $(\frac{1}{2}, \frac{155}{24})$. *[1 mark]*

Q2 a) Find the value of x that gives the minimum value of y — the stationary point of the curve $y = 2\sqrt{x} + \frac{27}{x}$.

Differentiate, and then solve $\frac{dy}{dx} = 0$:

$\frac{dy}{dx} = \frac{1}{\sqrt{x}} - \frac{27}{x^2}$ *[1 mark for each term]*

$\frac{1}{\sqrt{x}} - \frac{27}{x^2} = 0$ *[1 mark]* $\Rightarrow \frac{1}{\sqrt{x}} = \frac{27}{x^2}$

$\frac{x^2}{x^{\frac{1}{2}}} = 27 \Rightarrow x^{\frac{3}{2}} = 27$ *[1 mark]*

$x = \sqrt[3]{27^2} = 9$ *[1 mark]*. So 9 miles an hour gives the minimum rate of coal consumption.

b) $\frac{d^2y}{dx^2} = \frac{54}{x^3} - \frac{1}{2\sqrt{x^3}}$ *[1 mark]*

At the stationary point $x = 9$,

so $\frac{54}{9^3} - \frac{1}{2\sqrt{9^3}} = 0.05555...$ which is positive,

therefore $x = 9$ does give a minimum value for y *[1 mark]*.

c) $y = 2\sqrt{9} + \frac{27}{9} = 9$ *[1 mark]*.

So the minimum rate of coal consumption is 9 units of coal per hour.

Q3 a) Surface area of the tank = sum of the areas of all 5 sides = $x^2 + x^2 + xy + xy + xy = 2x^2 + 3xy$ *[1 mark]*.

Volume of the tank = length × width × height $= x^2y = 40\,000$ cm³ *[1 mark]* $\Rightarrow y = \frac{40\,000}{x^2}$

Putting this into formula for the area,

$A = 2x^2 + 3xy = 2x^2 + 3x(\frac{40\,000}{x^2})$ *[1 mark]*

$= 2x^2 + \frac{120\,000}{x}$ *[1 mark]*

b) To find stationary points, first find $\frac{dA}{dx}$ *[1 mark]*.

$\frac{dA}{dx} = 4x - \frac{120\,000}{x^2}$ *[1 mark]*

Then find the value of x where $\frac{dA}{dx} = 0$

$\frac{dA}{dx} = 4x - \frac{120\,000}{x^2} = 0$ *[1 mark]* $\Rightarrow x^3 = 30\,000$

$\Rightarrow x = 31.07... = 31.1$ cm (3 s.f.) *[1 mark]*

To check if it's a minimum, find $\frac{d^2A}{dx^2}$

$\frac{d^2A}{dx^2} = 4 + \frac{240\,000}{x^3} = 12$ *[1 mark]*.

Second derivative is positive, so it's a minimum *[1 mark]*

c) Put the value of x found in part b) into the formula for the area given in part a)

$A = 2(31.07...)^2 + \frac{120\,000}{31.07...}$ *[1 mark]* $= 5792.936...$

$= 5790$ cm² (3 s.f.) *[1 mark]*

Q4 a) Surface area

$= [2 \times (d \times x)] + [2 \times (d \times \frac{x}{2})] + [x \times \frac{x}{2}]$

$= 2dx + \frac{2dx}{2} + \frac{x^2}{2}$ *[1 mark]*

Surface area = 72, so $3dx + \frac{x^2}{2} = 72$

$\Rightarrow x^2 + 6dx = 144$ *[1 mark]*

$d = \frac{144 - x^2}{6x}$ *[1 mark]*

Volume = width × height × depth

$= \frac{x}{2} \times x \times d = \frac{x^2}{2} \times \frac{144 - x^2}{6x} = \frac{144x^2 - x^4}{12x}$

$= 12x - \frac{x^3}{12}$ *[1 mark]*

b) Differentiate V and then solve for when $\frac{dV}{dx} = 0$:

$\frac{dV}{dx} = 12 - \frac{x^2}{4}$ *[1 mark for each correct term]*

$12 - \frac{x^2}{4} = 0$ *[1 mark]* $\Rightarrow \frac{x^2}{4} = 12 \Rightarrow x = 4\sqrt{3}$ m *[1 mark]*

c) $\frac{d^2V}{dx^2} = -\frac{x}{2}$ *[1 mark]* so when $x = 4\sqrt{3}$,

$\frac{d^2V}{dx^2} = -2\sqrt{3}$ *[1 mark]*. $\frac{d^2V}{dx^2}$ is negative, so it's a maximum value. *[1 mark]*

$x = 4\sqrt{3}$ at V_{max}, so $V_{max} = (12 \times 4\sqrt{3}) - \frac{(4\sqrt{3})^3}{12}$

$= 32\sqrt{3}$ m³ (55.4 m³ to 3 s.f.) *[1 mark]*

Q5 a) $f'(x) = 2x^3 - 3 = 0$ at the stationary point. *[1 mark]*

$2x^3 = 3 \Rightarrow x = 1.1447... = 1.145$ (3 d.p.) *[1 mark]*, which gives:

$y = f(1.1447...) = \frac{1}{2}(1.1447...)^4 - 3(1.1447...)$

$= -2.5756... = -2.576$ (3 d.p.) *[1 mark]*.

So coordinates of the stationary point are (1.145, −2.576).

b) $f''(x) = 6x^2$ *[1 mark]* so at the stationary point: $f''(1.1447...) = 7.862... = 7.86$ (3 s.f.), which is positive, so it's a minimum *[1 mark]*.

c) (i) As the stationary point is a minimum, $f'(x) > 0$ to the right of the stationary point. So the function is increasing when $x > 1.145$ *[1 mark]*.

(ii) As the stationary point is a minimum, $f'(x) < 0$ to the left of the stationary point. So the function is decreasing when $x < 1.145$ *[1 mark]*.

d) The curve intersects the x-axis when:

$y = \frac{1}{2}x^4 - 3x = 0 \Rightarrow x(\frac{1}{2}x^3 - 3) = 0$

$\Rightarrow x = 0$ or $x = \sqrt[3]{6} = 1.817$ (3 d.p.) *[1 mark]*

So the graph looks like this.

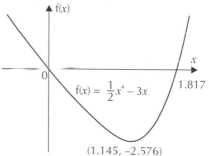

[1 mark]

Q6 a) First, multiply out the function to get

$y = 3x^3 - 8x^2 + 3x + 2$, so $\frac{dy}{dx} = 9x^2 - 16x + 3 = 0$ at the stationary point. *[1 mark]*

Solve using the quadratic formula:

$x = \frac{16 \pm \sqrt{(-16)^2 - (4 \times 9 \times 3)}}{2 \times 9} = \frac{16 \pm 2\sqrt{37}}{18}$

$x = 1.564...$ and $0.2130...$ *[1 mark]*. Substituting these values for x into the original equation for y gives: $y = -1.399...$ and $2.305...$

So the stationary points have coordinates (to 3 s.f.): (1.56, −1.40) *[1 mark]* and (0.213, 2.31) *[1 mark]*

b) $\dfrac{d^2y}{dx^2} = 18x - 16$ *[1 mark]*

At $x = 1.56$, $\dfrac{d^2y}{dx^2} = 12.1 > 0$, so it's a minimum *[1 mark]*

At $x = 0.213$, $\dfrac{d^2y}{dx^2} = -12.2 < 0$, so it's a maximum *[1 mark]*.

c) y is a positive cubic function, with stationary points as found in parts a) and b). The curve crosses the y-axis when $x = 0$, so $y = 2$. The initial cubic equation can be factorised to find where it intersects the x-axis:
$y = (x - 1)(3x^2 - 5x - 2)$
$y = (x - 1)(3x + 1)(x - 2)$
so $y = 0$ when $x = 1, -\dfrac{1}{3}$ and 2.
So the graph looks like this.

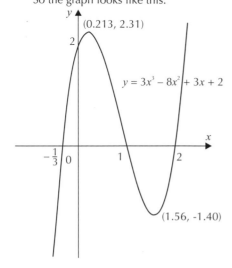

[3 marks available — 1 mark for shape and 1 mark for all x-intercepts, 1 mark for the y-intercept]

Chapter 7: Integration

1. Integration

Exercise 1.1 — Indefinite integrals

Q1 a) $\displaystyle\int 9x^8\,dx = \dfrac{9x^9}{9} + C = x^9 + C$

b) $\displaystyle\int 3x^2\,dx = \dfrac{3x^3}{3} + C = x^3 + C$

c) $\displaystyle\int (8x^7 + 2x)\,dx = \dfrac{8x^8}{8} + \dfrac{2x^2}{2} + C = x^8 + x^2 + C$

d) $\displaystyle\int (-8x^3 + 5x^2 - 3)\,dx = \dfrac{-8x^4}{4} + \dfrac{5x^3}{3} - 3x + C$
$= -2x^4 + \dfrac{5}{3}x^3 - 3x + C$

e) $\displaystyle\int (6x^2 - 2x^3)\,dx = \dfrac{6x^3}{3} - \dfrac{2x^4}{4} + C$
$= 2x^3 - \dfrac{x^4}{2} + C$

f) $\displaystyle\int (x^4 - x^2 - 3)\,dx = \dfrac{x^5}{5} - \dfrac{x^3}{3} - 3x + C$

g) $\displaystyle\int \left(\dfrac{3}{4}x^2 + 7x\right)dx = \left(\dfrac{3}{4} \times \dfrac{x^3}{3}\right) + \dfrac{7x^2}{2} + C$
$= \dfrac{x^3}{4} + \dfrac{7x^2}{2} + C$

h) $\displaystyle\int \left(\dfrac{1}{3}x^8 - \dfrac{1}{3}x^2 + 6\right)dx$
$= \left(\dfrac{1}{3} \times \dfrac{x^9}{9}\right) - \left(\dfrac{1}{3} \times \dfrac{x^3}{3}\right) + 6x + C$
$= \dfrac{x^9}{27} - \dfrac{x^3}{9} + 6x + C$

i) $\displaystyle\int -\dfrac{3}{x^4}\,dx = \displaystyle\int -3x^{-4}\,dx = \dfrac{-3x^{-3}}{-3} + C$
$= x^{-3} + C = \dfrac{1}{x^3} + C$

j) $\displaystyle\int \dfrac{1}{3}x^{-\frac{2}{3}}\,dx = \dfrac{\frac{1}{3}x^{\frac{1}{3}}}{\left(\frac{1}{3}\right)} + C = x^{\frac{1}{3}} + C$

k) $\displaystyle\int \sqrt{x}\,dx = \displaystyle\int x^{\frac{1}{2}}\,dx = \dfrac{x^{\frac{3}{2}}}{\left(\frac{3}{2}\right)} + C = \dfrac{2}{3}(\sqrt{x})^3 + C$

l) $\displaystyle\int \left(\dfrac{2}{x^3} + 2x\right)dx = \displaystyle\int (2x^{-3} + 2x)\,dx$
$= \dfrac{2x^{-2}}{-2} + \dfrac{2x^2}{2} + C$
$= -x^{-2} + x^2 + C$
$= -\dfrac{1}{x^2} + x^2 + C$

m) $\displaystyle\int (\sqrt{x})^3\,dx = \displaystyle\int x^{\frac{3}{2}}\,dx = \dfrac{x^{\frac{5}{2}}}{\left(\frac{5}{2}\right)} + C = \dfrac{2}{5}(\sqrt{x})^5 + C$

n) $\displaystyle\int (2x + 3)^2\,dx = \displaystyle\int (2x + 3)(2x + 3)\,dx$
$= \displaystyle\int (4x^2 + 12x + 9)\,dx$
$= \dfrac{4x^3}{3} + \dfrac{12x^2}{2} + 9x + C$
$= \dfrac{4}{3}x^3 + 6x^2 + 9x + C$

o) $\displaystyle\int \dfrac{x^5 + 7}{x^2}\,dx = \displaystyle\int \left(\dfrac{x^5}{x^2} + \dfrac{7}{x^2}\right)dx = \displaystyle\int (x^3 + 7x^{-2})\,dx$
$= \dfrac{x^4}{4} + \dfrac{7x^{-1}}{-1} + C = \dfrac{1}{4}x^4 - \dfrac{7}{x} + C$

p) $\displaystyle\int \dfrac{(3 - x)(2 + 3x)}{\sqrt{x}}\,dx$
$= \displaystyle\int \dfrac{6 + 7x - 3x^2}{x^{\frac{1}{2}}}\,dx$
$= \displaystyle\int (6x^{-\frac{1}{2}} + 7x^{\frac{1}{2}} - 3x^{\frac{3}{2}})\,dx$
$= \dfrac{6x^{\frac{1}{2}}}{\left(\frac{1}{2}\right)} + \dfrac{7x^{\frac{3}{2}}}{\left(\frac{3}{2}\right)} - \dfrac{3x^{\frac{5}{2}}}{\left(\frac{5}{2}\right)} + C$
$= 12\sqrt{x} + \dfrac{14}{3}(\sqrt{x})^3 - \dfrac{6}{5}(\sqrt{x})^5 + C$

Q2 $y = \displaystyle\int \dfrac{dy}{dx}\,dx = \displaystyle\int (5x^4 - 4x^3 + 5)\,dx$
$= \dfrac{5x^5}{5} - \dfrac{4x^4}{4} + 5x + C$
$= x^5 - x^4 + 5x + C$

The curve goes through $(-1, 1)$ so when $x = -1$, $y = 1$. Put these values into the equation above:
$y = x^5 - x^4 + 5x + C$
$\Rightarrow 1 = (-1)^5 - (-1)^4 + 5(-1) + C$
$\Rightarrow C = 8$
So $f(x) = x^5 - x^4 + 5x + 8$.

Q3 a) $y = \int \dfrac{dy}{dx}\, dx = \int 4\sqrt{x}\, dx = \int 4x^{\frac{1}{2}}\, dx$

$= \dfrac{4x^{\frac{3}{2}}}{\left(\frac{3}{2}\right)} + C = \dfrac{8}{3}x^{\frac{3}{2}} + C$

The curve goes through the origin, so when $x = 0$, $y = 0$.

$y = \dfrac{8}{3}x^{\frac{3}{2}} + C$

$\Rightarrow 0 = \dfrac{8}{3} \times 0^{\frac{3}{2}} + C$

$\Rightarrow C = 0$

So the equation of the curve is $y = \dfrac{8}{3}(\sqrt{x})^3$.

b) $y = \int \dfrac{dy}{dx}\, dx = \int \dfrac{4}{\sqrt[3]{x}}\, dx = \int 4x^{-\frac{1}{3}}\, dx$

$= \dfrac{4x^{\frac{2}{3}}}{\left(\frac{2}{3}\right)} + C = 6x^{\frac{2}{3}} + C$

The curve goes through (1, 2), so when $x = 1$, $y = 2$.

$y = 6x^{\frac{2}{3}} + C$

$\Rightarrow 2 = 6 \times 1^{\frac{2}{3}} + C$

$\Rightarrow C = -4$

So the equation of the curve is $y = 6(\sqrt[3]{x})^2 - 4$.

Exercise 1.2 — Definite integrals

Q1 a) $\displaystyle\int_1^3 3x^2\, dx = [x^3]_1^3 = (3^3) - (1^3) = 27 - 1 = 26$

b) $\displaystyle\int_{-2}^0 (4x^3 + 2x)\, dx = [x^4 + x^2]_{-2}^0$

$= (0^4 + 0^2) - ((-2)^4 + (-2)^2)$

$= -(16 + 4) = -20$

c) $\displaystyle\int_{-2}^5 (x^3 + x)\, dx = \left[\dfrac{x^4}{4} + \dfrac{x^2}{2}\right]_{-2}^5$

$= \left(\dfrac{5^4}{4} + \dfrac{5^2}{2}\right) - \left(\dfrac{(-2)^4}{4} + \dfrac{(-2)^2}{2}\right)$

$= \dfrac{625}{4} + \dfrac{25}{2} - \dfrac{16}{4} - \dfrac{4}{2} = \dfrac{651}{4}$

d) $\displaystyle\int_{-5}^{-2} (x + 1)^2\, dx = \int_{-5}^{-2} (x^2 + 2x + 1)\, dx$

$= \left[\dfrac{x^3}{3} + x^2 + x\right]_{-5}^{-2}$

$= \left(\dfrac{(-2)^3}{3} + (-2)^2 + (-2)\right)$

$\quad - \left(\dfrac{(-5)^3}{3} + (-5)^2 + (-5)\right)$

$= \left(\dfrac{-8}{3} + 4 - 2\right) - \left(\dfrac{-125}{3} + 25 - 5\right)$

$= 21$

e) $\displaystyle\int_1^4 x^{-2}\, dx = \left[\dfrac{x^{-1}}{-1}\right]_1^4 = \left[-\dfrac{1}{x}\right]_1^4$

$= \left(-\dfrac{1}{4}\right) - \left(-\dfrac{1}{1}\right) = \dfrac{3}{4}$

f) $\displaystyle\int_2^7 (x^{-3} + x)\, dx = \left[\dfrac{x^{-2}}{-2} + \dfrac{x^2}{2}\right]_2^7 = \left[-\dfrac{1}{2x^2} + \dfrac{x^2}{2}\right]_2^7$

$= \left(-\dfrac{1}{2(7^2)} + \dfrac{7^2}{2}\right) - \left(-\dfrac{1}{2(2^2)} + \dfrac{2^2}{2}\right)$

$= \left(-\dfrac{1}{98} + \dfrac{49}{2}\right) - \left(-\dfrac{1}{8} + \dfrac{4}{2}\right) = \dfrac{8865}{392}$

g) $\displaystyle\int_3^4 (6x^{-4} + x^{-2})\, dx = \left[\dfrac{6x^{-3}}{-3} + \dfrac{x^{-1}}{-1}\right]_3^4$

$= \left[-\dfrac{2}{x^3} - \dfrac{1}{x}\right]_3^4 = \left(-\dfrac{2}{4^3} - \dfrac{1}{4}\right) - \left(-\dfrac{2}{3^3} - \dfrac{1}{3}\right)$

$= -\dfrac{2}{64} - \dfrac{1}{4} + \dfrac{2}{27} + \dfrac{1}{3} = \dfrac{109}{864}$

h) $\displaystyle\int_1^2 \left(x^2 + \dfrac{1}{x^2}\right) dx = \int_1^2 (x^2 + x^{-2})\, dx = \left[\dfrac{x^3}{3} + \dfrac{x^{-1}}{-1}\right]_1^2$

$= \left[\dfrac{x^3}{3} - \dfrac{1}{x}\right]_1^2 = \left(\dfrac{2^3}{3} - \dfrac{1}{2}\right) - \left(\dfrac{1^3}{3} - \dfrac{1}{1}\right)$

$= \dfrac{8}{3} - \dfrac{1}{2} - \dfrac{1}{3} + 1 = \dfrac{17}{6}$

Q2 $\displaystyle\int_0^a x^3\, dx = \left[\dfrac{x^4}{4}\right]_0^a = \left(\dfrac{a^4}{4}\right) - \left(\dfrac{0^4}{4}\right) = \dfrac{a^4}{4}$

So $\dfrac{a^4}{4} = 64$

$\Rightarrow a^4 = 64 \times 4 = 256$

$\Rightarrow a = 4$

a can't be −4 since the question tells you that a > 0.

Q3 a) $\displaystyle\int_0^1 \sqrt{x}\, dx = \int_0^1 x^{\frac{1}{2}}\, dx = \left[\dfrac{x^{\frac{3}{2}}}{\left(\frac{3}{2}\right)}\right]_0^1 = \left[\dfrac{2}{3}(\sqrt{x})^3\right]_0^1$

$= \left(\dfrac{2}{3}(\sqrt{1})^3\right) - \left(\dfrac{2}{3}(\sqrt{0})^3\right) = \dfrac{2}{3} - 0 = \dfrac{2}{3}$

b) $\displaystyle\int_8^{27} \sqrt[3]{x}\, dx = \int_8^{27} x^{\frac{1}{3}}\, dx = \left[\dfrac{x^{\frac{4}{3}}}{\left(\frac{4}{3}\right)}\right]_8^{27} = \left[\dfrac{3}{4}(\sqrt[3]{x})^4\right]_8^{27}$

$= \left(\dfrac{3}{4}(\sqrt[3]{27})^4\right) - \left(\dfrac{3}{4}(\sqrt[3]{8})^4\right) = \left(\dfrac{3}{4} \times 3^4\right) - \left(\dfrac{3}{4} \times 2^4\right)$

$= \dfrac{195}{4}$

You can take the constants outside the square brackets if you find it easier to work with.

c) $\displaystyle\int_0^9 (x^2 + \sqrt{x})\, dx = \int_0^9 (x^2 + x^{\frac{1}{2}})\, dx = \left[\dfrac{x^3}{3} + \dfrac{x^{\frac{3}{2}}}{\left(\frac{3}{2}\right)}\right]_0^9$

$= \left[\dfrac{x^3}{3} + \dfrac{2}{3}(\sqrt{x})^3\right]_0^9$

$= \left(\dfrac{9^3}{3} + \dfrac{2}{3}(\sqrt{9})^3\right) - \left(\dfrac{0^3}{3} + \dfrac{2}{3}(\sqrt{0})^3\right)$

$= \left(\dfrac{729}{3} + \left(\dfrac{2}{3} \times 3^3\right)\right) - 0 = 261$

d) $\displaystyle\int_1^4 (3x^{-4} + \sqrt{x})\, dx = \int_1^4 (3x^{-4} + x^{\frac{1}{2}})\, dx$

$= \left[\dfrac{3x^{-3}}{-3} + \dfrac{x^{\frac{3}{2}}}{\left(\frac{3}{2}\right)}\right]_1^4 = \left[-\dfrac{1}{x^3} + \dfrac{2}{3}(\sqrt{x})^3\right]_1^4$

$= \left(-\dfrac{1}{4^3} + \dfrac{2}{3}(\sqrt{4})^3\right) - \left(-\dfrac{1}{1^3} + \dfrac{2}{3}(\sqrt{1})^3\right)$

$= \left(-\dfrac{1}{64} + \dfrac{2}{3} \times 2^3\right) - \left(-1 + \dfrac{2}{3}\right)$

$= \dfrac{1085}{192}$

e) $\displaystyle\int_0^1 (2x + 3)(x + 2)\, dx = \int_0^1 (2x^2 + 7x + 6)\, dx$

$= \left[\dfrac{2x^3}{3} + \dfrac{7x^2}{2} + 6x\right]_0^1$

$= \left(\dfrac{2 \times 1^3}{3} + \dfrac{7 \times 1^2}{2} + (6 \times 1)\right)$

$\quad - \left(\dfrac{2 \times 0^3}{3} + \dfrac{7 \times 0^2}{2} + (6 \times 0)\right)$

$= \left(\dfrac{2}{3} + \dfrac{7}{2} + 6\right) - 0 = \dfrac{61}{6}$

f) $\int_1^4 \frac{1}{\sqrt{x}} \, dx = \int_1^4 x^{-\frac{1}{2}} \, dx = \left[\frac{x^{\frac{1}{2}}}{\left(\frac{1}{2}\right)}\right]_1^4 = \left[2\sqrt{x}\right]_1^4$

$= (2\sqrt{4}) - (2\sqrt{1}) = 4 - 2 = 2$

Q4 a) $\int_1^4 \frac{x^2 + 2}{\sqrt{x}} \, dx = \int_1^4 (x^{\frac{3}{2}} + 2x^{-\frac{1}{2}}) \, dx$

$= \left[\frac{x^{\frac{5}{2}}}{\left(\frac{5}{2}\right)} + 2\frac{x^{\frac{1}{2}}}{\left(\frac{1}{2}\right)}\right]_1^4 = \left[\frac{2}{5}(\sqrt{x})^5 + 4\sqrt{x}\right]_1^4$

$= \left(\frac{2}{5}(\sqrt{4})^5 + 4\sqrt{4}\right) - \left(\frac{2}{5}(\sqrt{1})^5 + 4\sqrt{1}\right)$

$= \left(\frac{2}{5} \times 2^5 + 8\right) - \left(\frac{2}{5} + 4\right)$

$= \frac{64}{5} + 8 - \frac{2}{5} - 4 = \frac{82}{5}$

b) $\int_0^1 (\sqrt{x} + 1)^2 \, dx = \int_0^1 (x + 2\sqrt{x} + 1) \, dx$

$= \int_0^1 (x + 2x^{\frac{1}{2}} + 1) \, dx = \left[\frac{x^2}{2} + \frac{2x^{\frac{3}{2}}}{\left(\frac{3}{2}\right)} + x\right]_0^1$

$= \left[\frac{x^2}{2} + \frac{4}{3}(\sqrt{x})^3 + x\right]_0^1$

$= \left(\frac{1^2}{2} + \frac{4}{3}(\sqrt{1})^3 + 1\right) - \left(\frac{0^2}{2} + \frac{4}{3}(\sqrt{0})^3 + 0\right)$

$= \left(\frac{1}{2} + \frac{4}{3} + 1\right) - 0 = \frac{17}{6}$

c) $\int_4^9 \left(\frac{1}{x} + \sqrt{x}\right)^2 \, dx = \int_4^9 \left(\frac{1}{x^2} + 2\frac{\sqrt{x}}{x} + x\right) \, dx$

$= \int_4^9 (x^{-2} + 2x^{-\frac{1}{2}} + x) \, dx$

$= \left[\frac{x^{-1}}{-1} + \frac{2x^{\frac{1}{2}}}{\left(\frac{1}{2}\right)} + \frac{x^2}{2}\right]_4^9$

$= \left[-\frac{1}{x} + 4\sqrt{x} + \frac{x^2}{2}\right]_4^9$

$= \left(-\frac{1}{9} + 4\sqrt{9} + \frac{9^2}{2}\right) - \left(-\frac{1}{4} + 4\sqrt{4} + \frac{4^2}{2}\right)$

$= -\frac{1}{9} + 12 + \frac{81}{2} + \frac{1}{4} - 8 - 8$

$= \frac{1319}{36}$

Q5 a) The area is all above the x-axis so just integrate:

$\int_1^3 (x^3 + x) \, dx = \left[\frac{x^4}{4} + \frac{x^2}{2}\right]_1^3$

$= \left(\frac{3^4}{4} + \frac{3^2}{2}\right) - \left(\frac{1^4}{4} + \frac{1^2}{2}\right)$

$= \left(\frac{81}{4} + \frac{9}{2}\right) - \left(\frac{1}{4} + \frac{1}{2}\right)$

$= 24$

So the area is 24.

b) Again, the area is above the x-axis:

$\int_0^4 (x + \sqrt{x}) \, dx = \int_0^4 (x + x^{\frac{1}{2}}) \, dx$

$= \left[\frac{x^2}{2} + \frac{x^{\frac{3}{2}}}{\left(\frac{3}{2}\right)}\right]_0^4 = \left[\frac{1}{2}x^2 + \frac{2}{3}(\sqrt{x})^3\right]_0^4$

$= \left(\frac{1}{2}4^2 + \frac{2}{3}(\sqrt{4})^3\right) - \left(\frac{1}{2}0^2 + \frac{2}{3}(\sqrt{0})^3\right)$

$= \left(\frac{16}{2} + \frac{2}{3} \times 2^3\right) - 0$

$= \frac{40}{3}$

c) The limits aren't shown on the graph, but they are just the roots of the equation $0 = 4 - x^2$.

Set $y = 0$:

$4 - x^2 = 0 \Rightarrow x^2 = 4 \Rightarrow x = 2$ or -2.

So the limits of integration are -2 and 2:

$\int_{-2}^2 (4 - x^2) \, dx = \left[4x - \frac{x^3}{3}\right]_{-2}^2$

$= \left((4 \times 2) - \frac{2^3}{3}\right) - \left((4 \times (-2)) - \frac{(-2)^3}{3}\right)$

$= \left(8 - \frac{8}{3}\right) - \left(-8 - \frac{-8}{3}\right)$

$= 8 - \frac{8}{3} + 8 - \frac{8}{3}$

$= \frac{32}{3}$

d) This area lies above and below the x-axis so you'll have to integrate the bits above and below the axis separately.

First you need to find the points where the curve crosses the axis: $y = x(x - 1)(x - 3)$ is already factorised, so it's easy.

If $x(x - 1)(x - 3) = 0$ then either $x = 0$, $x = 1$ or $x = 3$. So these are the three points where the curve crosses the axis.

The area above the x-axis is between 0 and 1 so integrate:

$\int_0^1 x(x - 1)(x - 3) \, dx = \int_0^1 (x^3 - 4x^2 + 3x) \, dx$

$= \left[\frac{x^4}{4} - \frac{4x^3}{3} + \frac{3x^2}{2}\right]_0^1$

$= \left(\frac{1^4}{4} - \frac{4 \times 1^3}{3} + \frac{3 \times 1^2}{2}\right)$

$- \left(\frac{0^4}{4} - \frac{4 \times 0^3}{3} + \frac{3 \times 0^2}{2}\right)$

$= \frac{1}{4} - \frac{4}{3} + \frac{3}{2} - 0 = \frac{5}{12}$

The area below the x-axis is between 1 and 3 so integrate:

$\int_1^3 x(x - 1)(x - 3) \, dx = \int_1^3 (x^3 - 4x^2 + 3x) \, dx$

$= \left[\frac{x^4}{4} - \frac{4x^3}{3} + \frac{3x^2}{2}\right]_1^3$

$= \left(\frac{3^4}{4} - \frac{4 \times 3^3}{3} + \frac{3 \times 3^2}{2}\right) - \left(\frac{1}{4} - \frac{4}{3} + \frac{3}{2}\right)$

$= \left(\frac{81}{4} - \frac{108}{3} + \frac{27}{2}\right) - \frac{5}{12} = -\frac{8}{3}$

Areas cannot be negative so the area of the bit below the x-axis is $\frac{8}{3}$.

So the total area is $\frac{5}{12} + \frac{8}{3} = \frac{37}{12}$.

Q6 If x is positive then $y = x^2 + x$ is also positive so y is positive between 1 and 3 and so you can just integrate normally between the limits.

$\int_1^3 (x^2 + x) \, dx = \left[\frac{x^3}{3} + \frac{x^2}{2}\right]_1^3$

$= \left(\frac{3^3}{3} + \frac{3^2}{2}\right) - \left(\frac{1^3}{3} + \frac{1^2}{2}\right) = \left(\frac{27}{3} + \frac{9}{2}\right) - \left(\frac{1}{3} + \frac{1}{2}\right)$

$= \frac{38}{3}$

Q7 $y = 5x^3$ is negative for all negative values of x, and so it lies under the x-axis between $x = -5$ and $x = -3$.

So just integrate to find the 'negative' area and make it positive:

$$\int_{-5}^{-3} 5x^3 \, dx = \left[\frac{5x^4}{4}\right]_{-5}^{-3} = \left(\frac{5(-3)^4}{4}\right) - \left(\frac{5(-5)^4}{4}\right)$$

$$= \frac{405}{4} - \frac{3125}{4} = -680$$

So the area is 680.

Q8 The graph of $y = (x - 1)(3x + 9)$ crosses the x-axis at $x = 1$ and $x = -3$ so between $x = -2$ and $x = 2$, the graph crosses the axes at $x = 1$ which means the area lies both above and below the x-axis.

It might help to sketch a graph, but you don't really need to know which area is positive and which is negative — doing the integration will tell you which is which. Just integrate between $x = -2$ and $x = 1$ and then between $x = 1$ and $x = 2$.

Work out the area between $x = -2$ and $x = 1$:

$$\int_{-2}^{1}(x - 1)(3x + 9) \, dx = \int_{-2}^{1}(3x^2 + 6x - 9) \, dx$$

$$= [x^3 + 3x^2 - 9x]_{-2}^{1}$$

$$= ((1)^3 + 3(1)^2 - 9(1)) - ((-2)^3 + 3(-2)^2 - 9(-2))$$

$$= -5 - 22 = -27$$

So the area below the x-axis is 27.

Now work out the area between $x = 1$ and $x = 2$.

$$\int_{1}^{2}(x - 1)(3x + 9) \, dx = [x^3 + 3x^2 - 9x]_{1}^{2}$$

$$= ((2)^3 + 3(2)^2 - 9(2)) - ((1)^3 + 3(1)^2 - 9(1))$$

$$= 2 - (-5) = 7$$

So the area above the x-axis is 7.

Therefore the total area is $27 + 7 = 34$.

Q9 $y = \frac{20}{x^5}$ is always positive between $x = 1$ and $x = 2$ so just integrate:

$$\int_{1}^{2} \frac{20}{x^5} \, dx = \int_{1}^{2} 20x^{-5} \, dx = \left[\frac{20x^{-4}}{-4}\right]_{1}^{2} = \left[-\frac{5}{x^4}\right]_{1}^{2}$$

$$= \left(-\frac{5}{2^4}\right) - \left(-\frac{5}{1^4}\right) = -\frac{5}{16} + 5 = \frac{75}{16}$$

So the area is $\frac{75}{16} = 4.6875$

Q10 a) The area is the integral of $\frac{1}{x^2}$ between a and ∞: $\int_{a}^{\infty} \frac{1}{x^2} \, dx$.

To find this integral, replace ∞ with n:

$$\int_{a}^{n} \frac{1}{x^2} \, dx = \int_{a}^{n} x^{-2} \, dx = [-x^{-1}]_{a}^{n} = \left[-\frac{1}{x}\right]_{a}^{n}$$

$$= -\frac{1}{n} - \left(-\frac{1}{a}\right) = -\frac{1}{n} + \frac{1}{a}$$

As $n \to \infty$, $\frac{1}{n} \to 0$

so $-\frac{1}{n} + \frac{1}{a} \to 0 + \frac{1}{a} = \frac{1}{a}$ as $n \to \infty$.

So the shaded region has area $\int_{a}^{\infty} \frac{1}{x^2} \, dx = \frac{1}{a}$.

b) To find the shaded region you need to find the value of the integral of $18\left(\frac{2}{x^3} - \frac{1}{x^2}\right)$ between 3 and ∞: $\int_{3}^{\infty} 18\left(\frac{2}{x^3} - \frac{1}{x^2}\right) dx$.

Replace ∞ with n and integrate:

$$\int_{3}^{n} 18\left(\frac{2}{x^3} - \frac{1}{x^2}\right) dx = 18\int_{3}^{n}(2x^{-3} - x^{-2}) \, dx$$

$$= 18\left[\frac{2x^{-2}}{-2} - \frac{x^{-1}}{-1}\right]_{3}^{n} = 18\left[-\frac{1}{x^2} + \frac{1}{x}\right]_{3}^{n}$$

$$= 18\left(\left(-\frac{1}{n^2} + \frac{1}{n}\right) - \left(-\frac{1}{9} + \frac{1}{3}\right)\right)$$

$$= 18\left(-\frac{1}{n^2} + \frac{1}{n} - \frac{2}{9}\right)$$

As $n \to \infty$, $\frac{1}{n} \to 0$ and $\frac{1}{n^2} \to 0$

so $18\left(-\frac{1}{n^2} + \frac{1}{n} - \frac{2}{9}\right) \to 18\left(-0 + 0 - \frac{2}{9}\right) = -4$

as $n \to \infty$.

So the shaded region has area $\int_{3}^{\infty} 18\left(\frac{2}{x^3} - \frac{1}{x^2}\right) dx = -4$.

But area must be positive so the area is 4.

Q11 a) $\int_{1}^{n} \frac{3}{x^2} \, dx = \int_{1}^{n} 3x^{-2} \, dx = \left[\frac{3x^{-1}}{-1}\right]_{1}^{n} = \left[-\frac{3}{x}\right]_{1}^{n}$

$$= \left(-\frac{3}{n}\right) - \left(-\frac{3}{1}\right) = -\frac{3}{n} + 3$$

As $n \to \infty$, $\frac{3}{n} \to 0$ so $-\frac{3}{n} + 3 \to 0 + 3 = 3$

So $\int_{1}^{\infty} \frac{3}{x^2} \, dx = 3$

b) $\int_{3}^{n}\left(\frac{2}{x^2} + \frac{7}{x^3}\right) dx = \int_{3}^{n}(2x^{-2} + 7x^{-3}) \, dx$

$$= \left[\frac{2x^{-1}}{-1} + \frac{7x^{-2}}{-2}\right]_{3}^{n} = \left[-\frac{2}{x} - \frac{7}{2x^2}\right]_{3}^{n}$$

$$= \left(-\frac{2}{n} - \frac{7}{2n^2}\right) - \left(-\frac{2}{3} - \frac{7}{18}\right)$$

$$= -\frac{2}{n} - \frac{7}{2n^2} + \frac{19}{18}$$

As $n \to \infty$, $\frac{2}{n} \to 0$ and $\frac{7}{2n^2} \to 0$ so

$$-\frac{2}{n} - \frac{7}{2n^2} + \frac{19}{18} \to -0 - 0 + \frac{19}{18} = \frac{19}{18}$$

So the integral $\int_{3}^{\infty}\left(\frac{2}{x^2} + \frac{7}{x^3}\right) dx = \frac{19}{18}$

c) $\int_{-n}^{-2}\left(\frac{3}{x^4} + \frac{5}{x^3}\right) dx = \int_{-n}^{-2}(3x^{-4} + 5x^{-3}) \, dx$

$$= \left[\frac{3x^{-3}}{-3} + \frac{5x^{-2}}{-2}\right]_{-n}^{-2} = \left[-\frac{1}{x^3} - \frac{5}{2x^2}\right]_{-n}^{-2}$$

$$= \left(-\frac{1}{(-2)^3} - \frac{5}{2(-2)^2}\right) - \left(-\frac{1}{n^3} - \frac{5}{2n^2}\right)$$

$$= -\frac{1}{2} + \frac{1}{n^3} + \frac{5}{2n^2}$$

As $n \to -\infty$, $\frac{1}{n^3} \to 0$ and $\frac{5}{2n^2} \to 0$ so

$$-\frac{1}{2} + \frac{1}{n^3} + \frac{5}{2n^2} \to -\frac{1}{2} + 0 + 0 = -\frac{1}{2}$$

So the integral $\int_{-\infty}^{-2}\left(\frac{3}{x^4} + \frac{5}{x^3}\right) dx = -\frac{1}{2}$

Exercise 1.3 — Finding the area between a curve and a line

Q1 a) Start by finding the points where the curve and the line intersect. Solve $3x^2 + 4 = 16$:

$$\Rightarrow 3x^2 = 12 \Rightarrow x^2 = 4 \Rightarrow x = -2 \text{ or } 2.$$

So they intersect at $x = -2$ and $x = 2$.

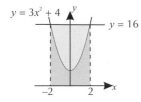

So the area is found by subtracting the integral of $3x^2 + 4$ between -2 and 2 from the integral of 16 between -2 and 2.

The area under the line is just a rectangle which is 16 by 4 so the area is $16 \times 4 = 64$.

$$\int_{-2}^{2} (3x^2 + 4)\,dx = [x^3 + 4x]_{-2}^{2}$$
$$= (2^3 + 4(2)) - ((-2)^3 + 4(-2))$$
$$= (8 + 8) - (-8 - 8)$$
$$= 16 + 16 = 32$$

So the area is $64 - 32 = 32$.

b) This might not look like a line and a curve at first, but if $x = 2$, $y = x^3 + 4 = 12$ so it's actually this area that you're after:

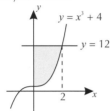

So you must find the area under the line from $x = 0$ and $x = 2$ minus the integral of the curve between from $x = 0$ and $x = 2$.

The first area is just a rectangle which is 12 by 2 so the area is $12 \times 2 = 24$.

Area under curve $= \int_{0}^{2} (x^3 + 4)\,dx$

$$= \left[\frac{x^4}{4} + 4x\right]_{0}^{2}$$
$$= \left(\frac{2^4}{4} + 4(2)\right) - \left(\frac{0^4}{4} + 4(0)\right)$$
$$= (4 + 8) - 0 = 12$$

So the area is $24 - 12 = 12$.

c) Solving to find the point of intersection of $y = 4$ and $y = \frac{1}{x^2}$: $4 = \frac{1}{x^2} \Rightarrow 4x^2 = 1 \Rightarrow x^2 = \frac{1}{4}$

$\Rightarrow x = \pm\frac{1}{2}$ so the intersection in the diagram is $x = \frac{1}{2}$:

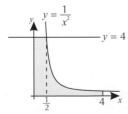

So required area is the area under the line $y = 4$ from $x = 0$ and $x = \frac{1}{2}$ added to the area under the curve $y = \frac{1}{x^2}$ from $x = \frac{1}{2}$ and $x = 4$.

The first area is just a rectangle which is $\frac{1}{2}$ by 4 so the area is $\frac{1}{2} \times 4 = 2$.

Area under curve $= \int_{\frac{1}{2}}^{4} \frac{1}{x^2}\,dx = \int_{\frac{1}{2}}^{4} x^{-2}\,dx$

$$= \left[\frac{x^{-1}}{-1}\right]_{\frac{1}{2}}^{4} = \left[-\frac{1}{x}\right]_{\frac{1}{2}}^{4}$$
$$= \left(-\frac{1}{4}\right) - \left(-\frac{1}{\left(\frac{1}{2}\right)}\right)$$
$$= -\frac{1}{4} + 2 = \frac{7}{4}$$

So the area is $2 + \frac{7}{4} = \frac{15}{4}$.

d) Solve $-1 - (x - 3)^2 = -5 \Rightarrow 4 = (x - 3)^2$
$\Rightarrow x - 3 = -2$ or $2 \Rightarrow x = 1$ or 5.
Be careful — this area is below the x-axis so the integrals will give negative areas.

So the area you want is the area above the line between 1 and 5 minus the area between the curve and the x-axis between 1 and 5.

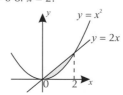

The first area is just a rectangle which is 5 by 4 so the area is $5 \times 4 = 20$.

So the area between the line and the x-axis between 0 and 5 is 20.

Area between curve and x–axis
$$= \int_{1}^{5} (-1 - (x - 3)^2)\,dx$$
$$= \int_{1}^{5} (-1 - (x^2 - 6x + 9))\,dx$$
$$= \int_{1}^{5} (6x - x^2 - 10)\,dx$$
$$= \left[\frac{6x^2}{2} - \frac{x^3}{3} - 10x\right]_{1}^{5} = \left[3x^2 - \frac{x^3}{3} - 10x\right]_{1}^{5}$$
$$= \left((3 \times 5^2) - \frac{5^3}{3} - (10 \times 5)\right)$$
$$\quad - \left((3 \times 1^2) - \frac{1^3}{3} - (10 \times 1)\right)$$
$$= \left(75 - \frac{125}{3} - 50\right) - \left(3 - \frac{1}{3} - 10\right)$$
$$= -\frac{28}{3}$$

So the area between the curve and the x-axis between 1 and 5 is $\frac{28}{3}$.

So the area is $20 - \frac{28}{3} = \frac{32}{3}$.

e) Solve $x^2 = 2x \Rightarrow x^2 - 2x = 0 \Rightarrow x(x - 2) = 0$
$\Rightarrow x = 0$ or $x = 2$.

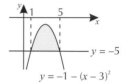

So the area is the area under the line between 0 and 2, minus the area under the curve between 0 and 2.

$$\int_0^2 2x \, dx = [x^2]_0^2 = (2^2) - (0^2) = 4$$

$$\int_0^2 x^2 \, dx = \left[\frac{x^3}{3}\right]_0^2 = \left(\frac{2^3}{3}\right) - \left(\frac{0^3}{3}\right) = \frac{8}{3}$$

So the area is $4 - \frac{8}{3} = \frac{4}{3}$.

f) Solve $x^3 = 4x \Rightarrow x^3 - 4x = 0 \Rightarrow x(x^2 - 4) = 0$

$\Rightarrow x(x + 2)(x - 2) = 0 \Rightarrow x = 0$ or $x = \pm 2$.

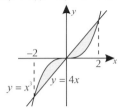

The area above the x-axis is the same as the area below the x-axis (from the symmetry of the graph).

For the area above the x-axis, find the area under the line from $x = 0$ to $x = 2$, minus the area under the curve from $x = 0$ to $x = 2$.

$$\int_0^2 4x \, dx = [2x^2]_0^2 = (2 \times 2^2) - (2 \times 0^2) = 8$$

$$\int_0^2 x^3 \, dx = \left[\frac{x^4}{4}\right]_0^2 = \left(\frac{2^4}{4}\right) - \left(\frac{0^2}{4}\right) = 4$$

So the area above the axis is $8 - 4 = 4$.

So the area below the x-axis is also 4.

The total area is $4 + 4 = 8$

For parts e) and f) you could have used the formula for finding the area of a triangle rather than integrating.

Q2 a) Solve $x^2 + 4 = x + 4 \Rightarrow x^2 - x = 0 \Rightarrow x(x - 1) = 0$
$\Rightarrow x = 0$ or $x = 1$.

Draw a diagram:

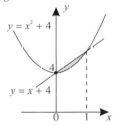

So the area is the area under the line from $x = 0$ and $x = 1$ minus the area under the curve from $x = 0$ and $x = 1$.

$$\int_0^1 (x + 4) \, dx = \left[\frac{x^2}{2} + 4x\right]_0^1$$

$$= \left(\frac{1^2}{2} + (4 \times 1)\right) - \left(\frac{0^2}{2} + (4 \times 0)\right)$$

$$= \frac{1}{2} + 4 = \frac{9}{2}$$

$$\int_0^1 (x^2 + 4) \, dx = \left[\frac{x^3}{3} + 4x\right]_0^1$$

$$= \left(\frac{1^3}{3} + (4 \times 1)\right) - \left(\frac{0^3}{3} + (4 \times 0)\right)$$

$$= \frac{1}{3} + 4 = \frac{13}{3}$$

So the area is $\frac{9}{2} - \frac{13}{3} = \frac{1}{6}$.

b) Solve $3x^2 + 11x + 6 = 9x + 6 \Rightarrow 3x^2 + 2x = 0$
$\Rightarrow x(3x + 2) = 0 \Rightarrow x = 0$ or $x = -\frac{2}{3}$.

Finding the coordinates of the points where the curve and the line meet will help you to sketch a diagram.

When $x = 0$, $y = 6$ and when $x = -\frac{2}{3}$, $y = 0$ so the curve meets the line at $(0, 6)$ and $(-\frac{2}{3}, 0)$.

Draw a diagram:

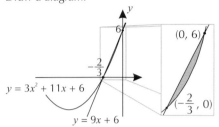

The shaded area is the area under the line from $x = -\frac{2}{3}$ and $x = 0$ minus the area under the curve from $x = -\frac{2}{3}$ and $x = 0$.

To find the area under the line, just find the area of the triangle with base $\frac{2}{3}$ and height 6:

$$A = \frac{1}{2} \times \frac{2}{3} \times 6 = 2$$

The area under the curve is given by:

$$\int_{-\frac{2}{3}}^{0} (3x^2 + 11x + 6) \, dx = \left[x^3 + \frac{11x^2}{2} + 6x\right]_{-\frac{2}{3}}^{0}$$

$$= \left(0^3 + \frac{11 \times 0^2}{2} + 6 \times 0\right)$$

$$- \left(\left(-\frac{2}{3}\right)^3 + \frac{11\left(-\frac{2}{3}\right)^2}{2} + 6\left(-\frac{2}{3}\right)\right)$$

$$= 0 - \left(-\frac{8}{27} + \frac{22}{9} - 4\right) = \frac{50}{27}$$

So the area is $2 - \frac{50}{27} = \frac{4}{27}$.

c) Solve $x^2 + 2x - 3 = 4x \Rightarrow x^2 - 2x - 3 = 0$
$\Rightarrow (x - 3)(x + 1) = 0 \Rightarrow x = -1$ or $x = 3$.

It'll also help to find where the curve meets the x-axis by solving $x^2 + 2x - 3 = 0 \Rightarrow (x + 3)(x - 1)$
$\Rightarrow x = 1$ or $x = -3$.

Draw a diagram:

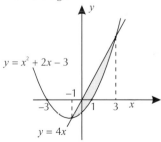

You'll need to find the areas above and below the x-axis separately.

For the area above the x-axis, integrate $4x$ from $x = 0$ to $x = 3$ and subtract the integral of $x^2 + 2x - 3$ from $x = 1$ to $x = 3$.

$$\int_0^3 4x\,dx = [2x^2]_0^3 = (2(3)^2) - (2(0)^2) = 18$$

$$\int_1^3 (x^2 + 2x - 3)\,dx = \left[\frac{x^3}{3} + x^2 - 3x\right]_1^3$$

$$= \left(\frac{(3)^3}{3} + (3)^2 - 3(3)\right) - \left(\frac{(1)^3}{3} + (1)^2 - 3(1)\right)$$

$$= \left(\frac{27}{3} + 9 - 9\right) - \left(\frac{1}{3} + 1 - 3\right) = \frac{32}{3}$$

So the area above the x-axis is $18 - \frac{32}{3} = \frac{22}{3}$

To find the area below the x-axis you need to find the positive area between the curve and the axis between -1 and 1 minus the positive area between the line and the axis between -1 and 0.

$$\int_{-1}^1 (x^2 + 2x - 3)\,dx = \left[\frac{x^3}{3} + x^2 - 3x\right]_{-1}^1$$

$$= \left(\frac{(1)^3}{3} + (1)^2 - 3(1)\right) - \left(\frac{(-1)^3}{3} + (-1)^2 - 3(-1)\right)$$

$$= \left(\frac{1}{3} + 1 - 3\right) - \left(\frac{-1}{3} + 1 + 3\right) = -\frac{16}{3}$$

So the area between the curve and the axis between -1 and 1 is $\frac{16}{3}$.

$$\int_{-1}^0 4x\,dx = [2x^2]_{-1}^0 = (2(0)^2) - (2(-1)^2) = -2$$

So the area between the line and the axis between -1 and 1 is 2.

So the area under the x-axis is $\frac{16}{3} - 2 = \frac{10}{3}$

So the total area enclosed by the curve and the line is $\frac{22}{3} + \frac{10}{3} = \frac{32}{3}$

Q3 Solving $x^2 = ax \Rightarrow x^2 - ax = 0 \Rightarrow x(x - a) = 0 \Rightarrow x = 0$ or $x = a$

Draw a diagram:

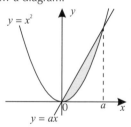

The area is given by the area under the line $y = ax$ between 0 and a, minus the area under the curve $y = x^2$ between 0 and a.

$$\int_0^a ax\,dx = \left[\frac{ax^2}{2}\right]_0^a = \left(\frac{a^3}{2}\right) - \left(\frac{a \times 0^2}{2}\right) = \frac{a^3}{2}$$

$$\int_0^a x^2\,dx = \left[\frac{x^3}{3}\right]_0^a = \left(\frac{a^3}{3}\right) - \left(\frac{0^3}{3}\right) = \frac{a^3}{3}$$

So the area is $\frac{a^3}{2} - \frac{a^3}{3} = \frac{a^3}{6}$.

The area is 36, so $\frac{a^3}{6} = 36 \Rightarrow a^3 = 216 \Rightarrow a = 6$.

2. The Trapezium Rule
Exercise 2.1 — The trapezium rule

Q1 a) You need to use 4 strips, so $n = 4$.

$$h = \frac{(b - a)}{n} = \frac{(3 - 1)}{4} = \frac{2}{4} = 0.5$$

So set up a table of x and y values:

x	$y = \frac{1}{x}$ (4 d.p.)
$x_0 = 1$	$y_0 = \frac{1}{1} = 1$
$x_1 = 1.5$	$y_1 = \frac{1}{1.5} = 0.6667$
$x_2 = 2$	$y_2 = \frac{1}{2} = 0.5$
$x_3 = 2.5$	$y_3 = \frac{1}{2.5} = 0.4$
$x_4 = 3$	$y_4 = \frac{1}{3} = 0.3333$

$$\int_1^3 \frac{1}{x}\,dx \approx \frac{0.5}{2}[1 + 2(0.6667 + 0.5 + 0.4) + 0.3333]$$

$$= 1.116675$$

$$= 1.12 \ (2\,d.p.)$$

b) You need to use 4 strips, so $n = 4$.

$$h = \frac{(b - a)}{n} = \frac{(2 - 0)}{4} = \frac{2}{4} = 0.5$$

So set up a table of x and y values:

x	$y = \sqrt{1 + 3^x}$ (4 d.p.)
$x_0 = 0$	$y_0 = \sqrt{1 + 3^0} = \sqrt{2} = 1.4142$
$x_1 = 0.5$	$y_1 = \sqrt{1 + 3^{0.5}} = 1.6529$
$x_2 = 1$	$y_2 = \sqrt{1 + 3^1} = \sqrt{4} = 2$
$x_3 = 1.5$	$y_3 = \sqrt{1 + 3^{1.5}} = 2.4892$
$x_4 = 2$	$y_4 = \sqrt{1 + 3^2} = 3.1623$

$$\int_0^2 \sqrt{1 + 3^x}\,dx \approx \frac{0.5}{2}[1.4142 + 2(1.6529 + 2 + 2.4892) + 3.1623]$$

$$= 4.215175$$

$$= 4.22 \ (2\,d.p.)$$

c) You need to use 4 strips, so $n = 4$.

$h = \dfrac{(b-a)}{n} = \dfrac{(6-1)}{4} = \dfrac{5}{4} = 1.25$

So set up a table of x and y values:

x	$y = \ln x$ (4 d.p.)
$x_0 = 1$	$y_0 = \ln 1 = 0$
$x_1 = 2.25$	$y_1 = \ln 2.25 = 0.8109$
$x_2 = 3.5$	$y_2 = \ln 3.5 = 1.2528$
$x_3 = 4.75$	$y_3 = \ln 4.75 = 1.5581$
$x_4 = 6$	$y_4 = \ln 6 = 1.7918$

$\displaystyle\int_1^6 \ln x \, dx \approx \dfrac{1.25}{2}[0 + 2(0.8109 + 1.2528$
$\qquad\qquad\qquad + 1.5581) + 1.7918]$
$\qquad = 5.647125$
$\qquad = 5.65 \ (2 \text{ d.p.})$

d) You need to use 4 strips, so $n = 4$.

$h = \dfrac{(b-a)}{n} = \dfrac{(3-0)}{4} = \dfrac{3}{4} = 0.75$

So set up a table of x and y values:

x	$y = 2^x$ (4 d.p.)
$x_0 = 0$	$y_0 = 2^0 = 1$
$x_1 = 0.75$	$y_1 = 2^{0.75} = 1.6818$
$x_2 = 1.5$	$y_2 = 2^{1.5} = 2.8284$
$x_3 = 2.25$	$y_3 = 2^{2.25} = 4.7568$
$x_4 = 3$	$y_4 = 2^3 = 8$

$\displaystyle\int_0^3 2^x \, dx \approx \dfrac{0.75}{2}[1 + 2(1.6818 + 2.8284$
$\qquad\qquad\qquad + 4.7568) + 8]$
$\qquad = 10.32525$
$\qquad = 10.33 \ (2 \text{ d.p.})$

Q2 a)

x	0	$\dfrac{\pi}{6}$	$\dfrac{\pi}{3}$	$\dfrac{\pi}{2}$
$y = \sin x$	0	$\dfrac{1}{2}$	$\dfrac{\sqrt{3}}{2}$	1

b) You just need to find the value of h. The x values increase by $\dfrac{\pi}{6}$ each time so $h = \dfrac{\pi}{6}$.

$\displaystyle\int_0^{\frac{\pi}{2}} \sin x \, dx \approx \dfrac{\left(\frac{\pi}{6}\right)}{2}\left[0 + 2\left(\dfrac{1}{2} + \dfrac{\sqrt{3}}{2}\right) + 1\right]$
$\qquad = 0.977048...$
$\qquad = 0.98 \ (2 \text{ d.p.})$

Q3 a) $\displaystyle\int_{-4}^0 (x^2 + 2x + 3) \, dx = \left[\dfrac{x^3}{3} + x^2 + 3x\right]_{-4}^0$

$= \left(\dfrac{0^3}{3} + 0^2 + 3 \times 0\right) - \left(\dfrac{(-4)^3}{3} + (-4)^2 + 3(-4)\right)$

$= 0 - \left(\dfrac{-64}{3} + 16 - 12\right)$

$= \dfrac{52}{3} = 17.33 \ (2 \text{ d.p.})$

b) 5 ordinates means 4 strips, so $n = 4$.

So $h = \dfrac{(b-a)}{n} = \dfrac{(0 - (-4))}{4} = \dfrac{4}{4} = 1$

So set up a table of x and y values:

x	$y = x^2 + 2x + 3$
$x_0 = -4$	$y_0 = 11$
$x_1 = -3$	$y_1 = 6$
$x_2 = -2$	$y_2 = 3$
$x_3 = -1$	$y_3 = 2$
$x_4 = 0$	$y_4 = 3$

$\displaystyle\int_{-4}^0 x^2 + 2x + 3 \, dx \approx \dfrac{1}{2}[11 + 2(6 + 3 + 2) + 3]$
$\qquad\qquad\qquad = 18$

c) The answer to part a), $\dfrac{52}{3} = 17.333...$ is the exact solution to the integral.
The answer to part b) is an estimate of the integral and is within 1 unit of the exact solution.

Q4 The only value you need to find now to use the trapezium rule is h. It is just the difference between the x values, so $h = 1$.

$\displaystyle\int_{-2}^2 f(x) \, dx \approx \dfrac{1}{2}[10 + 2(8 + 7 + 6.5) + 3]$
$\qquad\qquad = 28$

Q5 a) You need 4 strips so $n = 4$.

$h = \dfrac{(b-a)}{n} = \dfrac{(5-0)}{4} = \dfrac{5}{4} = 1.25$

So set up a table:

x	$y = 2^{-x}$ (4 d.p.)
$x_0 = 0$	$y_0 = 1$
$x_1 = 1.25$	$y_1 = 0.4204$
$x_2 = 2.5$	$y_2 = 0.1768$
$x_3 = 3.75$	$y_3 = 0.0743$
$x_4 = 5$	$y_4 = 0.0313$

$\displaystyle\int_0^5 2^{-x} \, dx \approx \dfrac{1.25}{2}[1 + 2(0.4204 + 0.1768$
$\qquad\qquad\qquad + 0.0743) + 0.0313]$
$\qquad = 1.4839375$
$\qquad = 1.484 \ (3 \text{ d.p.})$

b) You need 5 strips so $n = 5$.

$h = \dfrac{(b-a)}{n} = \dfrac{(5-0)}{5} = \dfrac{5}{5} = 1$

So set up a table:

x	$y = 2^{-x}$
$x_0 = 0$	$y_0 = 1$
$x_1 = 1$	$y_1 = 0.5$
$x_2 = 2$	$y_2 = 0.25$
$x_3 = 3$	$y_3 = 0.125$
$x_4 = 4$	$y_4 = 0.0625$
$x_5 = 5$	$y_5 = 0.03125$

$\displaystyle\int_0^5 2^{-x}\,dx \approx \frac{1}{2}[1 + 2(0.5 + 0.25 + 0.125$

$+\ 0.0625) + 0.03125]$

$= 1.453125$

$= 1.453\,(3\,\text{d.p.})$

c) The estimate in part b) should be more accurate because the area is split into more trapeziums, so they will be closer to the actual area under the curve.

Q6 a) 9 ordinates means 8 strips so $n = 8$.

$h = \dfrac{(b-a)}{n} = \dfrac{(3-1)}{8} = \dfrac{2}{8} = \dfrac{1}{4} = 0.25$

So set up a table:

x	$y = 3 - (2 - x)^2$
$x_0 = 1$	$y_0 = 2$
$x_1 = 1.25$	$y_1 = 2.4375$
$x_2 = 1.5$	$y_2 = 2.75$
$x_3 = 1.75$	$y_3 = 2.9375$
$x_4 = 2$	$y_4 = 3$
$x_5 = 2.25$	$y_5 = 2.9375$
$x_6 = 2.5$	$y_6 = 2.75$
$x_7 = 2.75$	$y_7 = 2.4375$
$x_8 = 3$	$y_8 = 2$

$\displaystyle\int_1^3 (3-(2-x)^2)\,dx \approx \frac{0.25}{2}[2 + 2(2.4375 + 2.75$

$+\ 2.9375 + 3 + 2.9375$

$+\ 2.75 + 2.4375) + 2]$

$= 5.3125$

b) The function $y = 3 - (2 - x)^2 = -x^2 + 4x - 1$ is a quadratic with a negative x^2 coefficient, so it is n-shaped. When $x = 0$, $y = -1$ so it intersects the y-axis at -1. Find the x-intercepts by letting $y = 0$, so $3 - (2 - x)^2 = 0 \Rightarrow 3 = (2 - x)^2$
$\Rightarrow \pm\sqrt{3} = 2 - x \Rightarrow x = 2 \pm \sqrt{3}$.

If you differentiate the curve $\dfrac{dy}{dx} = -2x + 4$ and let the derivative equal zero, you get that $x = 2$, so there is a maximum at $x = 2$.

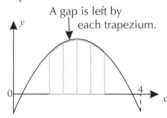

c) The approximation will be an underestimate because the graph curves in such a way that the trapeziums will lie underneath the curve.

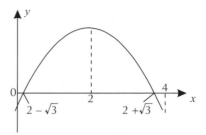

A gap is left by each trapezium.

Q7 a)

x	$-\dfrac{\pi}{2}$	$-\dfrac{\pi}{3}$	$-\dfrac{\pi}{6}$	0	$\dfrac{\pi}{6}$	$\dfrac{\pi}{3}$	$\dfrac{\pi}{2}$
y	0	0.5	$\dfrac{\sqrt{3}}{2}$	1	$\dfrac{\sqrt{3}}{2}$	0.5	0

b) Use the trapezium rule. You just need to find h, the difference between the x values.

So $h = \dfrac{\pi}{6}$

$\displaystyle\int_{-\frac{\pi}{2}}^{\frac{\pi}{2}} \cos x\,dx \approx \frac{\left(\frac{\pi}{6}\right)}{2}[0 + 2(0.5 + \frac{\sqrt{3}}{2} + 1$

$+\ \frac{\sqrt{3}}{2} + 0.5) + 0]$

$= \frac{\pi}{12}[0 + 2(2 + \sqrt{3}) + 0]$

$= \pi\left(\frac{2(2 + \sqrt{3})}{12}\right)$

$= \frac{\pi(2 + \sqrt{3})}{6}$

c) The graph of $y = \cos x$ looks likes this between the two limits:

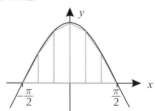

So the trapeziums will lie underneath the curve and so the approximation will be an underestimate.

Review Exercise — Chapter 7

Q1 To find y in terms of x, just integrate $\frac{dy}{dx}$ with respect to x.

a) $y = \int \frac{dy}{dx}\,dx = \int \left(\frac{5}{7}x^4 + \frac{2}{3}x + \frac{1}{4}\right)dx$

$= \frac{5x^5}{7 \times 5} + \frac{2x^2}{3 \times 2} + \frac{x}{4} + C$

$= \frac{x^5}{7} + \frac{x^2}{3} + \frac{x}{4} + C$

b) $y = \int \frac{dy}{dx}\,dx = \int \left(\frac{1}{\sqrt{x}} + \sqrt{x}\right)dx$

$= \int \left(x^{-\frac{1}{2}} + x^{\frac{1}{2}}\right)dx = \frac{x^{\frac{1}{2}}}{\left(\frac{1}{2}\right)} + \frac{x^{\frac{3}{2}}}{\left(\frac{3}{2}\right)} + C$

$= 2x^{\frac{1}{2}} + \frac{2x^{\frac{3}{2}}}{3} + C = 2\sqrt{x} + \frac{2(\sqrt{x})^3}{3} + C$

c) $y = \int \frac{dy}{dx}\,dx = \int \left(\frac{3}{x^2} + \frac{3}{\sqrt[3]{x}}\right)dx$

$= \int \left(3x^{-2} + 3x^{-\frac{1}{3}}\right)dx = \frac{3x^{-1}}{-1} + \frac{3x^{\frac{2}{3}}}{\left(\frac{2}{3}\right)} + C$

$= -\frac{3}{x} + \frac{9}{2}(\sqrt[3]{x})^2 + C$

Q2 **a)** $\int_0^1 (4x^3 + 3x^2 + 2x + 1)\,dx$

$- \left[x^4 + x^3 + x^2 + x\right]_0^1$

$= 4 - 0 = 4$

b) $\int_1^2 \left(\frac{8}{x^5} + \frac{3}{\sqrt{x}}\right)dx - \int_1^2 (8x^{-5} + 3x^{-\frac{1}{2}})dx$

$= \left[-\frac{2}{x^4} + 6\sqrt{x}\right]_1^2 = \left(-\frac{2}{16} + 6\sqrt{2}\right) - (-2 + 6)$

$= -\frac{33}{8} + 6\sqrt{2}$

c) $\int_1^6 \frac{3}{x^2}\,dx = \int_1^6 3x^{-2}\,dx = \left[\frac{-3}{x}\right]_1^6 = -\frac{1}{2} - (-3) = \frac{5}{2}$

Q3 **a)** **(i)** $\int_{-3}^3 (9 - x^2)\,dx = \left[9x - \frac{x^3}{3}\right]_{-3}^3$

$= 18 - (-18) = 36$

(ii) $\int_1^n \frac{3}{x^2}\,dx = \left[-\frac{3}{x}\right]_1^n$

$= -\frac{3}{n} - (-3) = -\frac{3}{n} + 3$

As $n \to \infty$, $\frac{3}{n} \to 0$ so $-\frac{3}{n} + 3 \to 3$

b) **(i)**

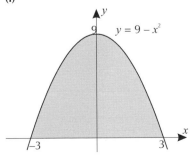

(ii)

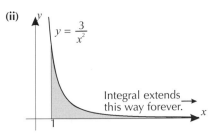

Q4 **a)** $\int_1^8 y\,dx = \int_1^8 x^{-\frac{1}{3}}\,dx = \left[\frac{3}{2}x^{\frac{2}{3}}\right]_1^8$

$= \left(\frac{3}{2} \times 8^{\frac{2}{3}}\right) - \left(\frac{3}{2} \times 1^{\frac{2}{3}}\right)$

$= \left(\frac{3}{2} \times 4\right) - \left(\frac{3}{2} \times 1\right) = \frac{9}{2}$

b) You need to find where the curve crosses the x-axis so set $x^2 + 8x + 7 = 0 \Rightarrow (x + 7)(x + 1) = 0$, so the graph crosses the axis at $x = -7$ and $x = -1$.

So the area is given by the area between the curve and the x-axis between $x = -7$ and $x = -1$ added to the area between the curve and the x-axis between $x = -1$ and $x = 0$.

$\int_{-7}^{-1} (x^2 + 8x + 7)\,dx = \left[\frac{x^3}{3} + 4x^2 + 7x\right]_{-7}^{-1}$

$= \left(\frac{(-1)^3}{3} + 4(-1)^2 + 7(-1)\right)$

$\quad - \left(\frac{(-7)^3}{3} + 4(-7)^2 + 7(-7)\right)$

$= \left(-\frac{1}{3} + 4 - 7\right) - \left(-\frac{343}{3} + 196 - 49\right)$

$= \quad 36$

So the area under the curve between -7 and -1 is 36.

$\int_{-1}^0 (x^2 + 8x + 7)\,dx = \left[\frac{x^3}{3} + 4x^2 + 7x\right]_{-1}^0$

$= \left(\frac{(0)^3}{3} + 4(0)^2 + 7(0)\right)$

$\quad - \left(\frac{(-1)^3}{3} + 4(-1)^2 + 7(-1)\right)$

$= 0 - \left(-\frac{1}{3} + 4 - 7\right)$

$= \frac{10}{3}$

So the total area is $36 + \frac{10}{3} = \frac{118}{3}$.

c) Again, find where the curve crosses the x-axis. Let $x^2 + x - 6 = 0 \Rightarrow (x + 3)(x - 2) = 0$ $\Rightarrow x = 2$ or $x = -3$.

So the area is given by the area between the curve and the x-axis between $x = -3$ and $x = -2$ added to the area between the curve and the x-axis between $x = 2$ and $x = 3$.

$\int_{-3}^2 (x^2 + x - 6)\,dx = \left[\frac{x^3}{3} + \frac{x^2}{2} - 6x\right]_{-3}^2$

$= \left(\frac{2^3}{3} + \frac{2^2}{2} - 6(2)\right) - \left(\frac{(-3)^3}{3} + \frac{(-3)^2}{2} - 6(-3)\right)$

$= \left(\frac{8}{3} + 2 - 12\right) - \left(-9 + \frac{9}{2} + 18\right)$

$= -\frac{125}{6}$

So the area between the curve and the x-axis between –3 and 2 is $\frac{125}{6}$.

$$\int_2^3 (x^2 + x - 6)\,dx = \left[\frac{x^3}{3} + \frac{x^2}{2} - 6x\right]_2^3$$

$$= \left(\frac{3^3}{3} + \frac{3^2}{2} - 6(3)\right) - \left(\frac{2^3}{3} + \frac{2^2}{2} - 6(2)\right)$$

$$= \left(9 + \frac{9}{2} - 18\right) - \left(\frac{8}{3} + 2 - 12\right) = \frac{17}{6}$$

So the total area is $\frac{125}{6} + \frac{17}{6} = \frac{71}{3}$.

Q5 a) $A = \int_0^2 (x^3 - 5x^2 + 6x)\,dx$

$$= \left[\frac{x^4}{4} - \frac{5}{3}x^3 + 3x^2\right]_0^2 = \frac{8}{3}$$

b) $A = \int_1^4 2\sqrt{x}\,dx = \int_1^4 2x^{\frac{1}{2}}dx = \left[\frac{4}{3}x^{\frac{3}{2}}\right]_1^4 = \frac{28}{3}$

c) $A = \int_0^2 2x^2\,dx + \int_2^6 (12 - 2x)\,dx$

$$= \left[\frac{2}{3}x^3\right]_0^2 + [12x - x^2]_2^6$$

$$= \frac{16}{3} + 16 = \frac{64}{3}$$

d) $A = \int_{-1}^4 (x + 3)\,dx - \int_{-1}^4 (x^2 - 4x + 7)\,dx$

$$= \left[\frac{x^2}{2} + 3x\right]_{-1}^4 - \left[\frac{x^3}{3} - 2x^2 + 7x\right]_{-1}^4$$

$$= \frac{33}{2} - 12 = \frac{9}{2}$$

Q6 a) $h = \frac{(3 - 0)}{3} = 1$

x	$y = \sqrt{9 - x^2}$
$x_0 = 0$	$y_0 = \sqrt{9} = 3$
$x_1 = 1$	$y_1 = \sqrt{8} = 2.82843$
$x_2 = 2$	$y_2 = \sqrt{5} = 2.23607$
$x_3 = 3$	$y_3 = \sqrt{0} = 0$

$$\int_a^b y\,dx \approx \frac{1}{2}[3 + 2(2.82843 + 2.23607) + 0]$$
$$= 6.5645 = 6.56 \ (3\text{ s.f.})$$

b) $h = \frac{(1.2 - 0.2)}{5} = 0.2$

x	$y = x^{x^2}$
$x_0 = 0.2$	$y_0 = 0.2^{0.04} = 0.93765$
$x_1 = 0.4$	$y_1 = 0.4^{0.16} = 0.86363$
$x_2 = 0.6$	$y_2 = 0.6^{0.36} = 0.83202$
$x_3 = 0.8$	$y_3 = 0.8^{0.64} = 0.86692$
$x_4 = 1$	$y_4 = 1^1 = 1$
$x_5 = 1.2$	$y_5 = 1.2^{1.44} = 1.30023$

$$\int_a^b y\,dx \approx \frac{0.2}{2}[0.93765 + 2(0.86363 + 0.83202$$
$$+ 0.86692 + 1) + 1.30023]$$
$$= 0.1 \times 9.36302 = 0.936 \ (3\text{ s.f.})$$

c) $h = \frac{(3 - 1)}{4} = \frac{2}{4} = 0.5$

x	$y = 2^{x^2}$
$x_0 = 1$	$y_0 = 2^1 = 2$
$x_1 = 1.5$	$y_1 = 2^{2.25} = 4.75683$
$x_2 = 2$	$y_2 = 2^4 = 16$
$x_3 = 2.5$	$y_3 = 2^{6.25} = 76.10926$
$x_4 = 3$	$y_4 = 2^9 = 512$

$$\int_a^b y\,dx \approx \frac{0.5}{2}[2 + 2(4.75683 + 16 + 76.10926)$$
$$+ 512]$$
$$= 176.933045$$
$$= 177 \ (3\text{ s.f.})$$

d) $h = \frac{(3 - 1)}{5} = \frac{2}{5} = 0.4$

x	$y = 2^{x^2}$
$x_0 = 1$	$y_0 = 2^1 = 2$
$x_1 = 1.4$	$y_1 = 2^{1.96} = 3.89062$
$x_2 = 1.8$	$y_2 = 2^{3.24} = 9.44794$
$x_3 = 2.2$	$y_3 = 2^{4.84} = 28.64080$
$x_4 = 2.6$	$y_4 = 2^{6.76} = 108.3834$
$x_5 = 3$	$y_5 = 2^9 = 512$

$$\int_a^b y\,dx \approx \frac{0.4}{2}[2 + 2(3.89062 + 9.44794$$
$$+ 28.64080 + 108.3834) + 512]$$
$$= 162.945104 = 163 \ (3\text{ s.f.})$$

Exam-Style Questions — Chapter 7

Q1 $\int_2^7 (2x - 6x^2 + \sqrt{x})\,dx = \left[x^2 - 2x^3 + \frac{2\sqrt{x^3}}{3}\right]_2^7$

[1 mark for each correct term]

$$= \left(7^2 - (2 \times 7^3) + \frac{2\sqrt{7^3}}{3}\right)$$

$$- \left(2^2 - (2 \times 2^3) + \frac{2\sqrt{2^3}}{3}\right)$$

[1 mark]

$$= -624.6531605 - (-10.11438192)$$
$$= -614.5387786 = -614.5388 \,(4\text{ d.p.})$$

[1 mark]

Q2 a) (i) $h = \dfrac{8-2}{3} = 2$ *[1 mark]*

$x_0 = 2 \quad y_0 = \sqrt{(3 \times 2^3)} + \dfrac{2}{\sqrt{2}} = 6.31319$

$x_1 = 4 \quad y_1 = \sqrt{(3 \times 4^3)} + \dfrac{2}{\sqrt{4}} = 14.85641$

$x_2 = 6 \quad y_2 = \sqrt{(3 \times 6^3)} + \dfrac{2}{\sqrt{6}} = 26.27234$

$x_3 = 8 \quad y_3 = \sqrt{(3 \times 8^3)} + \dfrac{2}{\sqrt{8}} = 39.89894$

[1 mark]

$\displaystyle\int_2^8 y\,dx \approx \dfrac{2}{2}[6.31319 + 2(14.85641$
$+ 26.27234) + 39.89894]$

[1 mark]

$= 128.46963 = 128.47$ to 2 d.p. *[1 mark]*

(ii) $h = \dfrac{5-1}{4} = 1$ *[1 mark]*

$x_0 = 1 \quad y_0 = \dfrac{1^3 - 2}{4} = -0.25$

$x_1 = 2 \quad y_1 = \dfrac{2^3 - 2}{4} = 1.5$

$x_2 = 3 \quad y_2 = \dfrac{3^3 - 2}{4} = 6.25$

$x_3 = 4 \quad y_3 = \dfrac{4^3 - 2}{4} = 15.5$

$x_4 = 5 \quad y_4 = \dfrac{5^3 - 2}{4} = 30.75$ *[1 mark]*

$\displaystyle\int_1^5 y\,dx \approx \dfrac{1}{2}[-0.25 + 2(1.5 + 6.25 + 15.5)$
$+ 30.75]$ *[1 mark]*

$= 38.5$ *[1 mark]*

b) Increase the number of intervals. *[1 mark]*

Q3 The limits are the x-values when $y = 0$, so first solve
$(x-3)^2(x+1) = 0$: *[1 mark]*

$x = 3$ *[1 mark]* and $x = -1$ *[1 mark]*

Hence, to find the area, calculate:

$\displaystyle\int_{-1}^3 (x-3)^2(x+1)\,dx$

$= \displaystyle\int_1^3 (x^3 - 5x^2 + 3x + 9)\,dx$ *[1 mark]*

$= \left[\dfrac{x^4}{4} - \dfrac{5}{3}x^3 + \dfrac{3}{2}x^2 + 9x\right]_{-1}^3$ *[1 mark]*

$-\left(\dfrac{3^4}{4} - \dfrac{5}{3}3^3 + \dfrac{3}{2}3^2 + (9 \times 3)\right)$

$\left.-\begin{array}{l}\dfrac{(-1)^4}{4} - \left(\dfrac{5}{3} \times (-1)^3\right)\\[2mm] + \left(\dfrac{3}{2} \times (-1)^2\right) + (9 \times (-1))\end{array}\right|$ *[1 mark]*

$= 15\dfrac{3}{4} - -5\dfrac{7}{12}$ *[1 mark]*

$= 21\dfrac{1}{3}$ *[1 mark]*

Q4 $n = 5$, $h = \dfrac{4 - 1.5}{5} = 0.5$ *[1 mark]*

$x_1 = 2.0$, $x_2 = 2.5$, $x_3 = 3.0$ *[1 mark]*

$y_0 = 2.8182$ *[1 mark]*, $y_3 = 6.1716$ *[1 mark]*,

$y_4 = 7.1364$ *[1 mark]*

$\displaystyle\int_{1.5}^4 y\,dx \approx \dfrac{0.5}{2}[2.8182 + 2(4 + 5.1216$
$+ 6.1716 + 7.1364) + 8]$ *[1 mark]*

$= 13.91935 = 13.9$ to 3 s.f. *[1 mark]*

Q5 a) $y_0 = \sqrt{1+0} = 1, \qquad y_1 = \sqrt{1+0.5} = 1.2247,$
$y_2 = \sqrt{1+1} = 1.4142, y_3 = \sqrt{1+1.5} = 1.5811,$
$y_4 = \sqrt{1+2} = 1.7321$

[3 marks available — 1 mark for 2 values,
2 marks for 4 values, 3 marks for all 5 values]

b) The value of h is 0.5. *[1 mark]*

$\displaystyle\int_0^2 \sqrt{1+x}\,dx \approx \dfrac{0.5}{2}[1 + 2(1.2247 + 1.4142$
$+ 1.5811) + 1.7321]$ *[1 mark]*

$= 2.793025$

$= 2.79$ (2 d.p.) *[1 mark]*

Q6 a) $m = \dfrac{y_2 - y_1}{x_2 - x_1} = \dfrac{0 - -5}{-1-4} = -1$ *[1 mark]*

$y - y_1 = m(x - x_1)$

$y - -5 = -1(x - 4)$ so $y + 5 = 4 - x$

$y = -x - 1$ *[1 mark]*

b) Multiply out the brackets and then integrate:

$(x+1)(x-5) = x^2 - 4x - 5$ *[1 mark]*

$\displaystyle\int_{-1}^4 (x^2 - 4x - 5)\,dx = \left[\dfrac{x^3}{3} - 2x^2 - 5x\right]_{-1}^4$ *[1 mark]*

$= \left(\dfrac{4^3}{3} - 2(4^2) - (5 \times 4)\right)$

$-\left(\dfrac{(-1)^3}{3} - 2(-1)^2 - (5 \times -1)\right)$ *[1 mark]*

$= -30\dfrac{2}{3} - 2\dfrac{2}{3}$ *[1 mark]* $= -33\dfrac{1}{3}$ *[1 mark]*

c) Subtract the area above the line from the area
above the curve to leave the area in between.
The area above the line is a triangle (where
$b = 5$ and $h = 5$), so use the formula for the area
of a triangle to calculate it *[1 mark]*:

$A = \dfrac{1}{2}bh = \dfrac{1}{2} \times 5 \times 5 = 12.5$ *[1 mark]*

Shaded area $= 33\dfrac{1}{3} - 12\dfrac{1}{2}$ *[1 mark]*

$= 20\dfrac{5}{6}$ *[1 mark]*

You could also have integrated $-x - 1$ from $x = -1$ to
$x = 4$ to find the area above the line — you'd have got
the same answer.

Glossary

A

Algebraic division
Dividing one algebraic expression by another.

Arc
The curved edge of a **sector** of a circle.

Asymptote
A line which a curve gets infinitely closer to, but never touches.

B

Binomial
A **polynomial** with only two terms e.g. $a + bx$.

Binomial expansion
The result of expanding a **binomial** raised to a power — e.g. $(a + bx)^n$.

Binomial formula
A formula which describes the general terms of a **binomial expansion**.

C

Chord
A line joining two points which lie on the circumference of a circle.

Coefficient
The constant multiplying the variable(s) in an algebraic term e.g. 4 in the term $4x^2y$.

Common ratio
The constant that you multiply by to get from one term to the next in a **geometric sequence** or **series**.

Constant of integration
A constant term coming from an indefinite **integration** representing any number.

Convergent sequence/series
A **sequence**/**series** that tends towards a **limit**.

D

Decreasing function
A function for which the **gradient** is always less than zero.

Definite integral
An **integral** which is evaluated over an interval given by two **limits**, representing the area under the curve between those limits.

Degree
The highest power of x in a **polynomial**.

Derivative
The result after **differentiating** a function.

Differentiation
A method of finding the rate of change of a function with respect to a variable.

Divergent sequence/series
A **sequence**/**series** that does not have a **limit**.

Divisor
The number or expression you're dividing by in a division.

E

Exponential decay
Exponential decay happens when the rate of decay gets slower and slower as the amount gets smaller (negative **exponential growth**).

Exponential function
A function of the form $y = a^x$.

Exponential growth
Exponential growth happens when the rate of growth gets faster and faster as the amount gets bigger.

F

Factor
A factor of a term or expression is something which divides into it.

Factorial
n factorial, written $n!$, is the product of all **integers** from 1 to n. So $n! = 1 \times 2 \times ... \times n$.

Factorising
The opposite of multiplying out brackets. Brackets are put in to write an expression as a product of its **factors**.

Factor Theorem
An extension of the **Remainder Theorem** that helps you factorise a polynomial. If $f(a) = 0$ then $(x - a)$ is a **factor** of $f(x)$.

G

Geometric sequence/series
A **sequence**/**series** in which you multiply by a **common ratio** to get from one term to the next.

Gradient
The gradient of a curve at a given point is how steep the curve is at that point.

I

Identity
An equation that is true for all values of a variable, usually denoted by the '$\equiv$' sign.

Increasing function
A function for which the **gradient** is always greater than zero.

Indefinite integral
An integral which includes a **constant of integration** that comes from integrating without **limits**.

Index
For a^n, n is the index and is often referred to as a power.

Integer
A positive or negative whole number (including 0).

Integral
The result you get when you integrate something.

Integration
Process for finding a function, given its **derivative** — the opposite of **differentiation**.

Limit (sequences and series)
The value that the individual terms in a **sequence**, or the sum of the terms in a **series**, tends towards.

Limits (integration)
The numbers between which you integrate to find a **definite integral**.

Linear factor
A **factor** of an algebraic expression of **degree** 1 — e.g. $ax + b$.

Logarithm
The logarithm to the base a of a number x (written $\log_a x$) is the power to which a must be raised to give that number.

Maximum
The highest point on a graph, or on a section of a graph (this is a local maximum).

Minimum
The lowest point on a graph or on a section of a graph (this is a local minimum).

Modulus
The modulus of a number is its positive numerical value.
The modulus of a function, $f(x)$, makes every value of $f(x)$ positive by removing any minus signs.

^nC_r
The **binomial coefficient** of x^r in the **binomial expansion** of $(1 + x)^n$.
Also written $\binom{n}{r}$.

Normal
A straight line passing through a curve that is perpendicular (at right angles) to the curve at the point where it crosses the curve.

Pascal's triangle
A triangle of numbers showing the **binomial coefficients**. Each term is the sum of the two above it.

Point of inflexion
A point on a graph where the curve briefly flattens out without changing direction. A type of **stationary point**.

Polynomial
An algebraic expression made up of the sum of constant terms and variables raised to positive **integer** powers.

Progression
Another word for **sequence**.

Quotient
The result when you divide one thing by another, not including the **remainder**.

Radian
A unit of measurement for angles. 1 radian is the angle in a **sector** of a circle with radius r that has an **arc** of length r.

Rational number
A number that can be written as the **quotient** (division) of two **integers**, where the denominator is non-zero.

Remainder
The expression that is left over following an **algebraic division** and that has a **degree** lower than the **divisor**.

Remainder Theorem
A method used to work out the remainder from an **algebraic division**, but without actually having to do the division. The **remainder** when $f(x)$ is divided by $(x - a)$ is $f(a)$.

Root
The roots of a function $f(x)$ are the values of x where $f(x) = 0$.

S

Second order derivative
The result of **differentiating** a function twice.

Sector
A section of a circle formed by two radii and part of the circumference.

Sequence
An ordered list of numbers (referred to as terms) that follow a set pattern. E.g. 2, 6, 10, ... or –4, 1, –4, 1, ...

Series
An ordered list of numbers, just like a **sequence**, but where the terms are being added together (to find a sum).

Sigma notation
Used for the sum of **series**. E.g. $\sum_{n=1}^{15} 3^n$ is the sum of the first 15 terms of the **geometric series** with **common ratio** 3 and **first term** 3.

Sine rule
A rule for finding missing sides or angles in a triangle. It can be used if you know any two angles and a side, and in some cases, if you know two sides and an angle that isn't between them.

Stationary point
A point on a curve where the **gradient** is zero.

Sum to infinity
The sum to infinity of a **series** is the value the sum tends towards as more and more terms are added.
Also known as the **limit** of a series.

T

Tangent
A straight line which just touches the curve at a point, without going through it and that has the same **gradient** as the curve at that point.

Trapezium rule
A way of estimating the area under a curve by dividing it into trapezium-shaped strips

Turning point
A **stationary point** that is a (local) **maximum** or **minimum** point of a curve.

Index

C2 Formula Sheet

These are the formulas you'll be given in the exam, but make sure you know exactly **when you need them** and **how to use them**. You might also need any formulas from the C1 formula sheet in C2.

Cosine Rule

$$a^2 = b^2 + c^2 - 2bc \cos A$$

Binomial Series

$$(a + b)^n = a^n + \binom{n}{1}a^{n-1}b + \binom{n}{2}a^{n-2}b^2 + \dots + \binom{n}{r}a^{n-r}b^r + \dots + b^n \quad (n \in \mathbb{N})$$

$$\text{where } \binom{n}{r} = {}^nC_r = \frac{n!}{r!(n-r)!}$$

$$(1 + x)^n = 1 + nx + \frac{n(n-1)}{1 \times 2}x^2 + \dots + \frac{n(n-1)\dots(n-r+1)}{1 \times 2 \times \dots \times r}x^r + \dots \quad (|x| < 1, \, n \in \mathbb{R})$$

Logs and Exponentials

$$\log_a x = \frac{\log_b x}{\log_b a}$$

Geometric Series

$$u_n = ar^{n-1}$$

$$S_n = \frac{a(1 - r^n)}{1 - r}$$

$$S_\infty = \frac{a}{1 - r} \text{ for } |r| < 1$$

Numerical Integration

The trapezium rule:

$$\int_a^b y \, dx \approx \frac{1}{2}h\{(y_0 + y_n) + 2(y_1 + y_2 + \dots + y_{n-1})\}, \text{ where } h = \frac{b - a}{n}$$

MEC2T51